Ecology and Conservation
of Butterflies

Edited by
Andrew S. Pullin

In Association with
The British Butterfly Conservation Society

CHAPMAN & HALL

London · Glasgow · Weinheim · New York · Tokyo · Melbourne · Madras

Published by Chapman & Hall, 2–6 Boundary Row, London SE1 8HN, UK

Chapman & Hall, 2–6 Boundary Row, London, SE1 8HN, UK

Blackie Academic & Professional, Wester Cleddens Road, Bishopbriggs, Glasgow G64 2NZ, UK

Chapman & Hall GmbH, Pappelallee 3, 69469 Weinheim, Germany

Chapman & Hall USA., One Penn Plaza, 41st Floor, New York NY 10119, USA

Chapman & Hall Japan, ITP-Japan, Kyowa Building, 3F, 2-2-1 Hirakawacho, Chiyoda-ku, Tokyo 102, Japan

Chapman & Hall Australia, Thomas Nelson Australia, 102 Dodds Street, South Melbourne, Victoria 3205, Australia

Chapman & Hall India, R. Seshadri, 32 Second Main Road, CIT East, Madras 600 035, India

First edition 1995
© 1995 Chapman & Hall

Typeset in 10/12pt Ehrhardt by Florencetype Ltd, Stoodleigh, Devon

Printed in Great Britain by St Edmundsbury Press, Bury St Edmunds, Suffolk

ISBN 0 412 56970 1

A catalogue record for this book is available from the British Library

Library of Congress Catalog Card Number: 94-70983

∞ Printed on acid-free text paper, manufactured in accordance with ANSI/NISO Z39 48–1992 and ANSI/NISO 239.48–1984 (Permanence of Paper)

Ecology and Conservation of Butterflies

Contents

Contributors

J. Asher, 24 Fettiplace Road, Marcham, Abingdon, Oxon OX13 6PL, UK

S.A. Clarke, Wessex Environmental Consultants, 4 Prospect Place, Grove Lane, Redlynch, Wilts SP5 2NT, UK

J.P. Dempster, Department of Zoology, University of Cambridge, Downing Street, Cambridge, UK

R.L.H. Dennis, Centre for Applied Entomology and Parasitology, Department of Biological Sciences, Keele University, Staffs ST5 5BG, UK

A. Erhardt, Botanisches Institut der Universität, Schönbeinstrasse 6, CH-4056 Basel, Switzerland

B.C. Eversham, Biological Records Centre, NERC Institute of Terrestrial Ecology, Monks Wood, Abbots Ripton, Huntingdon PE17 2LS, UK

R.E. Feber, Wildlife Conservation Research Unit, Department of Zoology, University of Oxford, South Parks Road, Oxford OX1 3PS, UK

A.P. Fowles, Countryside Council for Wales, Plas Penrhos, Ffordd Penrhos, Bangor, Gwynedd LL57 2LQ, UK

P.T. Harding, Biological Records Centre, NERC Institute of Terrestrial Ecology, Monks Wood, Abbots Ripton, Huntingdon PE17 2LS, UK

A.J. Kerr, Scottish Natural Heritage, 2 Anderson Place, Edinburgh EH6 5NP, UK

O. Kudrna, Ökologische Station, Universität Würzburg, Fabrikschleich, D-96181 Rauhenebrach, Germany

T.B. Larsen, 358 Coldharbour Lane, London SW9 8PL, UK

I.F.G. McLean, Species Conservation Branch, English Nature, Northminster House, Peterborough, PE1 1UA, UK

M.L. Munguira, Departmento de Biología (Zoología), Universidad Autónoma de Madrid, Cantoblanco, E-28049 Madrid, Spain

T.R. New, Department of Zoology, La Trobe University, Bundoora, Victoria 3083, Australia

M.R. Oates, The National Trust, 33 Sheep Street, Cirencester, Glos GL7 1QW, UK

P.A. Opler, US Fish and Wildlife Service, Office of Information Transfer, 1201 Oak Ridge Drive, Fort Collins, CO 80525, USA

E. Pollard, Springhill Farm, Benenden, Cranbrook, Kent TN17 4LA, UK

A.S. Pullin, Centre for Applied Entomology and Parasitology, Department of Biological Sciences, Keele University, Staffs ST5 5BG, UK

N.O.M. Ravenscroft, Research and Advisory Services Directorate, Scottish Natural Heritage, 2 Anderson Place, Edinburgh EH6 5NP, UK

P.A. Robertson, The Game Conservancy Trust, Fordingbridge, Hampshire SP6 1EF, UK

T.G. Shreeve, School of Biological and Molecular Sciences, Oxford Brookes University, Headington, Oxford OX3 0BP, UK

H. Smith, Wildlife Conservation Research Unit, Department of Zoology, University of Oxford, South Parks Road, Oxford OX1 3PS, UK

C.A.M. van Swaay, De Vlinderstichting, Postbus 506, NL-6700 AM Wageningen, The Netherlands

C.D. Thomas, School of Biological Sciences, University of Birmingham, Edgbaston, Birmingham B15 2TT, UK

J.A. Thomas, Institute of Terrestrial Ecology, Furzebrook Research Station, Furzebrook Road, Wareham, Dorset BH20 5AS, UK

M.L. Vickery, 3 The Deer Leap, Kenilworth, Warks CV8 2HW, UK

M.S. Warren, Butterfly Conservation, Conservation Office, PO Box 444, Dorchester, Dorset DT2 7YT, UK

M.R. Webb, Centre for Applied Entomology and Parasitology, Department of Biological Sciences, Keele University, Staffs ST5 5BG, UK

W.R. Williams, Computing Service, Durham University, Durham DH1 3LE, UK

T.J. Yates, Biological Records Centre, NERC Institute of Terrestrial Ecology, Monks Wood, Abbots Ripton, Huntingdon PE17 2LS, UK

M.R. Young, Department of Zoology, University of Aberdeen, Tillydrone Avenue, Aberdeen AB9 2TN, UK

Preface

This book was conceived to mark the Silver Jubilee of the British Butterfly Conservation Society. Interest in the conservation of butterflies has increased so rapidly that it is difficult to relate to the situation 25 years ago. Butterflies were on the decline in Britain, Europe and elsewhere but we lacked data on the extent of the decline and the underlying reasons, leaving us unable to implement effective conservation measures. An early recognition of the plight of British butterflies and moths led to the foundation of the society by a small group of conservationists in 1968. Today the society has over 10 000 members, owns a number of reserves and sponsors research, conservation and monitoring activities at the local and national level. As part of the Silver Jubilee celebrations an international symposium was held at Keele University in September 1993 entitled 'Ecology and Conservation of Butterflies'. This symposium clearly showed how much important work has been done in recent years and also gave me the impression that the subject had reached a watershed. This was not because the decline of butterflies has stopped or even slowed down, far from it, the threat to our butterflies continues to increase from habitat destruction and intensification of land use. The watershed is in our understanding of the relationship between butterflies and their habitat. For example, we now have monitoring schemes in place in Britain and some other European countries which have been running for a sufficient length of time to enable them to be used as tools to investigate changes in distribution and abundance; there is an increasing awareness of the importance of management of butterfly habitats; and, for the rarest species, detailed long-term research is uncovering complex habitat requirements that give us the chance of reversing decline and restoring populations after extinction. These advances in our understanding are represented in the first three sections of this book.

Another advance is in the increased communication and co-operation between butterfly conservation movements. Consequently, I am particularly pleased that this book includes contributions on European and global perspectives. These are of necessity increasingly general in nature as the areas considered get larger, but I hope they provide a comparison and a starting point for those interested in the wider view. We are perhaps guilty in Britain of considering only the British situation regardless of distribution and abundance in Europe. This is changing fast, and there are now many good examples of European collaboration. We have much to gain from exchange of information with our European colleagues and there can be no better time than the present when communication across Europe is becoming easier.

Britain, and indeed Europe as a whole, has relatively few species of butterflies compared with warmer continents. If butterfly conservation on a global scale is our concern, as I believe it should be, then we face real challenges in areas where tropical forest is being lost at alarming rates, and where the value of butterflies and other fauna to local people is low. A different kind of butterfly conservation is required in the Third World tropics, such as Africa and Central and South America. Even rich countries such as the USA and Australia do not have Europe's history of butterfly conservation activity, but, as the final chapters show, they are catching up fast and increased collaboration will benefit us all in the future.

A particular aim of this book is to try to bridge a number of gaps of relevance to butterfly conservation. For example, the communication gap between scientific research and practical conservation. Often nature reserve managers, at the sharp end of conservation, find results of scientific research inaccessible because of the form in which they are reported. More effort needs to be made by scientists to provide conservationists with information that they can act upon.

This book aims to give scientists, nature conservationists and amateur entomologists access to up-to-date information on a range of activities in butterfly conservation, representative of progress but by no means exhaustive. We have tried to present information in a form that will be easily understood by non-scientists. Although some chapters necessarily contain technical terms, we have tried to minimize them. The sections reflect, first, the different approaches of relevance to butterfly conservation – monitoring of distribution and abundance, habitat management in relation to land use and specific programmes for endangered species; and, secondly, the different spatial scales that must be considered when addressing national, continental and global conservation concerns.

Also, with accessibility of information in mind, we have used both Latin and English names (where they exist) of all species at their first mention in each chapter. Subsequently, the abbreviated Latin name is used, following normal convention. It is most important, particularly when the taxonomy is not stable, to state the author of each species name. This is often done at first mention of the species (or subspecies) in the text. However, this can be distracting and cumbersome in a book such as this, and therefore a taxonomic index lists the author of each species or subspecies. If a reader is uncertain of any scientific name used, he or she should consult this index for the author.

I must thank those who have contributed much to the production of this book. David Corke had the initial idea of the symposium and the book, and persuaded many of the authors to contribute. Andrew Phillips persuaded me to be the editor and to organize the Keele Symposium, and Roger Dennis provided constant support and advice. Special thanks go to Carol Jones, Stephanie Heath and Jayne Bratt for their assistance and patience in producing the manuscript.

ASP
Keele

PART ONE

Monitoring Distribution and Abundance

CHAPTER 1

Butterfly monitoring 1 – recording the changes

P.T. Harding, J. Asher and T.J. Yates

1.1 INTRODUCTION

To assess changes in the distribution and abundance of a butterfly fauna, reliable information on spatial and temporal occurrence is essential. This chapter examines present sources of information that could be used to assess changes, how those sources have been compiled and collated for Britain, how they have been and are being used, and how anticipated needs for information may be met in the future.

We discuss two distinct activities: recording (which is often equated with conducting surveys) and monitoring. These terms, together with an intermediate activity, surveillance, have been defined very precisely by Hellawell (1991), but for our purposes can be summarized as follows.

1. Survey: qualitative and/or quantitative observations, usually to a standard procedure and within a restricted time period, without preconceptions of what the results should be.
2. Surveillance: repeated surveys to provide a time series, to ascertain variability but without preconceptions of what the results should be.
3. Monitoring: intermittent surveillance to measure the extent of variation (or lack of it) from an established or expected norm.

The main areas of activity covered here are national and local butterfly recording schemes which are surveys (but the time series element of many schemes means that there is also some surveillance), and butterfly monitoring, using fixed transect routes.

Ecology and Conservation of Butterflies Edited by Andrew S. Pullin.
Published in 1995 by Chapman & Hall. ISBN 0 412 56970 1

1.2   AN OVERVIEW OF BUTTERFLY RECORDING

A comprehensive overview of recording could occupy several chapters. We have selected the most active and important elements, in particular, work that is co-ordinated in some way, wherever possible by reference to previously published work. We have concentrated on types of recording that offer opportunities to measure changes in the butterfly fauna of Britain. In addition to the work summarized below, other research and surveys carried out, mainly in the 1970s and 1980s, are listed by Harding and Green (1991).

1.2.1   The origins of butterfly recording in Britain

(a)   Why butterflies?

British butterflies are attractive and colourful, they fly in the daytime, they are usually most plentiful in good weather and two species are regarded as pests. Observing butterflies is relatively simple and can be done without collecting or killing specimens. It is not surprising that they have been the subject of attention by the general public, have a place in all the arts, have been collected as objects of beauty and scientific interest, and have been the subject of research and conservation practices. The sheer accessibility of butterflies has made them the most understood group of our fauna, apart from birds. We probably know as much about the autecology of British butterflies as about that of British mammals, a group with a similar number of species, and infinitely more than for any other invertebrate group. Like birds, butterflies are widely considered to be good indicators of the health of the environment. The true ecological importance of butterflies, as bioindicators, may be open to question when compared with, for example, chrysomelid beetles, spiders or molluscs, but detailed information on such groups of invertebrates is more difficult to obtain.

(b)   Collecting and recording

Butterfly recording has a long, if not always respectable, history in Britain. Collecting was the earliest form of recording, from which a list of the species occurring in Britain has been built up. Emmet (1989) suggested that butterfly collecting as a hobby began in Britain in the last quarter of the seventeenth century, but it is apparent that scholars, as early as the 1580s, were aware of upwards of 20 species of butterfly in Britain.

In these days of enlightened conservation, it is important to remember that the instinct to collect is not far below the surface. The earliest taxonomic and ecological experiences of many of our more senior and eminent entomologists and conservationists began with a butterfly net, a killing bottle and an improvised store box. This was at a time when collecting bugs was regarded as respectable (if slightly aberrant) behaviour, whereas collecting birds' eggs was considered to be unnecessary, although legal (until the passage of the Protection of Birds Act 1954).

Recording is, of course, a form of collecting; the collection of information rather than specimens.

Perceptions of the past ranges of British butterflies (see, for example, Ford, 1945; Heath, 1974; Heath, Pollard and Thomas, 1984; Shirt, 1987; Emmet and Heath, 1989; Warren, 1992b) have been based very largely on the records of collectors, either from publications or from data labels on specimens in museums and private collections. Because this information was originally generated without the intention of detecting changes and has been systematized only at the collation stage, it is inevitable that it is patchy, both spatially and temporally. None the less, it has proved possible to use such information to demonstrate apparent changes in the range and the national and regional status of species such as *Eurodryas aurinia* (marsh fritillary) (Warren, 1990a).

1.2.2 National butterfly recording

(a) Biological Records Centre 1967–82

The origins and history of the national Lepidoptera Recording Scheme (LRS), set up in 1967 by John Heath when he joined the Biological Records Centre (BRC) at Monks Wood, Cambridgeshire, are described by Heath, Pollard and Thomas (1984) and Harding and Sheail (1992). The primary intention of the scheme was to map the distribution of species at the level of 10 km squares of the National Grid. This limited objective was entirely consistent with the policy of BRC at that time. The popularity of this scheme was such that it occupied a significant part of John Heath's work between 1967 and 1982.

A site/species list record card for butterflies and macro-moths was designed by BRC in April 1967, and was widely used by volunteers to contribute records to the LRS. The butterfly component of the scheme was clearly more popular than the rest of the Lepidoptera so that a site/species record card for butterflies only, using both scientific and vernacular names, was designed in November 1968. Both cards were subsequently revised; RA8 in March 1971 and RA9 in November 1974, and records were also contributed on a variety of other cards for individual species.

By 1982, when the entire butterfly dataset (some 238 000 records) was to be computerized at BRC, it was decided to compile a database which would include as much detail as possible from the original record cards. Thus, full grid references, locality names, full dates and recorders' names were entered into the database (Harding and Greene, 1984), so that over 35% of the entries are localized to at least a 1 km square. This database has been used extensively during the past decade as a baseline of information for subsequent surveys and studies of individual species (see, for example, Warren, Thomas and Thomas, 1984; Morton, 1985; Read, 1985; Thomas *et al.*, 1986; Ravenscroft, 1991).

The original intention of the LRS was fulfilled by the *Atlas of Butterflies in Britain and Ireland* (Heath, Pollard and Thomas, 1984), which included 10 km square distribution maps of 62 species (e.g. Figure 1.1).

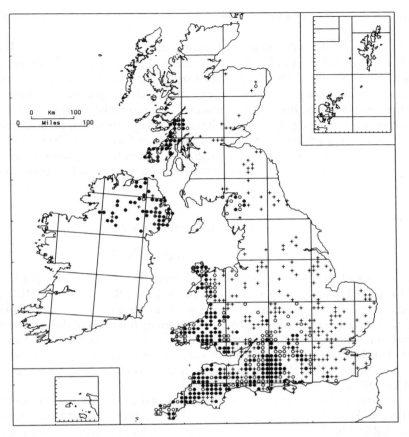

Figure 1.1 Distribution of *Eurodryas aurinia* in the UK. The recorded occurrence is mapped using the 10 km squares of the British and Irish national grids. ●, colonies recorded in 1990; ○, colonies recorded in the period 1970–89, but not since; +, colonies recorded before 1970, but not since.

The *Atlas* also drew on the growing body of ecological information resulting mainly from research by the Institute of Terrestrial Ecology (ITE). This *Atlas* clearly demonstrated substantial losses in the national range of about 30% of the butterfly species in Britain during the twentieth century.

(b)   National survey from 1982 onward

Following the completion of the collection of records for the *Atlas* in 1982, and the retirement of John Heath in January 1982, BRC combined, in 1983, with the British Butterfly Conservation Society (BC) to continue butterfly recording, but with a scheme organized by BC. A new recording card was designed (Figure 1.2) which designated 36 target species for which information about the site was required.

These species were those considered to be threatened or scarce within their range in Britain. A key objective of the BRC/BC scheme was to provide an early warning system, at the local level, on threats to sites with important butterfly populations (Harding, 1986).

This scheme was originally intended to run for 5 years, and although support for the scheme was considerably less than there had been for the LRS (1967–82), it was considered by BC to be fulfilling a useful function within the Society. The role of the BRC/BC scheme was reviewed in 1991, as part of the work of a joint working party which reported to BC in February 1992 (section 1.3).

(c)   *The Moths and Butterflies of Great Britain and Ireland*

Preparation of volume 7, part 1 of this series (Emmet and Heath, 1989), which covers the butterflies, began in 1985, soon after the publication of the *Atlas*. The original BRC maps from the *Atlas* were used by the editors and publisher to update the distribution maps for the volume. They collated information from a variety of sources, mostly from local and regional recorders and the BRC/BC scheme. Over 13 500 records were compiled up to the end of 1988 and these have subsequently been deposited, in a manuscript form, at BRC.

(d)   *The Irish dimension*

Some organized butterfly recording in Ireland was carried out in collaboration with the BRC scheme, until the Irish Biological Records Centre (IBRC) was set up in 1971. At that time all BRC's Irish records were copied to IBRC and the two centres continued to collaborate, until IBRC was closed in the late 1980s. IBRC published three editions of an atlas of butterflies in Ireland (Ni Lamhna, 1980). Copies of the Irish dataset held by BRC (as summarized by Heath, Pollard and Thomas, 1984) were deposited with IBRC in 1984 and with the Ulster Museum in 1989. The distribution of butterflies in Ireland was summarized as maps prepared by Tim Lavery in Hickin (1992), based almost entirely on Emmet and Heath (1989). In that publication it is stated that records of butterflies in Ireland should be sent to the Wildlife Service and/or the Irish Moth and Butterfly Society.

1.2.3   Regional butterfly recording

County lists, many of which include ecological or distribution information, have abounded for more than a century, and they continue to be published. The 1980s saw an unprecedented expansion of both county lists and county atlases (Corke and Harding, 1990). However, these recent publications under-represent the present scale of the activity, because most projects take at least a decade from inception to publication. A survey by BRC in 1988 showed that about 200 separate regional or local butterfly recording projects were operating in the UK. One county had at least six projects.

BUTTERFLIES 6446 | **LOCALITY** ELSWORTH | **V.C No.** | **Grid Reference** 2 9 5 2 3 2 2 6 2 7

Recorder	Determiner	Compiler	Vice-County name CAMBRIDGE	Date 2 4 0 9 1 9 8 9
P.T. &P.M. HARDING	P.T. &P.M. HARDING	P.T. HARDING	Source	Field ✓ Mus. Lit.

Altitude 3 0 metres

No. 2 7 8 2 | No. 2 7 8 2 | No. 2 7 8 2

BUTTERFLIES 6446

47502 * Adonis Blue
46502 * Black Hairstreak
45601 Brimstone
47301 * Brown Argus
46301 * Brown Hairstreak
47501 * Chalk-hill Blue
44201 * Chequered Skipper
45504 * Clouded Yellow
49301 Comma
47401 Common Blue
49605 * Dark Green Fritillary
44701 * Dingy Skipper
48101 * Duke of Burgundy Fritillary
44302 * Essex Skipper
50601 Gatekeeper/Hedge Brown
49802 * Glanville Fritillary
50401 * Grayling
46201 * Green Hairstreak
45803 Green-veined White
44901 * Grizzled Skipper
49901 * Heath Fritillary

*Target species

49604 * High Brown Fritillary
47901 Holly Blue
50802 * Large Heath
44601 Large Skipper
49101 * Large Tortoiseshell
45801 Large White
44303 * Lulworth Skipper
50301 * Marbled White
49701 * Marsh Fritillary
50701 Meadow Brown
50201 * Mountain Ringlet
47302 * Northern Brown Argus
46001 Orange Tip
48901 Painted Lady
49201 Peacock
49502 * Pearl-bordered Fritillary
48401 * Purple Emperor
46401 * Purple Hairstreak
48801 Red Admiral
50901 Ringlet
50202 * Scotch Argus

44401 * Silver-spotted Skipper
47201 * Silver-studded Blue
49606 * Silver-washed Fritillary
47001 * Small Blue
46701 Small Copper
50801 Small Heath
49501 * Small Pearl-bordered Fritillary
44301 Small Skipper
49001 Small Tortoiseshell
45802 Small White
50001 Speckled Wood
45201 * Swallowtail
50101 Wall Brown
48301 * White Admiral
46501 * White-letter Hairstreak
45401 * Wood White

OTHER SPECIES

Biological Records Centre   March 1983   RA 52

Figure 1.2 Species list/site recording card (RA52) designed for the BC/BRC recording scheme in 1983. (a) Front of card with species list, recorded species have been crossed through; (b) reverse of card with site information.

HABITAT DESCRIPTION

RURAL DOMESTIC GARDEN

GRAZED/MOWN MEADOW

DERELICT ARABLE FIELD

OWNER SEE MAP

a) Pr. & P.M. HARDING, ELSWORTH CB3 8TQ

b) J. SANDERCOCK, KNAPWELL, CAMBS

CONSERVATION (site status, management, etc.)

ALL PRIVATE LAND

DERELICT ARABLE MOWN IN JULY,

BUT FULL OF <u>RUMEX</u>, <u>SONCHUS</u>

& OTHER WASTELAND PLANTS

SPECIAL OBSERVATIONS

SMALL COPPER - FIRST RECORD IN 14 YEARS.

MANY SEEN IN GARDEN & DERELICT

ARABLE

OFFICE USE

SKETCH MAP (with scale)

This proliferation of local projects came about for several reasons, including:

1. the popularity of butterflies and increasing concern about the loss of sites and populations;
2. the development of local biological records centres;
3. the completion of recording for the BRC *Atlas* in 1982 – this left an experienced volunteer labour force with a desire to carry on recording, which seemed not to be satisfied by the revised BRC/BC scheme;
4. the publication of increasingly better and attractive identification guides throughout the 1980s.

The development of local projects has been almost completely without national co-ordination. BRC did not have the resources to support or advise local schemes, there was no co-ordinating body for local records centres and BC was primarily concerned with the national scheme. As a result, the range and quality of these projects vary greatly. The worst have acquired data which have subsequently become inaccessible, due to lack of planning, for example because a record centre ceased to operate or when an organizer died or moved out of the area. Some are concerned only with mapping distribution at ever-decreasing scales, such as 2×2 km squares (tetrads) or 1×1 km squares. Fortunately, many are closely associated with local records centres or conservation organizations (including BC) and are producing survey data with a wide range of uses.

However, the proliferation of uncoordinated local projects now makes it difficult to gain a comprehensive overview of the occurrence of species without making contact with 200 or more projects. This problem was already apparent in 1985 when the update of maps for Emmet and Heath (1989) was begun (section 1.2.2(c)).

A total of 21 local butterfly atlases had been published by September 1993 (Table 1.1). In addition, maps of individual species have been published as an interim stage in some local projects (e.g. Williams and Barker, 1993).

1.2.4   Fixed transect monitoring of butterfly abundance

(a)   *Butterfly monitoring scheme*

Fixed transect monitoring of butterflies was pioneered by Ernie Pollard and his team at Monk Wood in the 1970s through the Butterfly Monitoring Scheme (BMS). The scheme is intended to provide quantitative information on the fluctuations of butterfly numbers from year to year (Figure 1.3), so that any underlying trends can be assessed, and to detect changes from the overall trend at individual sites, caused by local factors such as management.

The scheme has been jointly funded throughout by ITE and the statutory conservation agency (since 1991 the Joint Nature Conservation Committee, JNCC). In the past 17 years the seasonal and annual abundances of butterflies have been quantified by BMS at up to 100 sites in the UK.

The methods used were described by Pollard (1977). At each site, the recorder

Table 1.1 British local butterfly atlases published before 1994

County	Author	Scale of maps (km²)	Period covered/ source of data
Berkshire, Buckinghamshire and Oxfordshire	Steel and Steel (1985)	10	1975–84
Carmarthenshire	Morgan (1989)	10 or 2	Literature
Cheshire	Rutherford (1983)	10	1961–82
Cornwall (SE only)	Frost and Madge (1991)	1	1985–90
Devon	Bristow, Mitchell and Bolton (1983)	2	1980–90
Dorset	Thomas and Webb (1984)	2	1978–84
Essex	Emmet and Pyman (1985)	10	1981–84, literature, British Museum (Natural History) and BRC files
Gloucestershire	Meredith (1989)	10 or 2	1975–88
Harrogate district	Barnham and Foggitt (1987)	1	1976–85
Hertfordshire	Sawford (1987)	2	1970– and 1984–86
Kent	Philp (1993)	2	1981–90
London area	Plant (1987)	2	Literature and BRC records
Norfolk	Hall (1991)	2	1984–88
Northumberland and Durham	Dunn and Parrack (1986)	2	–1985, literature and Rothamsted Experimental Station and Hancock Museum
Oxfordshire	Knight and Campbell (1986)	2	Oxfordshire Biological Recording Scheme
Scotland	Thomson (1980)	10	Literature and BRC files pre-1980
Sheffield	Garland (1981)	1	Sorby Natural History Society records and local collections
Shropshire	Riley (1991)	2	Mainly 1970–91
Staffordshire	R. G. Warren (1984)	10	Literature and BRC files
Suffolk	Mendel and Piotrowski (1986)	2	1983–85
Warwickshire	Smith and Brown (1987)	10	Warwickshire Biological Records Centre

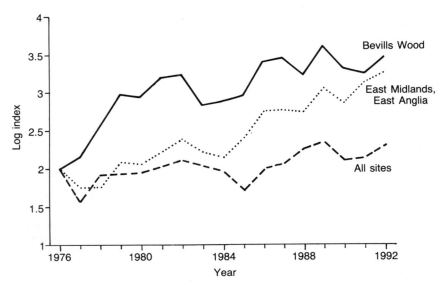

Figure 1.3 Fluctuations in the abundance of *Pararge aegeria* from data collected by the BMS. Data from a single site, a region, and all sites in the BMS are compared. This species is becoming generally more abundant and is spreading in the region. At Bevills Wood, a coniferized ancient wood, management conditions were particularly favourable for it in the late 1970s and early 1980s, but since then annual fluctuations have tended to mirror the national trends.

makes a series of counts, walking a fixed route, once a week from April to September, noting any butterflies seen within 5 m. To provide standardization, walks are carried out only between 10.45 and 15.45 (British Summer Time), and then only if the weather conditions meet specified criteria. The transect route is divided into sections, usually according to changes in habitat.

A series of reports and publications have summarized the results of the BMS, and the scheme and its methods are comprehensively reviewed by Pollard, Hall and Bibby (1986), Pollard (1991) and Pollard and Yates (1993a). Examples of short-term changes in butterfly abundance abound in the results of the BMS, many of which are summarized by Pollard and Yates (1993a) and in Pollard and Eversham (Chapter 2 this volume). In addition, similar patterns of fluctuations in abundance observed in the UK have begun to be detected in The Netherlands and Flanders, where comparable monitoring schemes have been set up (Pollard, van Swaay and Yates, 1993; van Swaay, Chapter 16 this volume).

(b) Other transect monitoring

As a result of publicity about the BMS and the publication of some early papers, growing interest led to the publication of the *Butterfly Monitoring Scheme – Instructions for Independent Recorders* (Hall, 1981). This small handbook has been widely distributed and many unofficial monitoring transects have been based on the principles of the BMS. A few of the more consistently monitored sites have subsequently been incorporated into the scheme, but most of the unofficial transects are not able to operate to the minimum standards (particularly regular weekly recording) regarded as essential for the scheme. A review in 1988 of unofficial transects (Harding and Green, 1989) showed that there was a strong concentration of sites in the south-east of England. Local co-ordination of transects outside the scheme continues to develop, for example in Dorset, Kent, the West Midlands and Hampshire/Isle of Wight (e.g. Williams and Barker, 1993).

1.2.5   Other co-ordinated projects

(a)   *JNCC Invertebrate Site Register*

The Invertebrate Site Register (ISR) was started in 1980 to identify, document and evaluate sites of importance for the conservation of terrestrial and freshwater invertebrates in Great Britain, and to provide a wide range of information and interpretation to the statutory conservation agencies (Ball, 1994). It draws on information from a variety of sources, including volunteers, so that there is inevitable overlap with local and national recording schemes. Although the ISR covers butterflies, it has been particularly directed to seek out and disseminate information on less well-studied groups.

(b)   *BC habitat survey*

BC has been running a habitat survey scheme for many years. Individual observers have been monitoring changes in habitat and butterfly populations on individual sites. This survey has followed its own methods and, although no comprehensive report on results has yet been produced, the scheme has been valuable in focusing the attention of non-professional observers on the problems of habitat change and the need for management.

(c)   *BC gardens survey*

This project collates and analyses information on the occurrence and numbers of mainly common species of butterflies in gardens from records contributed by volunteers (see Vickery, Chapter 9 this volume).

(d)   *Migrant butterflies*

The gathering of information on migrant Lepidoptera, including butterflies, was first co-ordinated in 1931 (Bretherton, 1983) and was undertaken by R. F. Bretherton and J. M. Chalmers-Hunt from 1969 until 1991. They published discursive annual summaries of records of migrants, including butterflies, mainly in the *Entomologist's Record* from 1982 to 1991. Since the death of Russell

Bretherton, the collation of information on migrant Lepidoptera has been taken over by B. Skinner.

1.2.6 Use and dissemination of data

(a) National datasets

The dataset of 238 000 records from the LRS (1967–82), is held in the BRC database using the ORACLE database management system. It is summarized in the *Atlas of Butterflies in Britain and Ireland* (Heath, Pollard and Thomas, 1984). This dataset has been widely consulted and used for conservation and research purposes (section 1.2.2(a)). Subsequent records, collected through the national BRC/BC survey and for the *Moths and Butterflies of Great Britain and Ireland* volume are held in the BRC archive, but have not yet been computerized due to lack of resources.

The annual data from the BMS are also held at BRC, where they are due to be incorporated into a fully relational ORACLE database in 1994. An annual report is circulated to all involved with the scheme. Extensive use has been made of data from the BMS for research, by ITE, JNCC and academic research workers, and is summarized in more than 30 scientific publications. This research is reviewed by Pollard and Yates (1993a) and further work is continuing.

(b) Local information

Attempts to trace local projects and their results have become increasingly difficult as the number of projects has increased in recent years. Even published results, such as annual or periodic reviews and local atlases, are often difficult to trace because of the lack of publicity outside the geographical area covered by the publication. Some enterprising specialist booksellers attempt to stock major publications, but valuable local reviews deserve a wider readership (e.g. Mabbett and Williams, 1993; Williams and Barker, 1993). Lack of conventional bibliographic details in these locally produced publications only increases their unwarranted obscurity.

The issue of access to the data from local projects is being addressed (section 1.3).

1.3   THE FUTURE NEED FOR RECORDING

Faced with the proliferation of local recording and monitoring projects during the 1980s, BRC and BC formed a joint working party, under the chairmanship of Jim Asher, to examine future policy for butterfly recording. The report of the working party (Asher, 1992) included detailed proposals for minimum standards for recording and for national co-ordination of recording effort. The report has formed a basis for subsequent discussions involving BC, BRC and the JNCC.

1.3.1 Redefining the objectives of butterfly recording

Butterfly recording of many types has been undertaken for a variety of narrow or poorly defined purposes, not the least of which being that it was seen as a good thing. In almost all cases, the results of recording have not been fully exploited for the scientific information that they contain. Even after some 10 years, new analyses are still being undertaken using the *Atlas* database at BRC which was compiled in 1982 (Prendergast *et al.*, 1993).

While acknowledging that volunteers have every right to pursue their own interests, most wish to be involved with projects which will lead to meaningful results and greater strength of knowledge about butterflies, their ecology and their protection.

As a contribution to attempts to rationalize butterfly recording, the main objectives may be defined as follows.

1. To provide knowledge of the distribution and rarity status of the butterfly fauna of the UK.
2. To identify sites of importance for butterflies and their conservation.
3. To monitor changes in the abundance and ranges of species, both at important sites and in the wider countryside.
4. To assess the factors that may be causing such changes, for example changes in environmental factors including land use, habitat management, local and national weather patterns and global climatic changes.
5. To provide a reliable and quantified basis from which to advance conservation decision-making and to advise planning and legislation at local, national and international levels.

1.3.2 Access to locally held data

(a) Changes in range

Greater access to locally held data is essential to provide regional and national overviews of changes in range. Collated data could be used, for example, to detect local changes in the ranges of species. For example, the spread of *Thymelicus sylvestris* (small skipper) in Yorkshire (Sutton and Beaumont, 1989) and *Pararge aegeria* (speckled wood) in many northern areas, such as Easter Ross and Sutherland (McAllister, 1993), have been published, but many local changes are often poorly documented and information percolates through to a national level in haphazard ways. Research on populations undergoing rapid range expansion should be undertaken to test ecological theories, and prospective research workers require up-to-date information in readily accessible forms.

(b) Development planning and site and species conservation

Few local recording initiatives are closely linked either with the local development planning system or with the statutory or voluntary conservation agencies. At least one local butterfly recording project is known to have actively withheld data from such organizations for fear of their misuse. Paranoia about the possible leakage of records to potential collectors is sadly misplaced if sites are then threatened by some form of development or change of management. Unless information is known to exist and is made accessible to those with a need to take wildlife interests into account, there must be little prospect of protecting sites and species. Safeguarding wildlife sites (even Sites of Special Scientific Interest) is difficult enough, so that all available information must be readily accessible to those with a responsibility or a commitment to protect wildlife.

1.4   RECOMMENDATIONS FOR FUTURE RECORDING

The principal recommendation of the BC/BRC Working Party (Asher, 1992) was that a new co-ordinated initiative for butterfly recording should be set up to cover Britain and Ireland, in which BC and BRC would take joint leading roles. In this section we summarize and explore the recommendations of the BC/BRC Working Party.

Future recording should be based on a regional structure for local recording, backed by local validation. It should draw on existing schemes where appropriate and be involved with developing new schemes in those areas where none currently exist.

Timely and regular feedback to individual recorders is essential if their enthusiasm is to be maintained and to enable efficient targeting of recording effort where it is most needed.

The need to co-ordinate regionally gathered data and to enable efficient feedback requires the use of computers, which are now, for the first time, widely available at modest cost. A further crucial requirement is to establish standards for recording.

1.4.1   Recommended standards for recording

Data can be interchanged and collated efficiently only if recording is carried out to a minimum set of standards. The following recommendations for a minimum set of data fields were made in the report:

1. name and address of recorder;
2. recording date, at least week and preferably day;
3. site or locality name and nearest town/village;
4. Ordnance Survey National Grid Reference – minimum 1 km square (four-figure, e.g. SP6011), but 100 m square (six-figure, e.g. SP605114) preferred;
5. habitat type(s) present on the site, using a simple agreed classification of basic land-use/vegetation types (the CORINE biotope codes provide a suitable scheme);
6. county;
7. numbers of each species observed using at least a simple scale of codes to represent numbers.

In addition, several other fields were advised:

1. evidence of breeding status from observation of early stages;
2. time at start of visit;
3. time spent at the site on the specified recording date(s);
4. weather conditions, expressed simply;
5. notes on significant changes in habitat, management or use;
6. site protection status;
7. absence of given species when searched for at an appropriate time of year under ideal weather conditions (particularly at sites where previously recorded).

The use of standardized recording forms that incorporate the minimum standards was recommended. In September 1993, 16 regional branches of BC, which cover 49 counties, were already using a common design of recording form, customized to meet their individual needs (Figure 1.4).

1.4.2 Management of data

Small but powerful microcomputers, which use standard operating systems such as MS DOS, are now widely available. The management of data has to be co-ordinated so that, even if different methods are being used to collect data, the data are able to be interchanged and transferred for wider use. All efforts should be made to establish interchange between formats and to minimize the proliferation of customized (home-brewed) data management systems.

Seven branch areas within BC, covering 21 counties, are now using the standardized software package LEVANA to enter and collate butterfly records made to the minimum set of recording standards. The wide take-up of the LEVANA package will lay the foundation within BC for a national network, called ButterflyNet, to co-ordinate and manage the collation of butterfly records originated at the BC branch level. Interfaces are being developed with other widely used software, in particular the larger RECORDER package, which is widely used by local record centres and wildlife trusts, to facilitate interchange of data in the future.

It is labour intensive to key-in data, and resources are limited at all levels. Data must be keyed-in only once; all subsequent data transfers should be by machine. The use of a limited number of data management packages, such as LEVANA and RECORDER, increases the potential for interchange of data. Clear working co-operation and data interchange between BC, BRC and other bodies, including local records centres, wildlife trusts and the conservation agencies must be strengthened and co-ordinated. However, there will have to be clear policies for the safeguard of data and their dissemination and use.

1.4.3 Standards for site monitoring

Recording at individual sites of importance for butterfly conservation usually includes transect monitoring to obtain quantitative information on species abun-

BUTTERFLY SITE RECORDING FORM

BBCS Upper Thames Branch
1992 edition

NAME: _Jim Asher_

ADDRESS: 24 Fettiplace Road
& Tel. No. Marcham, Abingdon OX13 6PL
0865 391727

YEAR: _1992_

BUTTERFLY CONSERVATION

Date(s) of visit(s): *please list overleaf*

SITE INFORMATION: If possible, please add a sketch map of the site and area visited

Site name:	SHABBINGTON WOOD / BERNWOOD FOREST	Grid ref:	SP612111
Nearest town:	OAKLEY	County:	BUCKS

HABITAT TYPES: Please tick box(es) that apply to the area visited and give any other details about habitat/land use:

Freshwater edges (lakes/rivers/canals)	2		Tree lines/ hedges	84	
Heath/ Scrub	31		Parks/ Gardens/ Churchyards	85	
Dry calcareous grassland/ Chalk down	34		Urban/ Industrial areas	86	
Meadow (unimproved)	38	✔	Fallow/ Waste land	87	
Broad-leaved deciduous woodland	41		Quarries / Chalk pits / Gravel pits / Clay pits	89	
Mixed woodland	43	✔	Road/rail verges/cuttings etc (active)	90A	
Bog/ Calcareous fen	54		Road/railway tracks (disused)	90D	
Fertilized/improved/reseeded grassland	81		Other details		
Crops	82		WOODLAND RIDES + MEADOW		
Orchards/ Plantations/ Commercial forestry	83		(BBONT)		

SUMMARY OF SPECIES SEEN

Enter details of the date and length (approx. minutes) of each visit overleaf, using the following scale for the number of each species seen: **A:** only 1 seen, **B:** 2-9, **C:** 10-29, **D:** 30-99, **E:** 100+. Give a summary below of the highest number of each species seen on any visit throughout the year (using the same codes), in column **X**. If any of the early stages (ova, larvae or pupae) of a particular species were seen, mark columns **O**, **L** or **P**. If mating is observed, mark column **M**.

	X	O	L	P	M		X	O	L	P	M
Small Skipper	C					Duke of Burgundy					
Essex Skipper						White Admiral	B				
Silver-spotted Skipper						Purple Emperor					
Large Skipper	C					Red Admiral	B				
Dingy Skipper						Painted Lady	B				
Grizzled Skipper	A					Small Tortoiseshell	B				
Wood White	B					Large Tortoiseshell					
Clouded Yellow						Peacock	C				
Brimstone						Comma	A				
Large White	D					Small Pearl-bordered Fritillary					
Small White	C					Pearl-bordered Fritillary					
Green-veined White	C					High Brown Fritillary					
Orange-tip			L			Dark Green Fritillary					
Green Hairstreak						Silver-washed Fritillary					
Brown Hairstreak	·	O				Marsh Fritillary					
Purple Hairstreak						Speckled Wood	B				
White-letter Hairstreak						Wall					
Black Hairstreak	B					Marbled White	C				
Small Copper	A					Grayling					
Small Blue						Gatekeeper	C				
Silver-studded Blue						Meadow Brown	C				
Brown Argus						Ringlet	C				
Common Blue	C					Small Heath	B				
Chalk Hill Blue						Other (specify):					
Adonis Blue											
Holly Blue											

Please return completed forms to:

Jim Asher BBCS/UTB, 24 Fettiplace Road, Marcham, Abingdon, Oxfordshire, OX13 6PL

Figure 1.4 Butterfly site recording form designed for use by the Upper Thames Branch of Butterfly Conservation. This type of form has been made available to many BC branches since 1992, through continued collaboration between BC and the BRC.

Recording Date:	Day:	21	27	16	31	2	6	20
	Month:	6	6	7	7	8	9	9
Length of visit (mins):		60	45	65	60	40	90	60
Weather conditions: (Poor/ Moderate/ Ideal)		M	M	I	M	I	P	P
Small Skipper		B	B	B	C	B		
Essex Skipper								
Silver-spotted Skipper								
Large Skipper		C	C	B	B	A		
Dingy Skipper								
Grizzled Skipper		A						
Wood White		B						
Clouded Yellow								
Brimstone								
Large White		A	B	B	B	D	B	
Small White				B	B	C		
Green-veined White			A	C	B	C	A	B
Orange-tip		L						
Green Hairstreak								
Brown Hairstreak					O			
Purple Hairstreak								
White-letter Hairstreak								
Black Hairstreak		B	A					
Small Copper			A					
Small Blue								
Silver-studded Blue								
Brown Argus								
Common Blue		B		C	B	B	B	A
Chalk Hill Blue								
Adonis Blue								
Holly Blue								
Duke of Burgundy								
White Admiral			B	B				
Purple Emperor								
Red Admiral		B	A	B	A			A
Painted Lady						B		A
Small Tortoiseshell			B	B	B	B		
Large Tortoiseshell								
Peacock					B	C		A
Comma			A					
Sm.Pearl-bordered Frit.								
Pearl-bordered Frit.								
High Brown Fritillary								
Dark Green Fritillary								
Silver-washed Fritillary								
Marsh Fritillary								
Speckled Wood		B		B	A	B		B
Wall								
Marbled White		B	C	B	B	B		
Grayling								
Gatekeeper				B	B	C		
Meadow Brown		C	C	C	B			
Ringlet		B	C	C	B	A		
Small Heath		A		B	A			

Comments:

Figure 1.4 Continued.

dance, as a means of assessing habitat change and the effectiveness of site management.

The BMS continues to be funded jointly by ITE and JNCC. In the current five-year programme (1992–97) the scheme is expected to increase the number of sites and the geographical spread of sites and coverage of habitats. Modified transect techniques for use in the wider countryside are being examined from 1993. Further work is being done on the analytical techniques used on BMS data, which may lead to some re-analysis of past data and re-examination of results.

The usual methods used for transect monitoring, normally based on the BMS techniques (Hall, 1981), already include the basic features of the new recording standards proposed in section 1.4.1. The main emphasis of the report, therefore, was to ensure that transect data can be made readily available to regional centres in a form that can be stored and converted for use with a minimum of additional work. Transect monitoring should be started only where a local recorder has the motivation and ability to maintain it over several years.

Transect monitoring has different methods and objectives from distribution recording. It may not record all species present at a site (but should be encouraged to do so where practicable, although transect recording cannot be applied, for example, to tree-canopy species) and the status of a species along a transect may not be representative of the site as a whole. It is, however, a good means of monitoring long-term changes. The resultant data also provide useful contributions to distribution mapping.

There is currently no means of handling all the data from all the sites which are being monitored outside the BMS, and data are often not truly comparable with BMS data, for a variety of reasons (e.g. short transect routes, selected species only, irregular recording). However, as BC branches increase their use of computers to store and analyse transect data, it is important to ensure that data from such transects can be brought into this wider recording initiative. A separate project may be required to co-ordinate these BMS-style transects and their data.

1.4.4 International collation of records

The geographical unit for data collation should cover Great Britain and Northern Ireland, together with the Isle of Man and the Channel Islands. The Republic of Ireland should be included if the co-operation and agreement of active lepidopterists in the country can be obtained.

There are already initiatives aimed at establishing wider European links for the collation and use of information on butterflies. Through the Standing Committee to the Berne Convention, the European Invertebrate Survey is actively involved with preparing data sheets for key species which are considered to be threatened throughout their continental range. Collaboration between the UK BMS and the *De Vlinderstichting* scheme in The Netherlands has already resulted in one publication (Pollard, van Swaay and Yates, 1993) and further work is planned for 1994.

There are individual projects to collect data and to map the distribution of butterflies, at national or provincial levels, in many European countries. No concerted attempt has yet been made to collate this information on a Europe-wide scale, as has been done for flowering plants, amphibians and reptiles, birds, mammals and myriapods. Without collated, Europe-wide information, there must always be some uncertainty about statements on the rarity and vulnerability of species, or on the spread of species. Perhaps the post-war spread, in continental Europe, of *Araschnia levana* (the aptly named European map butterfly) should be regarded as some form of sign to us to broaden our horizons!

1.4.5 Representation of data on maps

It is recognized that the use of simple dots on maps, to illustrate the presence of species, is inadequate. A range of symbols should be used to improve the information content of maps, for example to illustrate colony strength, and to avoid misleading conclusions about the status of species. The use of a variety of symbols in computer-generated maps has already been successfully demonstrated where data offer opportunities to analyse other factors, such as the number of colonies or proven breeding.

1.4.6 A national landmark project: the Millennium Atlas

Proposals are being considered for a new atlas of the butterflies of Britain and Ireland to mark the status of butterflies at the end of the twentieth century, based on recording effort over the last 3–5 years of the 1990s, and incorporating the recommended standards for data and new mapping concepts.

A major high-profile project of this kind will give impetus and focus to the initiative to co-ordinate butterfly recording and will lay foundations for a network to maintain a continuously updated butterfly database. It is anticipated that BC, jointly with BRC, will manage this Millennium Atlas project, with support from JNCC. Data collection will be organized on a regional basis, exploiting ButterflyNet (section 1.4.2) to collate data originating within BC and interchanging data with BRC on computer media. The active participation and co-operation of other recording centres will be strongly promoted, so that all sources of data will be tapped.

This project will aim for a very high level of coverage and will require the involvement of many recorders and encouragement for new recorders to become involved. A simple universal set of recording instructions is being developed, to be issued regionally. By collating data annually, it will be possible to track progress in widely issued annual reports and therefore to direct recording effort to under-recorded areas, both on a regional and on a national scale. A publicity drive to draw attention to the objectives and benefits of the project and a fund-raising campaign aimed at establishing sponsorship to meet project costs will be required. New recorders wishing to help with the project will be encouraged to contact BC or its local branches.

It is the firm long-term aim of BC and BRC that the momentum gained by the Millennium Atlas project will be used to maintain an up-to-date database of butterfly records into the future to meet the ever-increasing demand for data on species status.

1.5   CONCLUSIONS

The value of recording and monitoring in providing information essential to the development of conservation strategies and in setting priorities and evaluating results is increasing. The demand for data for environmental assessments and post-development land use is also growing rapidly, and, all too often, demand outstrips supply.

It is now widely recognized that there is a valuable resource of volunteer expertise available in British conservation organizations, and a willingness for co-operation between voluntary and professional bodies. A good start has been made in defining and disseminating standards for recording and monitoring butterflies, and in setting up a new basis for national data collation. Much work is still required to bring these early successes through to fruition.

Technology now exists for the collection, collation, analysis and interchange of data on a scale never before feasible. This brings with it the challenge of managing the rapid expansion of systems and maintaining compatibility.

The development, co-ordination and management of new initiatives which are badly needed requires a strong commitment by all parties and secure financial resources. It is recognized that government sources alone are unlikely to be able to provide all the financial backing required, and sponsorship will probably play a key role in the success of the schemes discussed here. A co-operative approach by JNCC and ITE to attract funding from government sources, jointly with a complementary approach by BC to seek industrial sponsorship, is seen as the only realistic way to secure the resources needed to promote and develop the Millennium Atlas project. Any potential funding body is going to want to know that the results will be useful.

Past experience has shown that once data and publications are available they are used frequently and widely. We are certain that the Millennium Atlas project offers the best method for updating regional and national information on the occurrence of butterflies in Britain, and for making the results widely available.

1.6   ACKNOWLEDGEMENTS

Almost all butterfly recording in Britain and Ireland is carried out by volunteers; without their contribution our present knowledge would be inadequate and any plans for future recording will be almost entirely dependent on their continued commitment.

CHAPTER 2

Butterfly monitoring
2 – interpreting the changes

E. Pollard and B.C. Eversham

2.1 INTRODUCTION

In Chapter 1 of this volume, Harding, Asher and Yates describe the methods that have been used to obtain information on changes in the distribution and abundance of British butterflies. In this chapter we use data mainly from the two major schemes, the Lepidoptera Recording Scheme (LRS) and the Butterfly Monitoring Scheme (BMS), to highlight the contrasting fortunes of two groups of butterflies.

The LRS has shown that many British butterflies have contracted sharply in their ranges in recent decades (Heath, Pollard and Thomas, 1984). These range contractions seem to have occurred most markedly from about 1960, but in many cases began much earlier. Heath, Pollard and Thomas listed 18 species that have undergone major contractions of range over the past 150 years. In addition to these 18 species, four species became extinct during the same period. This account relates largely to lowland Britain, where the decline of butterflies has been most severe.

Heath, Pollard and Thomas noted that a further six butterfly species had contracted in range during the nineteenth century, but subsequently spread again to occupy much of their previous ranges. More detailed evidence now available, especially from local (usually county) studies in the Midlands and northern England, shows that the number of species that contracted and later re-expanded their ranges was underestimated. In addition, the more recent local accounts suggest that the re-expansion of range of some of these butterflies has quickened during the 1980s.

Ecology and Conservation of Butterflies Edited by Andrew S. Pullin.
Published in 1995 by Chapman & Hall. ISBN 0 412 56970 1

2.2   WIDESPREAD AND LOCALIZED BUTTERFLIES: A DICHOTOMY IN THE BRITISH SPECIES

2.2.1   Evidence for the dichotomy

A dichotomy in distribution patterns of butterflies in lowland Britain is suggested by distribution data and also by the pattern of occurrence of butterflies at sites in the BMS. For example, a recent atlas of butterflies in Norfolk (Hall, 1991), a county with its land use dominated by agriculture but also with semi-natural areas, such as coastal dunes, heaths and wetlands, clearly shows two groups of species (Figure 2.1a).

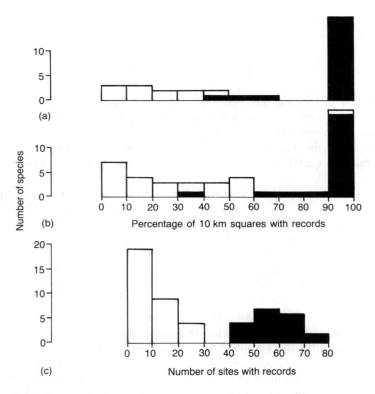

Figure 2.1   Number of 10 km squares in which resident butterfly species occur in (a) Norfolk and (b) Berkshire, Buckinghamshire and Oxfordshire. (c) Number of sites in the BMS at which species occurred in 1990 (results for other years are similar). Three regular immigrant species are excluded. Solid area: species categorized as widespread in Table 2.1. Open area: species categorized as localized in Table 2.1. The distribution patterns suggest a clear dichotomy between the two categories; the species that do not occur in their expected histogram columns are *Thymelicus lineola*, which is relatively sparsely recorded in both (a) and (b), and is probably under-recorded, and *Melanargia galathea*, which is widespread in (b) and occurs in the right-hand column. In (c), *T. lineola* records are not separated from those of *T. sylvestris*.

One group of 17 resident species was recorded in virtually all of the 69 10 km squares in the county, while 12 of the 14 other species have been recorded in fewer than 30 squares. In Berkshire, Buckinghamshire and Oxfordshire (considered together in one account: Steel and Steel, 1985) the dichotomy is also clearly evident (Figure 2.1b), although the rarer species occupy more 10 km squares than is the case in Norfolk.

Local studies where recording has been intensive over relatively short periods are suited to analyses of this type, as the distribution of most butterflies has changed little. In addition, over a long period, occasional sightings of rare species tend to obscure the small number of 10 km squares with breeding populations. Nevertheless, even over the very long period (*c.* 1850–1982) in which data are available from the LRS, a similar dichotomy emerges in areas of intensive agriculture (Figure 2.2).

For example, in the East Midlands 16 common and widespread species have been recorded in more than 80% of 10 km squares, while all of the other 34 species have been recorded in fewer than 60% of squares.

The occurrence of species at sites in the BMS shows a similar separation into those that are present at most sites within their ranges and those that are much rarer and have localized distributions (Figure 2.1c). The division is evident at the monitored sites in spite of the fact that they are predominantly nature reserves, with a disproportionate number of the rarer species.

Of the resident butterflies of lowland Britain, 29 species can be categorized as having localized distributions (Table 2.1); that is, they usually breed in discrete areas of semi-natural biotopes and are rarely seen away from their breeding areas.

A further 20 resident butterfly species can be regarded as characteristic of the farmed landscape of lowland Britain. In several cases the ranges of these 20 species are restricted to more southerly areas of Britain, but, within their ranges, they are almost invariably widespread. Any new sites that become available within the ranges are likely to be colonized.

Few, if any, ecological groupings are completely clear-cut. This suggested division of butterflies in lowland Britain is no exception, although most species fall readily into one of the two categories. Exceptions which are less clear include species such as *Pararge aegeria* (speckled wood), *Anthocharis cardamines* (orange tip), *Melanargia galathea* (marbled white) and *Quercusia quercus* (purple hairstreak), which are widespread in parts of their ranges, but more localized in others. Some species classed here as localized, such as *Melitaea cinxia* (Glanville fritillary), appear to move relatively frequently between populations in a restricted area (Thomas and Lewington, 1991). In effect, this species and several other species classified here as localized may occur as interconnected populations within restricted parts of their ranges. It may also be argued that species such as *Lycaena phlaeas* (small copper) and *Polyommatus icarus* (common blue) are no longer widespread in areas of highly intensive arable cultivation. However, these last species seem able to colonize virtually any patches of countryside that become suitable for them.

2.2.2 Causes of the dichotomy

The presence of a breeding population of a species at a site requires, as a first essential, the presence of the food-plant. Thomas (1984a, 1991) has emphasized that the food-plant must be in the particular conditions required by the butterfly

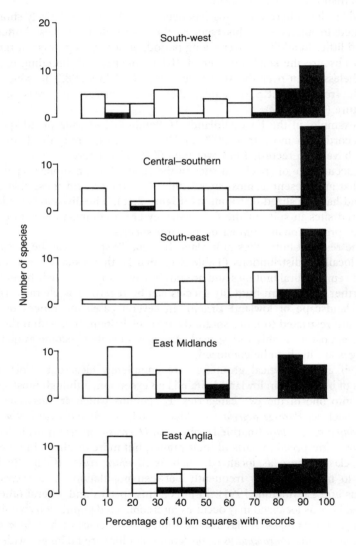

Figure 2.2 Percentage of 10 km squares in which resident butterfly species occur in five regions of England. Solid and open areas are as in Figure 2.1. The dichotomy is most pronounced in the more intensively agricultural areas of eastern England. The 'widespread species' occur in relatively few squares. In most areas these are either *Thymelicus lineola*, which is probably under-recorded and has an easterly distribution, or *Pararge aegeria*, which is widespread in the west of England but less so in the east.

Table 2.1 Almost all of the resident British butterflies of lowland Britain can be categorized as either common and widespread in the general countryside or rare and localized, restricted to discrete areas of semi-natural biotopes such as woods, heaths, wetlands or downs

Common and widespread species	Rare and localized species
Thymelicus sylvestris*	Thymelicus acteon
Thymelicus lineola*	Hesperia comma[†]
Ochlodes venata*	Erynnis tages
Gonepteryx rhamni	Pyrgus malvae
Pieris brassicae	Papilio machaon
Pieris rapae	Leptidea sinapis[†]
Pieris napi	Callophrys rubi
Anthocharis cardamines*	Thecla betulae[†]
Lycaena phlaeas	(Quercusia quercus)
Polyommatus icarus	Satyrium w-album
Celastrina argiolus*	Satyrium pruni
Aglais urticae	Cupido minimus[†]
Inachis io*	Plebejus argus[†]
(Polygonia c-album)*	Aricia agestis
(Pararge aegeria)*	Lysandra coridon
Lasiommata megera*	Lysandra bellargus[†]
Pyronia tithonus*	Hamearis lucina[†]
Maniola jurtina	Ladoga camilla*
Coenonympha pamphilu	Apatura iris[†]
Aphantopus hyperantus*	Boloria selene[†]
	Boloria euphrosyne[†]
	Argynnis adippe[†]
	Argynnis aglaja[†]
	Argynnis paphia[†]
	Eurodryas aurinia[†]
	Melitaea cinxia
	Mellicta athalia[†]
	(Melanargia galathea)[†]
	Hipparchia semele

In only a few cases (in parentheses) is the appropriate category doubtful. In general the common and widespread butterflies have flourished in recent decades, and many have expanded in range (indicated by *). In sharp contrast, many of the rare and localized butterflies have contracted in range and become rarer (indicated by [†]). Omitted from the lists are *Lycaena dispar* and *Maculinea arion*, which occur as reintroduced populations; *Carterocephalus palaemon*, which is extinct in England but occurs in Scotland; and *Nymphalis polychloros*, the status of which as a resident British species is doubtful. Four British species with northerly distributions are also omitted. Evidence on range changes is from a variety of sources, as discussed in the text.

and that the range of such conditions is often very narrow. Thus a possible explanation for the dichotomy is that the localized butterflies find their requirements only in few isolated patches, while the requirements of common butterflies are widely available.

Localized and widespread butterflies may also differ to some extent in other respects. The butterflies we have categorized as common and widespread tend to have more larval food-plants, more generations in a year and greater mobility than the localized butterflies (Bink, 1992; Hodgson, 1993). Nevertheless, in the intensively managed landscape of Britain, the distribution of a species seems to depend primarily on the way in which man has modified the vegetation and provided food-plants in suitable conditions. For example, the present abundance and widespread distribution of butterflies that feed on *Urtica dioica* (stinging nettle) is possible because this plant is favoured by disturbance and nutrient enrichment in uncultivated corners of farmland. These butterflies would almost certainly have been very much rarer before humans dominated the countryside.

The dichotomy may indicate that there is a critical point at which, as isolation of habitats increases, free movement through the countryside becomes too hazardous and so mobility is selected against. Thus the individuals of isolated populations may become increasingly sedentary (Dempster, King and Lakhani, 1976; Dempster, 1991). This possibility merits further study.

In summary, there is a division of butterflies in lowland Britain into two groups, either highly localized or widely distributed; this division is in large part a reflection of land-use patterns. The large majority of the land area of the lowlands comprises agricultural land, typically of arable and grass fields with field boundaries, which are often hedges, road verges, copses and other small uncultivated fragments. Within this agricultural matrix, and occupying a small proportion of the land area, are the island biotopes. The division of British butterflies is into those that can find the resources they require in the matrix (and often also on the islands) and those which need resources which are restricted to the 'islands'.

2.3 CONTRASTING FORTUNES OF WIDESPREAD AND LOCALIZED BUTTERFLIES

The records of the LRS span some 130 years, and for most of this period they were not collected for the specific purpose of recording distributions. Thus there is geographical and temporal variation in the intensity of recording and, in the early years, a bias in museum and literature records in favour of recording the rarer species (for a general review of the problems and some solutions, see Eversham *et al.*, 1993). Thus, the assessment of changes in status of common butterflies is not straightforward and, in the context of this study, the data are not adequate to establish the timing of the retreat and subsequent expansion of some of the widespread species.

Nearly all of the 18 species listed by Heath, Pollard and Thomas (1984) as showing severe declines come into our category of localized butterflies (Table 2.1). A possible exception is *Nymphalis polychloros* (large tortoiseshell), the present status of which as a resident British species is doubtful. *Carterocephalus palaemon* (chequered skipper) formerly occurred in localized populations in England, but is now extinct there. A few species, such as *Hesperia comma* (silver-spotted skipper) (Thomas and Jones, 1993), have shown limited recoveries in the past few years, but in general the declines continue (Warren, 1993a).

2.3.1   Evidence from local distribution records

The most striking evidence on recent expansion of the range of butterflies is from local studies (e.g. Harrison and Sterling, 1985; Dunn and Parrack, 1986; Sutton and Beaumont, 1989; Rimington, 1992; and others listed by Harding, Asher and Yates in Chapter 1 of this volume). This evidence is the simple documentation of the arrival and spread of a species in a region, or regions, of Britain. Eleven of the 12 species which these local studies show to have expanded in range in recent decades are categorized as widespread (Table 2.1). The twelfth species, *Ladoga camilla* (white admiral), is a localized woodland butterfly.

2.3.2   Evidence from the Butterfly Monitoring Scheme

Evidence from the BMS on changes in abundance at individual sites suggests that trends in the abundance of the two categories of butterflies have continued in very recent years. Nearly 1300 populations have been monitored for 8 or more years between 1976 and 1992. If the species are divided as in Table 2.1, the widespread species have shown a preponderance of significant increases and the localized species a preponderance of significant declines (Pollard and Yates, 1993a) and the difference between the categories is highly significant. Not all of the widespread species which have increased at monitored sites have the potential for expansion of range, because their ranges already occupy virtually the whole of Britain.

These data also show that increases at monitored sites have been largely concentrated in eastern England, where colonization of many new sites has occurred (Pollard, Moss and Yates, 1994). This easterly bias is shown most markedly by *Polygonia c-album* (comma), *P. aegeria*, *Maniola jurtina* (meadow brown) and *Aphantopus hyperantus* (ringlet), but occurs to a lesser extent in several other species. In the example illustrated, that of *M. jurtina* (Figure 2.3), 80 sites have data for periods of 8 or more years.

Of these sites, 18 show significant population increases and all except one are in eastern England. Only five sites show significant declines. *Maniola jurtina* is one of the species which appear to be flourishing, but has little scope in Britain for expansion of range.

2.4   CAUSES OF THE DECLINE OF LOCALIZED BUTTERFLIES

There is a consensus of views that the decline of the localized butterflies in Britain has been caused largely by the loss of areas with suitable food-plants, or by changes in the character of such areas (e.g. Heath, Pollard and Thomas, 1984; Thomas, 1984a, 1991; Warren, 1992b, 1993a). There is little doubt that this widely accepted view explains a large part of the decline of many British butterflies.

There are now a number of autecological studies (e.g. Thomas, 1974, 1983b; Warren, 1987b,c; Warren, Pollard and Bibby, 1986), including several described in this volume, which provide details of the particular conditions required by individual species. Frequently, the cause of losses has been shown to be the cessation of traditional types of land use, such as coppicing of woodland and grazing of downland. The food-plants often survive on sites where species have disappeared, but the specific conditions required by particular butterflies have been lost.

The rate of decline of some butterfly species has been spectacular. For example *Argynnis adippe* (high brown fritillary) was described as 'not a rarity' in a field guide to butterflies in the late 1970s (Goodden, 1978), but is now considered a vulnerable species in Britain (Warren, Chapter 14 this volume). The decline of this butterfly, and many others, in Britain and in other European countries (Pavlicek-van Beek, Ovaa and van der Made, 1992) is a major cause of concern.

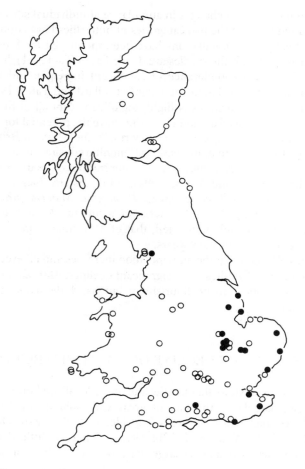

Figure 2.3 Sites in the BMS where populations of *Maniola jurtina* have been monitored for 8 or more years. Closed circles indicate significant increases in abundance. As with several other species, there is a strong tendency for populations in the south-east and east of England to have increased in abundance.

Their survival depends to a large extent on the appropriate management of areas where populations survive. In Britain, many of these sites are nature reserves. Although the degree of success in managing nature reserves for rare butterflies has in the past been poor (e.g. Thomas, 1984a), the success rate has improved in recent years.

The expansion of the woodland butterfly *L. camilla* is consistent with known changes in woodland, as well as with warm summers (Pollard, 1979), and this exception tends to 'prove the rule' that changes in biotopes have been of great importance. *L. camilla* thrives in relatively shady conditions, such as occur when unmanaged coppice grows tall, and this type of woodland has become much more widespread as coppicing has declined.

In spite of some recent successes, the conservation of rare butterflies on nature reserves presents many problems. Research is needed to establish the requirements of species, and, even then, management may be expensive and the desired vegetation difficult to create or maintain. Monitoring is necessary to ensure that a population is not in slow decline. Even where short-term success is achieved, long-term commitment is required. Once a species becomes restricted to isolated sites, further extinctions, perhaps caused by extreme weather, are unlikely to be offset by natural colonizations.

There is some evidence that the mobility of some butterfly species has declined as populations have become more isolated (Dempster, King and Lakhani, 1976; Dempster, 1991), thus exacerbating the difficulty of conservation and strengthening the division of butterflies into widespread and localized species. While little is known of the expected rates of extinction of localized populations, occasional re-introductions are likely to be necessary, and these also require considerable organization.

In summary, there is no doubt that a substantial proportion of the British butterfly species must, in the longer term if not immediately, be regarded as threatened. These butterflies characteristically occur in isolated populations and management of their biotopes is usually required for their conservation.

2.5 THE EARLY CONTRACTION AND RECENT EXPANSION OF RANGE OF WIDESPREAD BUTTERFLIES

2.5.1 Summary of the changes

As discussed above, during a period when a number of rare butterflies in Britain have declined sharply, some 11 species, mostly already widespread within their ranges, have flourished and spread. Most of these species underwent a period of range contraction during the nineteenth century.

The timing and pattern of the changes varies in detail from species to species, and in most cases has not been well documented. However, some exceptions follow.

The most extreme example is that of *P. c-album* documented, by Pratt (1986–87). This species was widespread, reaching as far as southern Scotland in the early nineteenth century. After a long period of decline, with temporary minor recoveries, it reached a low point in the early years of the twentieth century when it was restricted to an area on the borders of England and Wales. Since then, it has expanded its range again, this time with periods of minor withdrawals, and now occupies most of England and Wales as far north as Yorkshire. Its recent spread has been monitored in particular by some of the regional studies and by the BMS. In the latter it has increased significantly in abundance, especially in eastern England, and has colonized several sites close to the current northern edge of its range (Pollard and Yates, 1992).

The contraction and re-expansion of *P. aegeria* has also been well documented (Downes, 1948; Chalmers-Hunt and Owen, 1952, Barbour, 1986b; McAllister, 1993). This species was probably continuously distributed in suitable biotopes throughout England and Wales in the early nineteenth century. During the latter years of that century and the early years of the twentieth century it declined dramatically and became virtually restricted to two areas, one in central southern and south-west England and Wales and the other in western Scotland. The re-expansion seems to have started just prior to 1940 and the most recent information suggests that it is still continuing. *Pararge aegeria* has colonized several monitored sites in eastern England in the late 1980s and early 1990s (Pollard and Yates, 1993a).

The recent, very rapid spread of *Celastrina argiolus* (holly blue) may be of a different character from the other cases considered. *Celastrina argiolus* is univoltine in restricted northern areas of Britain and these populations seem to be relatively stable and permanent. In the south, the species is bivoltine and these butterflies have spread to much of the north of England in the early 1990s. The expansion has been over a very short period (1989–92) and is an extreme example of numerous 'explosions' of abundance which seem to be characteristic of *C. argiolus*. Since the start of the monitoring scheme in 1976, there have been three such periods of abundance, which in each case were evident first in the south and south-west of Britain and were followed by an apparent, temporary, spread north and east (Pollard and Yates, 1993b). Whether or not the present range will be maintained is, of course, uncertain, but precedents suggest that it will soon be lost from those northern and midland areas where it has not had long-standing univoltine populations.

2.5.2 Causes of the changes

The early contraction and later re-expansion of ranges of many of the widespread butterflies may have been caused by a single fluctuating factor, but equally it is possible that different factors have been involved during the two periods. In most cases, the timing of the changes is not known with precision, but they seem to follow a broadly similar chronology. Declines typically occurred during the middle to late part of the nineteenth century and may have continued in the early years of the twentieth century; expansions of some species are reported as having begun as early as 1910–20 (*P. c-album*), but more usually the period of expansion seems to have been from about 1940 to the present, with perhaps an acceleration in the past decade.

(a) Changes in the countryside

We have emphasized loss of and changes in particular biotopes as the likely cause of the declines in abundance and contractions of range of the rare and localized butterflies. Is there evidence that changes in the farming countryside have had equivalent effects on the abundance and ranges of the butterflies that breed there?

Over the past 200 years, the farming landscape has undergone profound changes, including the main period of the Georgian enclosures and the recent losses of many of the hedges that were planted at that time. More generally, in prosperous times, the area under cultivation has increased, drainage improved, stocking rates increased and land use in general has become more intensive; in periods of depression the converse has occurred. Underlying this periodic flux of agricultural prosperity and consequent landscape change, there has been the increasing application of science and technology to farming. It is difficult to imagine that such changes as hedgerow loss, improved drainage, and the increased use of herbicides and pesticides could have been other than detrimental to most butterflies.

It is striking, therefore, that the major intensification of farming during and after the Second World War, and particularly during the 1960s to 1980s, has seen expansion of the ranges of some farmland butterflies. It is possible that a few species, such as the nettle-feeders, have benefited from the increased use of fertilizers, as mentioned above, but in general it is difficult to reconcile the pattern of change in agriculture with that of butterfly ranges. For example, species of hedges, ditches and other marginal areas, such as *Thymelicus sylvestris* (small skipper), *Pyronia tithonus* (gatekeeper) and *A. hyperantus*, seem to have prospered in spite of changes in farmland rather than as a result of them. Indeed, it is certain that populations have been lost within the ranges of such butterflies, for example when hedgerows have been removed, during a period in which expansion of range has been in progress.

In summary, unlike the rare, localized butterflies, it does not seem that the changing status of the common, widespread butterflies is related in a simple and direct way to changes in their biotopes.

(b) Weather

Several authors (e.g. Dennis, 1977; Heath, Pollard and Thomas, 1984; Turner, Gatehouse and Corey, 1987) have drawn attention to the southerly distributions of many butterflies in Britain, and have suggested that summer temperature is likely to be the factor responsible for this restriction of range. Recent analysis of results from the BMS (Pollard, 1988; Pollard and Yates, 1993a) also shows a strong correlation between summer warmth and increase in numbers of several butterfly species. Of the 'expanding' butterflies, such significant relationships with summer weather have been shown for *T. sylvestris*, *P. tithonus*, *C. argiolus*, *P. c-album*, *Lasiommata megera* (wall brown) and *P. aegeria*.

These analyses do not suggest that winter weather is of great importance, although dry winters may be beneficial. However, for some species, such as

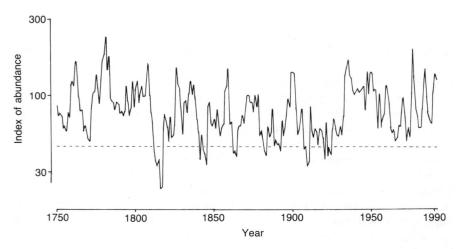

Figure 2.4 Reconstruction of changes in abundance of *Pyronia tithonus* since 1750, based on the predicted response to temperature. Many equations were calculated using data from the BMS and monthly weather records over the period 1976–91. The best-fitting equation included the previous butterfly index value and mean June, July, August temperature. This equation was then run with historical temperature data. The dotted line is used to highlight periods of predicted low numbers. The pattern of abundance is in broad accord with known changes in distribution; *P. tithonus* declined through the nineteenth century and began a gradual recovery from about 1940 (note log scale).

P. aegeria, high rainfall in spring and early summer is associated with an increase in numbers. This, and other species, seem to be adversely affected by drought, presumably because of desiccation of food-plants.

Although the BMS has now been in progress for 18 years, this is a short period in which to establish the influence of weather on butterflies with sufficient precision to allow the confident prediction of effects. It is likely that the relationships are complex, and several different factors may need to be included in realistic mathematical models. Nevertheless, we can use the simple models currently available to make a start with such studies.

Here we consider one broad measure of summer temperature (mean June–August temperature), and use it with historical weather data (Manley, 1974; Met. Office, 1976–92) to 'reconstruct' past changes in abundance of *P. tithonus*, one of the species that has retreated and advanced over the past 150 years. Over the monitoring period, the fluctuations of this species have been similar to those of several other butterflies (Pollard and Yates, 1993a); thus the results of this modelling may have wider application.

The preliminary results presented here (Figure 2.4) are, as might be expected, not conclusive, but do show features of interest.

Our model suggests that weather conditions for the 'national' *P. tithonus* population were generally good in the late eighteenth century, intermittently poor or very poor from about 1810 to 1930, and then generally good until the present.

There is thus a very broad agreement between the fluctuation of the *P. tithonus* population and summer temperatures. Given the known limitations of the model, that the enormous changes in character of the farming countryside are ignored and that the historical weather data are themselves collations and subject to error, this agreement is encouraging; suggesting that summer temperatures have played some part in the changes for this species and, by analogy, others that have shown the same general trends. The results do not simply reflect average summer temperatures over the period; extreme highs and lows of abundance depend on sequences of warm and cool summers, respectively. The occurrence of series of cool summers in the nineteenth century may have been the cause of sharp contractions of range.

In summary, this limited evidence suggests that weather may have played a role in the distributional changes of some widespread British butterflies. Whether it has played the major role is unclear; as further data accrue in the BMS, more ambitious analyses will become possible and the role of weather should be clarified further.

(c) Other factors

From time to time, changes in butterfly numbers have been attributed to a wide variety of causes. In the 1960s, pesticides were often considered to be important causes of decline of farmland butterflies; however, no evidence was then available on the nature and extent of any such declines, and subsequent experimental studies suggested that an effect of pesticides was unlikely (Moriarty, 1969). As has been described, several of the farmland butterflies, which would be expected to be most affected, have in fact expanded in range.

Atmospheric pollution has been considered, solely on circumstantial evidence, to be of potential importance (Barbour, 1986a). Similarly, Pollard and Yates (1993a) have speculated that raised levels of carbon dioxide in the atmosphere, together with deposition of atmospheric nitrogen compounds (originating mainly from vehicle exhausts and from agriculture), might affect butterflies via their food-plants. Again, there is no direct evidence for such effects, although the tendency for increases in the abundance of widespread butterflies to be greatest at monitored sites in the east of England may suggest a role for some form of pollution.

2.6 CONCLUSIONS

We have discussed two very different trends in the distribution and abundance of British butterflies. On the one hand there is a group of localized butterflies, generally restricted in lowland Britain to areas of semi-natural countryside; these butterflies have been in severe decline. It seems likely that, except on some nature reserves managed for their conservation, these declines will continue. On the other hand there are widespread butterflies, generally able to breed within or on the fringes of the farmed landscape, which either show little sign of decline, or have, in many cases, been expanding in range in recent years.

The decline of the localized butterflies is thought to be largely the result of loss of areas that provide suitable food resources for the larvae. The reason for the success of the widespread species is not clear, although it seems likely that weather may have played a role both in the recent expansion and in an earlier retreat to more southerly areas during the nineteenth century.

The expansion of the widespread species in recent years suggests that, where suitable resources have been available, other conditions have been favourable for butterflies. Thus the decline of the localized butterflies may be even more serious than has been thought; in most cases they have failed to respond to conditions which seem to have been favourable for other butterflies. It is possible that the potential for expansion has occurred, but suitable sites have been too distant from existing populations for individuals to locate them. Thomas and Jones (1993) have described such a limitation to the spread of *H. comma* on the chalk in southern England. This species has recently benefited from an increase in grazing of downland by rabbits (and perhaps also from favourable weather), but suitable sites separated by more than 10 km from existing sites have not been colonized. Thomas and Jones also emphasized the importance of groups of populations within a limited area, so that sites where populations become extinct have a good chance of recolonization when they become suitable again.

Climatic warming is likely to cause changes in the distribution and abundance of butterflies. Such changes might be of an order far greater than has been experienced since records have been kept. Conservation measures must, in such circumstances, be very flexible and must be adjusted as changes in the status of butterflies become evident. The role of distribution recording and monitoring will then be even more important than at present.

2.7 ACKNOWLEDGEMENTS

We are grateful to the Joint Nature Conservation Committee for their support of the Biological Records Centre and the Butterfly Monitoring Scheme. In particular we thank Henry Arnold (BRC), Tina Yates (BMS) and the very many individuals who have contributed to the results by field recording.

CHAPTER 3

Butterfly mobility

T.G. Shreeve

3.1 INTRODUCTION

Of the 55 species of butterfly which are resident within the British Isles, the majority have undergone changes in range (overall geographical occupation) and distribution (occurrence within range) within the past 150 years. Almost all have reduced distributions, but, more importantly, five species have become extinct and 18 have undergone major range contraction. In only eight species have there been consistent patterns of range expansion during the past 50 years (Pollard, 1991; Warren, 1992b). This general pattern of decline is matched in other areas of the Palaearctic (Kudrna, 1986; van Swaay, 1990). It is also the general trend for the Nearctic, where many formally common species are becoming increasingly scarce, especially in the region of the Atlantic and near Atlantic coast. These declines, which have been more rapid within the past 50 years than previously, are generally ascribed to the loss of habitat associated with changes of land use compounded by habitat fragmentation and isolation (Warren, 1992b). Other factors, such as climate change (Pratt, 1986) and pollution (Barbour, 1986a), may also be implicated but are less easily identified as causal factors. However, for those species which have recently expanded their range, or continue to do so – *Anthocharis cardamines* (orange tip), *Ladoga camilla* (white admiral), *Polygonia c-album* (comma), *Pararge aegeria* (speckled wood) and *Pyronia tithonus* (gatekeeper) – habitat, and especially climate, have key roles in their range expansion (Pollard, 1979, 1991; Barbour, 1986b; Pratt, 1986; Pollard and Eversham, Chapter 2 this volume).

Given the general decline of most species, much conservation effort is placed on maintaining species *in situ* with considerable debate about the merits of reintroduction. This last is a potentially contentious issue. The rationale for reintroduction is usually based on one of two premises that are not necessarily exclusive. The first is that, for certain species, existing populations are now so isolated and/or fragmented that natural recolonization of suitable habitats will be such a slow process that recolonization rates will not match extinction rates. The second is that candidate species for reintroduction have such poor mobility that natural recolonization will not occur. In Britain reintroductions are currently being carried out and

Ecology and Conservation of Butterflies Edited by Andrew S. Pullin.
Published in 1995 by Chapman & Hall. ISBN 0 412 56970 1

planned for a number of scarce or threatened species. Some introductions have been successful, but the majority have failed (Warren, 1992b). Fundamental to the role of reintroductions and to understanding the effects that habitat loss and fragmentation have on the long-term viability of butterfly species within the British landscape must be an understanding of their mobility, especially of the dispersal of individuals from existing locations. In particular, the mechanisms of dispersal and colonization, and the role of habitat features and population characteristics in maintaining or promoting mobility, must be understood.

3.2   THE PROBLEMS OF STUDYING MOBILITY

Some species of butterfly are obviously mobile. The most mobile are those that are either migratory or are partial migrants, e.g. *Colias croceus* (clouded yellow), *Vanessa atalanta* (red admiral) and *Pieris brassicae* (large white). Information on the mobility of other species is much less clear. Those that have recently expanded their ranges must have some degree of mobility (e.g. *L. camilla*, *P. c-album*, *P. aegeria* and *P. tithonus*), but for the majority of other species there is little direct evidence of mobility patterns. Within the literature, those species that are scarce or are restricted in distribution are often assumed to be poorly mobile. Scarcity and restriction are frequently equated with the possession of a closed population structure, which, in turn, is taken to be indicative of poor mobility (e.g. Hodgson, 1993). It is equally likely that a lack of suitable habitats and isolation impose a closed population structure independent of mobility.

It is easier to study within-site movements than those movements between sites, but the latter will result in distribution and range changes. Furthermore, data for within-site movements are not appropriate to the study of mobility and can lead to misinterpretation if they are so applied. For example, studies of the within-site movements of *Maniola jurtina* (meadow brown) (Brakefield, 1984) and *P. aegeria* (Shreeve, 1985) could be interpreted to propose that both these species are relatively sedentary and have closed population structures. The primary aims of such studies are usually to reveal patterns of habitat use and gather information on population size. They will not detect movements away from particular sites. In typical mark–release–recapture programmes, the movement of individuals away from sites can only be ascribed to loss, it cannot be distinguished from natural mortality and predation. For both *M. jurtina* and *P. aegeria* there is evidence for considerable mobility. For *M. jurtina* individuals have been tracked making movements away from habitats in excess of 600 m, and there is evidence for lifetime mobility and displacement in excess of this (Dennis, personal observations). Data from North Wales on the lack of between-site variation of morphological characteristics, expected on the basis of differing selection parameters in different habitat types, are also most easily explained by invoking between-population movements (Shreeve and Dennis, personal observations). Recent expansion of the range of *P. aegeria* in north-eastern Scotland and eastern England points to high mobility in this species; recent occupation of woodland habitats is indicative of individuals crossing unsuitable habitat areas (e.g. arable fields) to colonize new sites.

Data from studies of colonization events of a limited range of species are indicative that species may be mobile. Maximum colonizing distances of 0.6–1.0 km for *Plebejus argus* (silver-studded blue), 8.65 km for *Hesperia comma* (silver-spotted skipper), 2.25 km for *Thymelicus acteon* (Lulworth skipper), 0.65–2.50 km for *Mellicta athalia* (heath fritillary) and 1.40 km for *Satyrium pruni* (black hairstreak) are given by Thomas, Thomas and Warren (1992). These figures are the distances travelled in association with colonizing events, they are not the recorded maximum flight distances of individuals of these species. For other species there is anecdotal evidence for individual mobility. This includes documented observations of individuals of *Cupido minimus* (small blue), *Lysandra coridon* (chalkhill blue), *Aricia agestis* (brown argus) and *Satyrium w-album* (white-letter hairstreak) at considerable distances from known colonies (Horton, 1977; Emmet and Heath, 1989), together with casual observations of other species in unsuitable habitats, such as *S. pruni*, *Callophrys rubi* (green hairstreak), *Lysandra bellargus* (adonis blue), *Boloria euphrosyne* (pearl-bordered fritillary) and *Hipparchia semele* (grayling) (Shreeve, personal observation). Such observations do not equate with estimates of mobility obtained from studies of within-site mobility. They also point to the potential for colonization by a number of species and may also indicate that individuals of apparently sedentary species have the potential to move between existing populations. Whether this potential is fulfilled is obviously dependent on both the availability of suitable habitats and the spatial arrangements of such habitats in the landscape.

3.3   A HISTORICAL PERSPECTIVE

Historical and biogeographical considerations indicate that the British butterfly fauna has the potential for greater mobility than is generally accepted. The majority of species that are currently resident in Britain would have arrived and established populations since the Loch Lomond Stadial or Younger Dryas, approximately 10 000 BP (Dennis, 1992). Conditions during this cold period of the current interglacial may have permitted a few cold-tolerant species, such as *Aricia artaxerxes* (northern brown argus), that arrived during the earlier part of the interglacial to persist in a few south-facing slopes in southern England, but the end of this period marks the beginning of reinvasion of the land mass by most species of butterfly now resident in the British Isles (Dennis, 1977, 1992). Because of conditions during the height of the last glacial, the majority of these species would have been restricted to the Mediterranean fringe. It is from these areas that species have spread to their current locations. Such spread points to past mobility, including that of species which are currently classified as of low mobility, such as *C. minimus*, *P. argus* and *Thecla betulae* (brown hairstreak). Accompanying their establishment within the British Isles has been some adjustment of morphological characteristics, current variation correlates with current climate (Dennis and Shreeve, 1989). Whether there has also been adjustment of mobility is a matter of conjecture. That flight capacity and hence mobility can change is demonstrated by the reduction of wing expanse and its relationship to flight activity in *Papilio machaon* (swallowtail) at Wicken Fen (Dempster King and Lakhani, 1976). Changes of wing expanse occurred during the period 1820–1960 and it is suggested that this change was a direct product of the combination of isolation and the lack of alternative habitats, strong selective pressures against the maintenance of mobility.

The range of habitats occupied by species may vary on a geographical scale. For some species the range occupied in Britain is less than in southern parts of their geographical distributions. For example, *Leptidea sinapis* (wood white) is generally restricted to woodland in Britain, but in Continental Europe it additionally occurs in meadows. If species have intrinsic mobility, it is possible that their poor ability to colonize vacant sites in Britain is influenced by their highly specialized habitat requirements towards the limits of their ranges. Thus an understanding of habitat requirements may be essential to understanding mobility.

3.4   A CLASSIFICATION OF MOBILITY

The habitat occupied by any species must include locations where host-plants grow in conditions that will facilitate larval growth and development and are accessible to egg-laying females. There must also be pupation sites and sites for mate location, as well as places for resting, roosting and adult feeding. Together these factors determine the suitability of particular habitats. Species of butterfly can be classified by the types of habitat that they occupy. The habitats occupied by the butterflies of Britain range from short-lived early seral stages to long-lived climax communities (Table 3.1).

This classification differs from the more conventional biotopic schemes in that species which are traditionally placed as woodland species (e.g. *B. euphrosyne*) are, in fact, more closely associated with temporary features within that biotope.

Classification by seral stages more closely associates species with their essential habitat components and leads to expectations of mobility and also of population structure in relation to the stability of their habitats (Southwood, 1962, 1977). Species that are most closely associated with short-lived habitats should have good dispersal properties to facilitate colonization of new habitat patches before existing patches become unsuitable through succession. By contrast, intrinsic mobility is not such a necessary characteristic of species associated with long-lived habitats.

More than half of the British butterfly fauna is associated with early seral stages (Table 3.1). In the absence of human intervention, these seral stages are only temporary, changing to woodland and eventually to climax forest. Permanent early seral stages are maintained only in a few locations, such as scree slopes and some sand-dunes where geomorphological processes limit succession. In addition to consequences for mobility, habitat stability and longevity may have consequences for population structure. Those species associated with short-lived habitats are expected to form short-lived populations from colonizing events. To be effective, these colonizing events must be viable if the founding population size is low.

Table 3.1   The seral habitats occupied by the resident species of butterfly in the British Isles. Species of early seral stages are expected to be more mobile than those of late seral stages. All species are classified by the earliest seral stage that they inhabit but seral stages 1 and 2 are not distinguished

Species	Seral stage	Species	Seral stage
Carterocephalus palaemon	3	*Celastrina argiolus*	5
Thymelicus sylvestris	3	*Maculinea arion*	1/2
Thymelicus lineola	3	*Hamearis lucina*	3
Thymelicus acteon	3	*Ladoga camilla*	5
Hesperia comma	1/2	*Apatura iris*	6
Ochlodes venata	3	*Aglais urticae*	3
Erynnis tages	1/2	*Inachis io*	3
Pyrgus malvae	1/2	*Polygonia c-album*	3
Papilio machaon	3	*Boloria selene*	1/2
Leptidea sinapis	3	*Boloria euphrosyne*	1/2
Gonepteryx rhamni	4	*Argynnis adippe*	1/2
Pieris brassicae	3	*Argynnis aglaja*	1/2
Pieris rapae	3	*Argynnis paphia*	5
Pieris napi	1/2	*Eurodryas aurinia*	3
Anthocharis cardamines	3	*Melitaea cinxia*	1/2
Callophrys rubi	3	*Mellicta athalia*	3
Thecla betulae	5	*Pararge aegeria*	4
Quercusia quercus	6	*Lasiommata megera*	1/2
Satyrium w-album	6	*Erebia epiphron*	1/2
Satyrium pruni	5	*Erebia aethiops*	3
Lycaena phlaeas	1/2	*Melanargia galathea*	3
Lycaena dispar	3*	*Hipparchia semele*	1/2
Cupido minimus	1/2	*Pyronia tithonus*	3
Plebejus argus	1/2	*Maniola jurtina*	3
Aricia agestis	1/2	*Aphantopus hyperantus*	3
Aricia artaxerxes	1/2	*Coenonympha pamphilus*	1/2
Polyommatus icarus	1/2	*Coenonympha tullia*	3
Lysandra coridon	1/2		
Lysandra bellargus	1/2		

1/2   Seral stages 1 and 2 (bare ground, short herbs and grasses).
3   Seral stage 3 (shrubs, tall herbs and grasses).
4   Seral stage 4 (trees and shrubs).
5   Seral stage 5 (pre-climax forest).
6   Seral stage 6 (climax forest).
3*   Plagioclimactic bog (= seral stage 3 in structure).

Unfortunately, there is scant information about this important aspect of butterfly biology, despite its significance to the planning of reintroduction programmes.

Even if mobility and the likelihood of individuals finding vacant habitat patches is poorly documented, the potential for successful establishment of populations once such patches are found is recognized. The potential reproductive output of females, measured as eggs/female, is large. Estimates range from *c.* 150 for *A. cardamines* (Courtney and Duggan, 1983) to *c.* 600 for *Lycaena dispar* (large copper) (Duffey, 1968, 1977). Actual fecundity may be much less, with a reduction of approximately one order of magnitude. However, the potential for the rapid establishment of populations from small numbers of founding individuals is evident, even if actual fecundity is as low as conservative estimates.

3.5 FACTORS AFFECTING MOBILITY

Although much understudied, a variety of factors can be identified which influence the mobility of species. These can be divided into factors associated with the habitat, with population parameters and with environmental factors which influence individual activity.

Theoretically, habitat stability has an important influence on mobility. As well as influencing overall patterns of mobility, habitat suitability can influence individual flight activity. *Mellicta athalia* is associated with temporary habitats and, in woodlands in Kent, these are short-lived clearings created by coppicing. When habitat patches are suitable for egg-laying and are large, the majority of individual females remain within such habitat patches, but when they become less suitable, such as when shade causes a decline in the number of egg-laying sites and reduces adult activity, or when these clearings are small, an increased proportion of females leave existing patches (Warren, 1987b). How females of this species detect overall suitability is not precisely known, but the data are a clear indicator that habitat quality can affect mobility.

The suitability of a habitat for an individual may also be influenced by the activity of other members of the population. Baker (1969, 1984) argues that every individual has a particular habitat suitability threshold (= migratory threshold), which may be influenced by physiological condition. This will influence the probability of an individual leaving an existing habitat patch. For colonization the factors that influence female movement are important, and for population persistence within particular locations the combined effects of male and female mobility and residence are important.

Within any habitat patch there are finite resources, including host-plants and potential mates. For egg-laying females the crucial resource is the availability of suitable host-plants where the predictability of larval survival is high. When particular host-plants are repeatedly exploited by females, their value to any individual female declines. If their value falls below that dictated by the migratory threshold (*sensu* Baker) of any particular female, then alternative resource locations must be sought by that female. These may be within the habitat if the habitat is large or population size is small, or in alternative habitat patches if the habitat is small or population size is large. Different individuals will have different thresholds, determined by previous experience (= physiological state) and by genotype. Baker argues that young females may have lower thresholds than older females because their potentially longer future life span may facilitate their success in finding alternative habitat patches. Alternatively, older females may take the risk of moving away from an existing habitat once they have laid some eggs; such a strategy ensures that some eggs are laid, provided suitable host-plants are available in the original habitat. Associated with finding new habitat patches are the risks associated with leaving an existing patch. These may range from the absolute risk of not finding an alternative to those associated with the chance of predation before finding an alternative. As the quality of an existing patch changes with time, so does the relative cost of moving. With decline in resource quality the relative cost of moving also declines. For males the availability of females is critical. When mates are scarce, either through absolute female scarcity or through the activity of competing males, migratory thresholds will be exceeded.

According to the model of Baker (1969, 1984), population density and environmental conditions will have an effect on mobility. When population density is high, mobility should increase because the migratory thresholds of an increasing proportion of individuals will be exceeded, because the activity of the population reduces habitat quality. If habitat quality declines, for example through successional changes, or via unpredictable events such as drought, then mobility should also increase.

This model can be used to explain the mobility and distribution of some species. The movements of *M. athalia* in relation to habitat-patch quality can be fitted to this model, as can variation in the distribution of *A. cardamines*. When populations of this species are at high abundance, individuals are widely distributed and widely disperse their eggs, in comparison to years when they are at low abundance (Dennis, 1982a). Females can assess the suitability of host-plants by the presence of eggs of conspecifics; newly laid eggs are detected by the presence of a pheromone (Dempster, 1992) and older eggs by their red colour (Shapiro, 1981). Where the quality of resources is low, that is when they have been previously used, egg-laying females will disperse to alternative host-plants. Whether the model of Baker offers a general explanation of the mobility of species is as yet unclear, but it undoubtedly deserves further investigation.

In addition to habitat features and the activity of other members of the population, flight activity is influenced by environmental conditions, particularly temperature and solar radiation. Warm, sunny conditions generally facilitate activity, especially flight activity (see Shreeve, 1992). It is interesting that most observations of individuals of species away from known habitats are made during, or immediately after, periods of warm, sunny weather. In addition, the probability of individuals locating new habitats, independent of the factors that determine whether they leave an area, will be increased by conditions which minimize resting and basking but facilitate flight.

Weather conditions that facilitate flight activity vary on a geographical gradient within Britain. Conditions that facilitate flight occur more predictably and for longer periods in southern than in northern parts. Thus in southern parts of Britain the thermally determined risks of mobility are less than in northern parts. Given an equal availability of habitat patches, colonizing ability and individual mobility should therefore be greater in southern than in northern parts. The only exceptions may be *Erebia aethiops* (Scotch argus), *E. epiphron* (mountain ringlet) and *Coenonympha tullia* (large heath) which are cold adapted and have their southern range limits in northern and central parts of Britain. There is some evidence that the mobility of *A. cardamines* is greater in southern than in northern parts of its range in Britain (Courtney, 1980; Dennis, 1982b; Dempster, 1989). In addition, it has been suggested that there may be a genetic component to differing mobility rates of this species in different geographical areas (Dennis, 1982a).

If mobility is influenced by thermal conditions, the risks associated with finding new habitat patches are greater in northern than in southern areas because of limited time for flight activity. If there is a genetic component to mobility, then the absolute abundance of habitats should additionally influence overall mobility. A scarcity of habitats should reduce mobility but an abundance promote mobility. Whether individual species or populations can track the abundance of habitats given the rate of habitat change over much of Britain is a matter of speculation. The genetic component of mobility has received little attention, but warrants investigation.

3.6 MOBILITY AND THE LANDSCAPE

Within the British Isles, habitats that are suitable for most species of butterfly are continuing to decline, both in quantity and quality. This is leading to increasing fragmentation and reducing the chance that the movement of individuals from existing populations will lead to the establishment of new populations. For recolonization, continuous habitats are probably not required. Colonization does not require the movement of a single invasion front, it can be achieved by the movement of isolated individuals to suitable habitats some distance from the main region of established populations (Hengeveld, 1988). There is an element of chance in the frequency and distance of movement of individuals away from populations. The frequency of potential colonizing movements may depend on factors within existing habitats, but the probability of a suitable area being colonized will decline with its distance from existing locations.

Habitats differ in their suitability for particular species. Some sites may always be marginal and only able to support small or temporary populations while others support larger or more long-lived populations. The suitability of any particular location will also vary between years, either as a response to vegetation changes and management or because of climate. If species have the ability periodically to use habitats which are marginal in suitability, then the probability of them reaching and colonizing more favourable sites is enhanced if these suboptimal sites act as 'stepping-stones'. The use of such sites may even be part of the general opportunistic strategy of many species, playing an important role in population structure (see Thomas, Chapter 4 this volume), specifically by reducing the isolation of existing populations in optimal sites. This aspect of population dynamics is receiving increasing attention. It is probable that *M. athalia* and *Eurodryas aurinia* (marsh fritillary) have long-lived populations in optimal habitat areas and form temporary populations in more marginal areas (Warren, 1990a, 1991). Likewise, the structure of populations of *A. cardamines* in southern England and *Pieris napi* (green-veined white) in Scotland and southern England may be similar, with temporary occupation of suboptimal habitat patches (Dennis, 1982b; Shreeve, personal observation).

To cater for declining species and to facilitate their ability to recolonize sites using their intrinsic properties of dispersal will require more than the protection of existing colonies *in situ*. Consideration must be given to the provision of alternative sites which are at least suboptimal. This requires an emphasis of the value of non-designated reserve areas. Equally important should be the realization that the temporary loss of populations of most species in Britain is to be expected. If mobility can be facilitated, recolonization should also be an expected event.

Currently, temporary areas of suboptimal habitat which are suitable for the majority of species are scarce. Even areas that could support such common species as *M. jurtina*, *P. tithonus* and *P. napi* are frequently unsympathetically managed. There is also debate about the value of such areas as aids to the mobility of species, and their role as areas that may support persistent but small populations. Much of this debate has focused around the role of roadside verges, and also of the role of major landscape features, such as motorways, as barriers to dispersal. Munguira and Thomas (1992) documented the presence of 27 species of butterfly on roadside verges in Dorset and Hampshire, with individual verges containing a maximum of 12 species. From their studies they concluded that roads may not be an effective barrier to dispersal, individuals of 21 species were observed flying across roads, were detected as crossing roads by mark–release–recapture or were seen killed on roads. This is in contrast to the findings of Dennis (1986), that part of the M56 motorway in Cheshire was an effective barrier to the movement of *A. cardamines*. Individuals of this species were observed to fly towards the embankment of this motorway but to change direction and not to cross the motorway. Clearly, there are differences of scale, of time and of location in these two studies. It is likely that large roads can limit mobility. Whether they do so will depend on local factors that influence flight activity and flight direction. These include temperature, sunshine, windspeed, aspect, verge type and traffic density. No general conclusions can be drawn.

3.7 CONCLUSIONS

Although strong theoretical arguments can be made to support the hypothesis that species are generally more mobile than currently described, specific rather than anecdotal evidence is lacking. This lack of data is partly because between-site mobility is not easily studied in non-migratory species. If only a few individuals are needed to establish new colonies, then systematic studies designed to identify such events may be both extremely slow and time consuming if conventional approaches to study are used. It is hoped that the available information will improve with time. Observations of individuals occurring away from recognized populations need to be carefully documented. With adequate design and monitoring, planned reintroductions could also be used to monitor mobility and colonization events arising from the reintroduction itself. Studies of isozyme variation, the introduction of DNA studies into butterfly biology and the use of genetic markers all offer the opportunity to study the consequences of mobility, even if the event of mobility is inaccessible to systematic study.

CHAPTER 4

Ecology and conservation of butterfly metapopulations in the fragmented British landscape

C.D. Thomas

4.1 INTRODUCTION

Over the past 25 years, efforts to conserve British butterflies have concentrated on local processes. Researchers have identified the habitat requirements of many species, and used this information to make recommendations for their conservation (this volume). Conservation agencies have then attempted to maintain or re-create these conditions on reserves, in some cases with great success (Thomas, 1991; Warren, 1991). However, at a national scale many species have continued to decline. Agricultural and forestry practices have changed so rapidly and so widely that many species no longer find suitable habitats in most of their former British range: conservation organizations have been unable to protect and maintain more than a scattering of local populations in small reserves. Unfortunately, rare species often have such subtle habitat requirements that many populations have been lost even from reserves and from other fragments of 'semi-natural' vegetation as a result of apparently minor habitat changes (Thomas, 1991; Warren, 1993a,b). Good management can reduce the rate of local extinction, sometimes to a very low level, but extinctions cannot be stopped completely. If reserves are completely isolated from one another, population after population will be lost with no opportunity for recolonization. In the long term, the persistence of rare species will depend on the rate of recolonization as well as on the rate of local extinction.

The local and regional distributions of many species of insect are not static (Andrewartha and Birch, 1954), and butterflies are no exception (e.g. Ehrlich, 1984; Heath, Pollard and Thomas, 1984; Dennis and Williams, Chapter 15 this volume). It is essential that changing patterns of distribution or abundance be

Ecology and Conservation of Butterflies Edited by Andrew S. Pullin.
Published in 1995 by Chapman & Hall. ISBN 0 412 56970 1

understood if effective recovery programmes are to be designed and implemented. The purpose of this chapter is to assess the need to expand the conservation perspective from local processes to the dynamics of populations over wider areas, and to identify ways in which long-term persistence might be achieved.

4.2 HABITAT CONTINUITY

4.2.1 Successional species

Many British butterflies breed in some kind of successional or plagioclimax vegetation (Thomas, 1991; Warren, 1993b; Shreeve, Chapter 3 this volume). Successional species occupy habitats which inevitably become unsuitable in time, and managers must provide continuity of habitat close enough for a species to colonize each new succession. Good examples are several woodland fritillaries, *Mellicta athalia* (heath fritillary), *Boloria selene* (small pearl-bordered fritillary) and *B. euphrosyne* (pearl-bordered fritillary), which frequently occupy woodland clearings in the first few years of succession (Warren, 1991; Warren and Thomas, 1992), and *Plebejus argus* (silver-studded blue), which occupies successional habitats within heathland (Thomas and Harrison, 1992).

In south-east England, *M. athalia* lays its eggs on *Melampyrum pratense* (common cow wheat) in the 5 or so years after an area of woodland has been coppiced. To persist, *M. athalia* must shift from clearing to clearing. Although some traditional coppicing was carried out in Blean Woods National Nature Reserve (NNR) in Kent in the early and mid-1970s, management was not systematic and little attention was paid to the ability of the species to disperse between clearings (M.S. Warren, personal communication): by 1980 the population had dwindled to a handful of individuals (Warren, Thomas and Thomas, 1984). Systematic coppicing was restored in the late 1970s, since when the butterfly has recovered, with more than 1500 present each year since 1984 (Warren, 1991; Warren and Thomas, 1992). Not only has habitat continuity been ensured, but managers have created open rides and glades that allow the adult butterflies to disperse from one clearing to another (Figure 4.1).

Blean Woods NNR can be considered as a model for *M. athalia* management in coppiced woodland, provided that the management is replicated over 5–10 times the area shown in Figure 4.1: in 1989, the emerging population was apparently restricted to one small clearing that was created in 1988 and a path along one side of a 1987 clearing (adults at that point would not have moved into the 1989 clearings, and there is little breeding in the permanent glades and rides). In fact, Blean Woods NNR is adjacent to other reserves where similar forms of management are being carried out, so the area available is larger than that shown.

Plebejus argus occupies successional habitats on heathland, following cutting, grazing, fire or disturbance, and shows a similar pattern of local extinctions and colonizations (Thomas and Harrison, 1992). Figure 4.2 shows the butterfly's localized and spatially dynamic distribution in an essentially uninterrupted heathland at the South Stack reserve of the Royal Society for the Protection of Birds (RSPB) in North Wales.

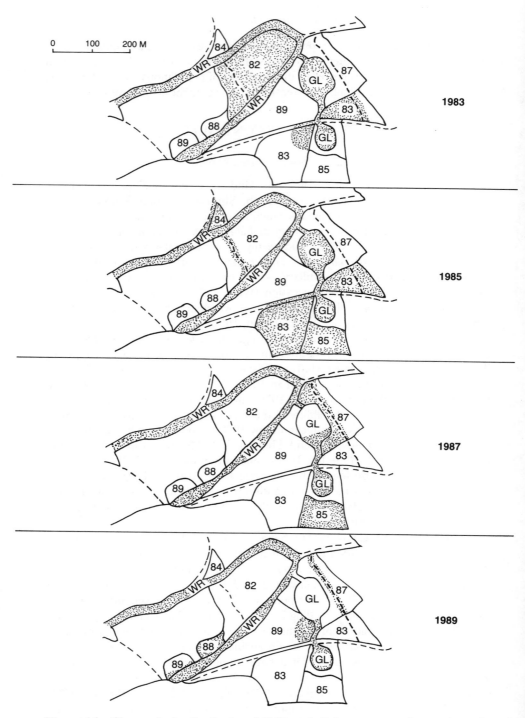

Figure 4.1 Changes in the distribution of *Mellicta athalia* in response to the coppice cycle in Blean Woods NNR, 1983–89. Numbers show the year in which each area was cut. Shaded areas show the distribution of adult butterflies. Wide rides (WR) and glades (GL) were rarely used for breeding. (Reprinted from Warren (1991), courtesy of Elsevier Science Publishers.)

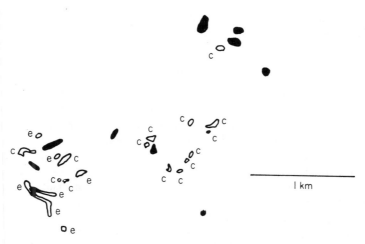

Figure 4.2 Distribution of *Plebejus argus* in the South Stack Cliffs metapopulation in 1983 and 1990. Solid = *P. argus* present in 1983 and 1990. Open = *P. argus* present in 1983 or 1990, but not both: e = extinction (1983 only); c = colonization (1990 only). (Reprinted from Thomas and Harrison 1992, courtesy of Blackwell Scientific Publications.)

For the species to persist in any area in isolation, a population must be present continuously. Now superimpose in your mind's eye, 1 ha (100 × 100 m), 5 ha (*c.* 225 × 225 m), 25 ha (500 × 500 m) and 100 ha (1 × 1 km) squares over Figure 4.2. It immediately becomes obvious why *P. argus* has become extinct from most small, isolated heaths; only large areas are likely to provide the continuity of seral conditions required by this species. Deliberate conservation management may be able to provide habitat continuity in heathlands as small as, say, 5–10 ha, but with difficulty.

4.2.2 Plagioclimax species

A plagioclimax is a community in which succession, usually to woodland, is halted more or less permanently because of continued management. For example, most permanent grasslands in Britain are maintained by cutting or grazing, and succession to woodland may be prevented indefinitely. For butterflies inhabiting plagioclimax vegetation, the habitat is potentially stable provided that management is stable, and it may therefore be possible to maintain butterfly populations in small reserves. Only if management is discontinued will the vegetation undergo succession. However, the empirical dynamics of most rare butterflies in plagioclimax vegetation (grasslands, heathlands) show major changes in abundances and distributions (see below). Most plagioclimax vegetation consists of many different microhabitats and microclimates, such that very subtle changes in the vegetation can quickly render an area suitable for one species and unsuitable for another, even though the habitat appears unchanged to a casual observer. Many extinctions of local and rare species from plagioclimax vegetation fragments are testimony to this (Thomas, 1991; Warren, 1993a).

4.2.3   Habitat heterogeneity

For some species, different host-plants, habitats or microhabitats may be required in different years or seasons, e.g. *Celastrina argiolus* (holly blue). For species with mobile larvae, a variety of microhabitats might be required to ensure survival within a generation (Weiss, Murphy and White, 1988).

In 1982, eggs of *Hesperia comma* (silver-spotted skipper) were laid on small *Festuca ovina* (sheep's fescue) plants that were growing next to bare ground (Thomas *et al.*, 1986). The likelihood of finding an egg in a small quadrat and the abundance of adults in entire habitat patches increased with the percentage bare ground and percentage *F. ovina*. Heavily grazed, south-facing hillsides were ideal. The summer of 1990 was very dry and, combined with heavy rabbit grazing, some of the sites that had appeared ideal in 1982 were parched, with hardly a green grass blade to be seen. In a repeat survey the following year, 1991, I predicted that *H. comma* numbers would be exceptionally low. To my surprise, average densities were exactly the same in 1991 as in 1982 (Thomas and Jones, 1993), but with a shift in local distribution to longer vegetation. In 1991, *H. comma* was scarce (but rarely extinct) on the most heavily grazed, south-facing slopes, and was more abundant where the percentage bare ground was relatively low and in sites which did not face due south. There was no longer any correlation between the 1982 measure of habitat quality and the density of adult butterflies, although the skippers were still restricted to sites with short vegetation, *F. ovina* and some bare ground. After very wet summers, or if rabbit grazing was reduced, the areas of highest density would presumably revert to sites that appeared to be overgrazed in 1991. Because the best breeding habitat cannot be guaranteed to remain in exactly the same place, effective long-term conservation may require much larger areas than are at first apparent.

4.3   POPULATION STRUCTURE

British butterflies have been placed in two dispersal categories, 'open/migratory' and 'closed' (Thomas, 1984a; Warren, 1992a; Pollard and Eversham, Chapter 2 this volume). For species with closed populations, there is some dispersal but local birth and death processes are presumed to be the major determinants of local abundance, whereas for open populations local abundances are determined more by immigration and emigration than by local breeding success. It may eventually be possible to replace these two categories with a continuum of mobility, or with many more categories, but the classification 'open' or 'closed' has been useful. For species with closed populations, local management of a small area, say one field, can increase local breeding success and thereby increase local abundance. For species with open or migratory population structures, increased breeding success in one field would be dissipated over, say, hundreds of km^2, so management would have to be carried out at a much broader scale.

4.3.1 Metapopulations

Most of the three-quarters of British butterfly species that are regarded as having closed local populations exist as metapopulations. A metapopulation is a collection of local populations (like those in Figures 4.1 and 4.2), connected by occasional dispersal, in which there are local extinctions and colonizations (Gilpin and Hanski, 1991). A local population is defined as a group of individuals in a distinct habitat patch (a colony): most individuals will remain within one habitat patch throughout their adult life, but some will disperse. A proportion of these will reach other local populations or colonize fresh habitats. Metapopulations may persist indefinitely, even though each local population will eventually become extinct, provided that, in the long run, the number of colonizations equals the number of extinctions.

What is the evidence that rare species exist as metapopulations in the British landscape rather than as single, isolated populations? Metapopulations are characterized by:

1. occasional movements between local populations;
2. colonizations and extinctions; and
3. local populations occur in groups rather than as single isolates.

To assess the evidence, I focus on *P. argus, M. athalia, H. comma, Thymelicus acteon* (Lulworth skipper) and *Melitaea cinxia* (Glanville fritillary). Evidence for local extinction is reviewed elsewhere (Thomas, 1991; C. Thomas, 1993; Warren, 1993a).

(a)   Dispersal between local populations

Individual movements between local populations have been detected in four of the species, and have not been examined in the fifth. Estimated rates of exchange are:

1. *Plebejus argus*: 0% over 1–3 days in North Wales (C. Thomas, 1983), 1.4% over 1–6 days for Martlesham Heath in Suffolk and 10% over the same period for two patches that were separated by just 15 m at another site in Suffolk (Ravenscroft, 1986);
2. *Mellicta athalia*: 1.4% over ≤17 days in Kent (Warren, 1987b);
3. *Hesperia comma*: 14% in 2 days for two patches separated by just 18 m in Surrey (Thomas *et al.* 1986); and
4. *Melitaea cinxia*: 15% for males and 30% for females measured over most of the flight period in a series of habitat patches in Finland (Hanski, Kuussaari and Nieminen, 1994).

For all four species, local populations are connected by some dispersal, the percentage exchange depending on the species, distance between patches and duration of study.

(b)   Colonization

Colonization has been observed in all five species.

1. *Plebejus argus* colonizations are shown in Figures 4.2 and 4.6. Other examples
    are given in Thomas and Harrison (1992). The maximum known single-step
    colonization distance observed in North Wales is 600 m, although colonization
    may occasionally take place up to about 1 km from an existing population
    source (N.O.M. Ravenscroft (personal communication) has detected one
    female movement of *c.* 1 km in Suffolk).
2. *Mellicta athalia* colonization of fresh habitat is shown in Figure 4.1. In Kent,
    where *M. athalia* habitat is so ephemeral, colonization has only been observed
    over distances up to 650 m, but in south-west England colonization has been
    observed up to 2.5 km from a source population (Warren, 1987b).
3. It can be deduced that *T. acteon* was once restricted to the coast in Dorset and
    has recently colonized inland, establishing new local populations (J. Thomas,
    1983a). The longest single-step colonization was 2.25 km (Thomas, Thomas
    and Warren, 1992).
4. *Hesperia comma* was observed to colonize 29 new patches of habitat in the
    North and South Downs between 1982 and 1991 (Figure 4.3) – all but five
    were within 1 km of an existing source population, and the remainder were
    within 10 km (maximum 8.65 km) (Thomas and Jones, 1993).
5. *Melitaea cinxia*'s reliance on recent cliff-falls on the Isle of Wight implies that
    colonizations must take place regularly (Simcox and Thomas, 1979).

(c)   The distributions of occupied habitat patches

If these species exist as metapopulations, then single, isolated local populations
should be unusual, and the most isolated populations should be at the limits of
each species' colonization ability.

1. The most isolated local population of *P. argus* in North Wales is 800 m from
    its nearest neighbour, similar to the maximum colonization distance (Figure
    4.4; Thomas and Harrison, 1992).
2. All but two local *M. athalia* populations in Kent are within 1 km of their
    nearest neighbour, and the most isolated is 2.6 km, similar to the maximum
    colonization distance (2.5 km) observed in south-west England (Thomas,
    Thomas and Warren, 1992).
3. All but one local *T. acteon* population in Dorset is within 2.25 km of another,
    the deduced maximum colonization distance (Thomas, Thomas and Warren,
    1992). One local population at 15.5 km from its nearest neighbour can be
    regarded as truly isolated.
4. All but six local *H. comma* populations in south-east England are within 1 km
    of another local population, and the remaining six are within 10 km (Thomas
    and Jones, 1993). This would be expected from the distribution of coloniz-
    ation distances (Figure 4.3).

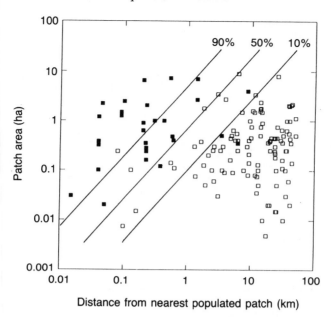

Figure 4.3 Colonization of empty habitat patches by *Hesperia comma* between 1982 and 1991 in relation to patch area and isolation. All patches were vacant in 1982. Solid = patches colonized by 1991; open = patches still vacant in 1991. Lines show combinations of area and isolation which give 90%, 50% and 10% chances of colonization in 9 years, fitted from logistic regression. (Reprinted from Thomas and Jones (1993), courtesy of Blackwell Scientific Publications.)

5. The most isolated local *M. cinxia* population out of 42 studied in Finland by Hanski, Kuussaari and Nieminen (1994) was only 1.6 km from its nearest neighbour, within the colonization ability of this species.

These five species vary by a factor of perhaps 10 in dispersal capacity, yet they all have characteristics of metapopulations. Only one really isolated local population was found out of all those examined in the five species studied. This implies that single, isolated local populations rarely survive for long: they either become extinct or new habitat patches are colonized and they become metapopulations.

4.3.2 The importance of different local populations to persistence

Even though rare species may exist as metapopulations, this does not imply that each local population is equally important to long-term persistence (Harrison, 1991, 1994). Large populations in large, high-quality habitat patches are much less likely to become extinct than are small local populations in small or poor habitat patches.

The potential importance of population refuges is well illustrated by the dynamics of *H. comma* over the past 50 years. This species requires close-cropped chalk turf, where it lays its eggs on the grass *F. ovina* (above). When calcareous

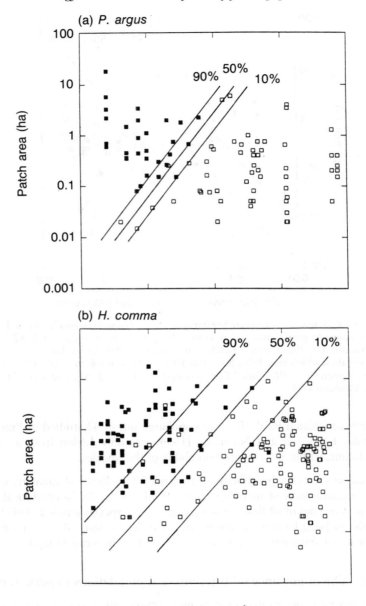

Figure 4.4   Distributions of occupied (solid) and vacant (open) habitat patches for four species. Lines give combinations of area and isolation which give 90%, 50% and 10% probabilities of occupancy, fitted from logistic regression. *Plebejus argus* on limestone grassland in North Wales, *Hesperia comma* on chalk grassland in south-east England, *Thymelicus acteon* on calcareous grassland in Dorset, *Mellicta athalia* in woodland clearings in Kent. (Reprinted from Thomas, Thomas and Warren (1992), courtesy of Springer-Verlag.)

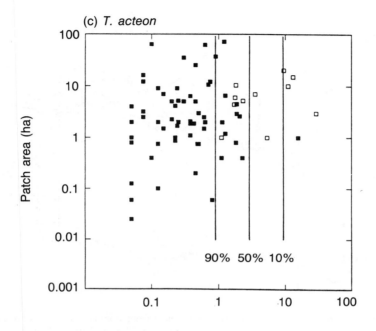

(c) *T. acteon*

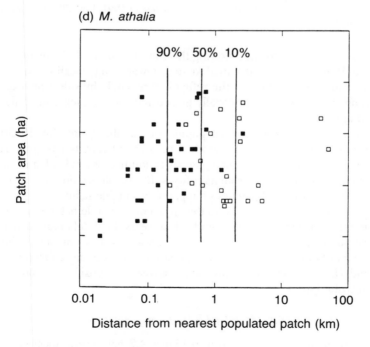

(d) *M. athalia*

Distance from nearest populated patch (km)

Figure 4.4 Continued

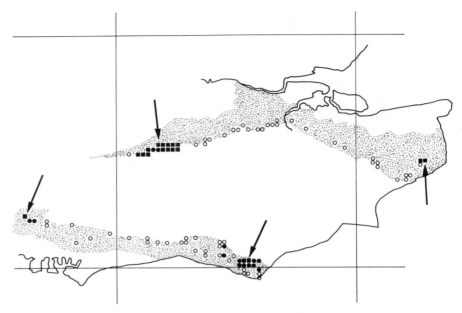

Figure 4.5 Limited expansion by *Hesperia comma* from population refuges (arrowed) in the North and South Downs. Distribution mapped in 2 km × 2 km grid squares, 100 km grid lines are shown. Filled squares = refuges; filled circles = recolonization between 1975 and 1991; open circles = habitat suitable for *H. comma* in 1991, but still unoccupied. (After Thomas and Jones,1993.)

grasslands became overgrown following the elimination of rabbits by myxomatosis in the mid-1950s, *H. comma* declined to 46 or fewer refuge localities in 10 regions (Thomas *et al.*, 1986). Many of the refuges were single hillsides; for example, it survived on single, large commercially grazed hillsides at each end of the South Downs (Figure 4.5).

The species survived as isolated populations in these sites for 20–30 years. Conservation management and a gradual recovery of rabbits has since restored habitat quality to at least 144 habitat patches in south-east England, and *H. comma* has begun to recolonize away from its refuges. But recolonization has been limited (Figure 4.5), and over 100 patches of apparently suitable habitat remained unoccupied in 1991. Because so much of the apparently suitable habitat is so isolated, complete recolonization may take hundreds of years. Had deliberate management for *H. comma* provided extra habitat refuges in reserves at several points along both the North and South Downs, there would have been many more foci for recolonization and *H. comma* could already have recovered a considerable portion of its former distribution.

(a)   Mainland–island metapopulations

The *P. argus* metapopulation shown in Figure 4.2 has several local populations, each of which is fairly similar in size to each other local population and, given the

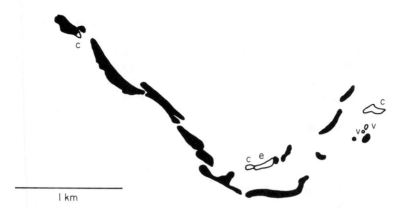

Figure 4.6 Distribution of *Plebejus argus* in the Great Orme metapopulation in 1983 and 1990. Symbols and letters as in Figure 4.2; v = vacant habitat in 1990. (Reprinted from Thomas and Harrison (1992), courtesy of Blackwell Scientific Publications.)

ephemeral nature of the heathland habitat (even large local populations eventually become extinct), all local populations are likely to be of comparable importance. This is close to the original, theoretical concept of a metapopulation (Levins, 1970). The *P. argus* metapopulation on the limestone grassland of the Great Orme in North Wales shows a very different structure. It exists as several large local populations which occur more or less continuously along one hillside, with local colonization, extinction and vacant habitat found in smaller, peripheral patches (Figure 4.6).

This hillside can be regarded as a 'mainland' population of *P. argus*, with most of the extinction and colonization restricted to the peripheral 'island' habitat patches. It is possible for a single habitat patch to dominate an entire metapopulation (Harrison, Murphy and Ehrlich, 1988). Population systems of this type can still be called metapopulations (they fit the definition given above), but conservation efforts should focus on the main patch(es). Other *P. argus* metapopulations are intermediate between the extremes shown in Figures 4.2 and 4.6 (Thomas and Harrison, 1992): in reality there is a continuum between 'all patches are equal' (Levins-type) and 'mainland–island' metapopulations.

(b)   Sources and sinks

Dispersing individuals which emigrate from a good breeding habitat (source) may arrive in a poor-quality habitat (sink) which is populated because of repeated immigration rather than because the butterfly is able to reproduce successfully therein. The most likely example among British butterflies is *Eurodryas aurinia* (marsh fritillary). Many local populations have existed in their present locations for as long as records have been available, but other sites have supported temporary local populations in an apparently poor habitat that was heavily grazed or where host-plants grew at low density (Warren, 1994). *Eurodryas aurinia* definitely bred in these habitat patches for a while, but local populations were short lived. To be quite certain that a metapopulation is operating in this way, good data are required on immigration, emigration, birth and death at a range of densities in each habitat type. This is a tall order, so we have little idea whether it is a widespread phenomenon among British butterflies (cf. Shreeve, Chapter 3 this volume). However, it is a serious issue. For some species, we may need to target conservation towards habitat patches where breeding success is highest.

4.3.3 Patchy populations

The remaining quarter of British butterflies have open or migratory populations. Although these species also exploit patchy resources, individuals fly in and out of habitat patches readily, each butterfly visiting many patches during its adult lifetime (Harrison, 1991, 1994). These species exploit small patches of habitat over such large areas that they are almost impossible to study or conserve when rare. The problem is exemplified by *Nymphalis polychloros* (large tortoiseshell) which is presumed to have an open population structure; we don't even know if it has been resident in Britain in recent years, and it is now considered extinct.

4.4 CONSERVATION IN A FRAGMENTED LANDSCAPE

Only rarely will an entire viable metapopulation be present within a single British nature reserve. Most reserves are too small. Nevertheless, reserves can protect important parts of metapopulations (e.g. habitat 'mainlands' and population sources), and provide refuges of suitable habitat when conditions are temporarily unfavourable in the wider (non-reserved) countryside. The potential of refuges to provide foci for recolonization is clearly illustrated by the dynamics of *H. comma* (above). In parts of the world where habitats are less fragmented than in Britain, most effort should probably be directed at the preservation of extensive blocks of each habitat. Conservation of networks of habitat fragments is often a last-ditch attempt, once a particular biome is all but obliterated.

An understanding of metapopulations forces us to consider regional as well as local conservation issues. We should focus on ways of increasing rates of colonization as well as on trying to minimize rates of extinction, and implement programmes which encompass several reserves as well as non-reserved habitats. Management will still be carried out patch by patch, reserve by reserve, but local management must be carried out in the context of the wider countryside. Single local populations on reserves may provide important refuges during periods when conditions are unsuitable for a particular species in the wider countryside, but isolated populations are unlikely to survive for ever. Recovery programmes should actively aim to produce conditions that will allow species to re-expand from temporary refuges.

4.4.1 Metapopulation dimensions

Networks of habitat patches are needed for long-term persistence, but how many patches and how far apart? There is little good empirical evidence on the number of patches needed by most species. For *H. comma* and *P. argus*, 10 local populations per metapopulation is likely to be too few and 20 enough for persistence (C. Thomas, 1993); the larger each patch, the fewer patches will be needed. Metapopulations of these two species typically occupy tens of hectares, with several times this area of similar vegetation also present (the remainder is currently unsuitable, but may become suitable following a relatively minor change in habitat management). Viable metapopulations apparently require many times the minimum areas required by individual populations (cf. Thomas, 1984a; Warren, 1992a).

In the absence of clear empirical data on the minimum areas required by metapopulations, simulation models can be used to assess the probability of metapopulation persistence and the possible effects of some patches being destroyed (Hanski, 1994; Hanski and Thomas, 1994). These models specify the exact location of each patch of habitat and the area of each patch, with rules governing dispersal between patches and probabilities of local extinction. Simulation outputs show which habitat patches are occupied at a particular time and which are not, and give measures of population density in each patch. These patterns match those seen in real metapopulations quite satisfactorily. Simulation models have predicted persistence in existing metapopulations and eventual extinction in areas where habitat patches are widely scattered. They have also successfully predicted colonization of networks of habitat patches from population refuges or sites of introduction. With present models, the habitat distribution is static, but, in principle, habitat patches can be added and lost in specific locations and at various rates to assess the potential consequences of different management options.

We must be able to identify which local populations belong to a single metapopulation if we are to plan conservation management at this scale. A first approximation can be made by plotting 'incidence functions', as in Figure 4.4. All local populations of *P. argus* are within 1 km of another local population, whereas there are many empty habitat patches beyond this distance. This implies that *P. argus* rarely, if ever, colonizes further than 1 km in North Wales; local populations that are less than 1 km apart can be regarded as belonging to a single metapopulation (Thomas and Harrison, 1992). The local *P. argus* populations in Figure 4.2 belong to a single metapopulation because it is possible to move throughout the system of patches in steps of less than 1 km: they are connected over a number of generations even though there will be no direct movement of individuals between the two patches that are farthest apart. To treat two groups of local populations that are divided by the specified distance as separate metapopulations, there should be little possibility that the intervening land will become suitable in the future. In 1983, the South Stack *P. argus* metapopulation (Figure 4.2) existed as two groups of local populations just over 1 km apart, but changes in the intervening heathland vegetation allowed *P. argus* to colonize the gap between 1983 and 1990. The distance required to separate metapopulations of the other species considered here is greater than 1 km. It is approximately 10 km for *H. comma*.

4.4.2 Colonization and the creation of new habitat

Encouraging colonization by producing new habitat has an essential part to play in the conservation of metapopulations. The probability that a fresh patch of habitat will be colonized in a given length of time is likely to depend on its isolation from all possible population sources, and on the population size in each potential source. One way to approach the problem is to use simulation models, adding possible new patches of various areas in several feasible locations. Simulations can then be run to estimate the likely time to colonization, and the long-term probability that a patch will be occupied. The trouble is that considerable knowledge of population parameters is needed, and some of them are extremely difficult to estimate (Hanski, 1994; Hanski and Thomas, 1994). Future developments will make models of this type easier to use and more realistic, and they will undoubtedly be used widely in conservation planning for critically endangered species.

It will probably be some years before models of this type are in routine use by reserve managers and local conservation agencies for species that are not internationally endangered. In the meantime, information of the type given in section 4.3.1 can be used to provide guidelines. Figure 4.3 shows colonization by *H. comma*, and the lines show the probability that a formerly empty habitat patch will be colonized over a 9 year period. Potentially, a manager could read off such a graph to make decisions about habitat management, aiming perhaps to create a habitat that has greater than 50% probability of being colonized within 9 years.

In the absence of direct information on colonization, which is slow to acquire, can existing distributions be used to predict colonization? For *H. comma*, colonization (Figure 4.3) and occupancy (Figure 4.4) graphs are extremely similar to one another, although one should bear in mind that the two sets of data are not independent. This implies that non-dynamic (single-year) distributions could be used to predict which new habitat patches are likely to be colonized, although they cannot predict a time scale for colonization. None the less, in periods of up to 10 years the most isolated local populations of *H. comma* and *M. athalia* are the same as the maximum observed colonization distances, and only 2 out of 99 local *P. argus* populations were (slightly) more isolated than the maximum observed colonization distance. This suggests that a manager could reasonably expect colonization within 5–10 years if a sufficiently large area (see Figure 4.4 and Thomas, 1984a; Warren, 1992a) of new breeding habitat was created within the range of existing nearest-neighbour distances. It may be advisable to exclude perhaps the most isolated 5% of nearest-neighbour distances, because these sites might have been colonized at a time when they were not so isolated, or they may have remained as empty habitat for decades before colonization. This simplistic approach is imperfect, but it is likely to work most of the time and has the advantage of being extremely easy to use. Suppose that a reserve manager is interested in establishing a population of a rare species in a reserve. All the manager has to do is measure the distance from the nearest local population (not every other population) to different parts of the reserve, and read off the graph to see whether any feasible combination of area and distance would give, say, a greater than 50% probability that the patch would eventually be populated. If not, the manager could then look at the intervening land and examine the possibility of providing habitat stepping-stones, either on another reserve or by agreement with a private landowner. Similarly, the potential for linking reserves which are currently isolated could be explored.

Separate incidence functions will be needed for a further 25–30 species, but they will (usually) only need to be plotted once for each species. This can be achieved relatively easily by carrying out a thorough survey of all local populations and unoccupied habitat in a small fraction of a species' range. The region selected must include areas where maximum isolation values can be large.

4.5   INTRODUCTIONS

Increasingly, managers and enthusiasts are turning to re-establishment as a means of conserving butterflies in fragmented landscapes. The logic is good. Humans have fragmented natural habitats such that many butterfly species are no longer capable of dispersing through the British countryside. Why should we not restore connectivity by moving insects from patch to patch? This is an important conservation option, but it should always take third place to:

1. maintaining existing habitat connectivity, and
2. restoring connectivity by management which provides new habitat stepping-stones (C. Thomas, 1992).

The first two options will favour a wide variety of invertebrates, not just butterflies.

A hierarchy of questions should be asked before carrying out an introduction to unoccupied, but apparently suitable, habitat. First, 'Is the species likely to colonize naturally?' If an area is unlikely to be recolonized naturally in the next 10–100 years, the appropriate second question is 'Could any of the intervening vegetation be managed to create new habitat and increase the rate of spread?' Only if the answer is again 'no' should we proceed to 'Is establishment viable?' The answer to this question should be 'yes' only when a potential release site is considerably larger than the minimum area required by single local populations (as a very rough guide, multiply by $\geqslant 5$ the minimum local areas given by Thomas (1984a) and Warren (1992a)) or when a new metapopulation can be established. Establishments of small, isolated local populations are doomed to eventual failure, and their temporary success lulls us into a false sense of security (C. Thomas, 1992). By far the most successful introductions have been where butterflies have been transferred across tens of kilometres of unsuitable habitat (much further than natural colonization distances) to a whole series of empty habitat patches in which it has been possible to establish a new metapopulation (Oates and Warren, 1990; C. Thomas, 1992). Yet, because the species in question is unlikely to have been recorded there in the past (if it had, it would probably still be present), these introductions would not be encouraged by existing guidelines on (re)establishment

(Anon, 1986). Simulation models can be used to assess whether an introduced metapopulation will spread and whether it is likely to persist (Hanski, 1994; Hanski and Thomas, 1994).

4.6   PRIORITY REGIONS

Because most British butterflies are widespread European species, conservation in Britain has been based principally on status in Britain. There is a tendency to concentrate on species that are rare in Britain, and to put the greatest effort into preserving them in parts of the country where they are particularly rare. This has some value because refuge populations may eventually act as foci for recolonization, but the fate of most refuge populations is to become extinct rather than to initiate a phase of recolonization. A metapopulation approach suggests that we should target regions where rare species are still relatively common, regions which still represent a connected landscape for a particular species.

Warren's (1994) research on *E. aurinia* highlights the plight of a particularly important species. Declining throughout Europe, western Britain apparently remains one of its strongholds. But much more concern seems to be generated by the loss of a local population in eastern or central Britain, where it is nearly extinct, than by the loss of local populations in the west. *Eurodryas aurinia* has been the subject of many attempts to re-establish it in isolated habitat patches in central and eastern England, in areas where there is little hope of persistence. The priority must be to ensure the continued persistence of viable metapopulations in the west. Other British species clearly have a much better chance of persisting in some regions than in others. Examples include *Hamearis lucina* (Duke of Burgundy) on Salisbury Plain and in the Cotswold Hills (Oates and Emmet, 1990), and *P. argus* on relatively extensive heathlands in Surrey/Hampshire, Dorset, and in the New Forest.

4.7   LANDSCAPE CONSERVATION

The interconnected nature of the landscape allows metapopulations to persist and allows them to track changes in the spatial distribution of suitable habitats. Most legal battles over proposed developments are fought over the importance of a single site: the presence of similar examples of the same habitat nearby that support populations of the same rare species is often used as an excuse for allowing development. A metapopulation approach to conservation could be used to argue that these are the very areas we should target for conservation. The approach means that we should regard every patch of habitat as potentially important, not just the best examples.

Unfortunately, we will not succeed in saving every remaining habitat fragment in Britain, and conservation organizations will never own the bulk of the landscape. To maintain and expand metapopulations of rare species will require increasing co-operation with other land owners. Rather than fight every development, there is much more scope for biologists and conservation organizations to influence development so that damage to existing systems is minimized, and benefits from new habitats are maximized. Conservation organizations need to be involved proactively when there is potential to create new habitats, and not just become involved when existing habitats are threatened. This is already happening to some extent, but not enough. For each rare species, we should be asking how we can increase local abundances, and how we can protect, manage and create habitats that will allow distributions to expand. We should not be content simply to slow the rate of decline.

4.8 ACKNOWLEDGEMENTS

I thank Susan Harrison, Ilkka Hanski, Neil Ravenscroft, Jeremy Thomas and Martin Warren for access to unpublished manuscripts and information.

PART TWO

Butterflies and Land Use in Britain

CHAPTER 5

Butterflies on nature reserves in Britain

*I.F.G. McLean, A.P. Fowles, A.J. Kerr, M.R. Young
and T.J. Yates*

5.1 INTRODUCTION

Nature reserves in Britain have been established by statutory conservation agencies, local authorities, voluntary conservation organizations, commercial companies and private individuals. While the majority of these sites have been selected for acquisition or lease because they represent good examples of particular biotopes, a few have achieved nature reserve status due to the presence of significant populations of scarce or threatened species, including butterflies. There is no comprehensive guide to all Britain's nature reserves, although several books summarize the location and interest of many of them. Hywel-Davies, Thom and Bennett (1986) give a general account of a wide selection of nature reserves; Smith (1982) gives a general account of a selection of Royal Society for Nature Conservation (RSNC) Trusts' reserves; Chapman (1987) and Newbery (1983) review Royal Society for the Protection of Birds (RSPB) reserves. The role of nature reserves in conserving British butterflies, and information on the representation of butterflies on nature reserves, has been reviewed previously in publications and unpublished reports: for National Nature Reserves (NNRs) and other statutory sites see Morris (1967) and Peachey (1982); Pollard, Hall and Bibby (1986) include accounts of butterflies on NNRs with Butterfly Monitoring Scheme (BMS) transects; for reserves administered by the RSNC Trusts see Steel and Parsons (1989); for RSPB reserves see Cadbury (1990); for Butterfly Conservation (BC) reserves see Butterfly Conservation (1992). Ratcliffe (1977) includes a general review of the conservation of butterflies and moths for the series of Nature Conservation Review (NCR) key sites. Thomas (1984a) and Warren (1992b) present broader reviews which include analyses of the roles of nature reserves and examples of the failures and successes of conserving butterflies on these sites.

Ecology and Conservation of Butterflies Edited by Andrew S. Pullin.
Published in 1995 by Chapman & Hall. ISBN 0 412 56970 1

Table 5.1 National Nature Reserves, RSNC and RSPB reserves in Great Britain in 1990

	Number of reserves	Total area (ha)
NNRs		
England	121	41 312
Scotland	68	112 241
Wales	46	12 798
Great Britain	235	166 351
Other reserves		
RSNC Wildlife Trust Partnership	About 2000	52 000
RSPB	118	74 700

In 1990, at the start of the period for this review, there were 235 NNRs in Great Britain (Table 5.1). This is a substantial and welcome increase from the 170 NNRs declared when Peachey conducted the last review of butterflies on NNRs (Peachey, 1982). In the same year the RSNC and RSPB reserves extended to nearly 127 000 ha (Table 5.1). Many of the RSNC Trusts' reserves are small, but nevertheless they include many significant sites for butterflies.

This chapter discusses the results of a recent review of butterflies on NNRs, and also includes brief accounts of RSNC and RSPB reserves drawn from publications and unpublished reports. The various ways nature reserves contribute towards butterfly conservation are illustrated here by reference to examples of projects undertaken on nature reserves in England, Scotland and Wales. Our aim is to indicate what has been achieved so far and to consider briefly future developments that can increase the contributions made by nature reserves to butterfly conservation, as well as suggesting how butterfly conservation itself may refine and assist conservation of other wildlife.

5.2   METHODS

This review draws on new data for butterflies on NNRs obtained via a question-naire sent to wardens and site managers in 1993. The questionnaire (Figure 5.1) asked whether a butterfly transect is recorded, within or outside the BMS co-ordinated by the Institute of Terrestrial Ecology (ITE).

It also sought information on whether specific management for butterflies had been carried out in the 3 year period 1990–92. For 53 species of resident British butterflies (those listed in Table 5.3), excluding migrant species, the extinct *Maculinea arion* (large blue) and *Nymphalis polychloros* (large tortoiseshell) and the pests of cultivated crops, *Pieris rapae* (small white) and *P. brassicae* (large white), respondents were asked to assess the status of butterflies recorded within each reserve over the period 1990–92. The subjective categories chosen (vagrant, strong, moderate, weak and present but of unknown status) are the same as those used in a similar questionnaire survey by Peachey (1982), allowing direct comparison with that review.

The guidance notes accompanying the questionnaire included the following definitions of the status categories.

1. Vagrant: observed on the NNR but not a breeding resident.
2. Strong: a large breeding population present each year.
3. Moderate: species breeds on the NNR but not in large numbers.
4. Weak: only a small breeding population present.
5. Present unknown: known to be resident but population size unknown.

Respondents were also asked to state the last year in which species were observed for those species that had become extinct since 1980, and the year in which species were first observed for those butterflies that had colonized the reserve since 1980.

The results from the NNR questionnaire were also compared with quantitative results from a sample of 16 NNRs participating in the BMS during the period 1990–92. The assessment of the status of butterflies on NNRs from BMS transects was undertaken by one of us (TJY) without reference to the returns from site managers and wardens. While some species are known to be under-recorded by the BMS method (notably Theclinae, the hairstreaks), others may be under-represented if the transect misses out areas where the adults occur; conversely some butterflies may appear to be relatively more abundant than they really are if the transect passes through breeding or nectaring areas where adults spend much of their time. Nevertheless, the comparison between the subjective assessment of status from the questionnaire, and an interpretation of results from the transects, gives a simple means of examining the level of agreement between the two approaches. Neither method is completely reliable for all sites and species; the numerical assessments of the BMS are inherently more precise, but these are available for fewer reserves than have records of butterflies. If these two methods give similar results when compared on the same sites, then it is more likely that the subjective scoring of the status of butterflies (as used in previous and present questionnaire surveys) agrees with the more quantitative approach of the BMS, at least when the latter is translated into an equivalent subjective assessment.

5.3 RESULTS

The results of the questionnaire survey for NNRs are presented in Tables 5.2 and 5.3.

Returns were received from 127 sites in England, 41 in Scotland and 49 in Wales, giving a total of 217 NNRs. Of these returns, 22 NNRs had no records of butterflies for the period 1990–92.

Of the 217 NNRs, 53 (24% of all NNRs in the survey) had contributed to the national BMS during the period 1990–92 and there were a further 28 recorded transects which were not part of this scheme, but used the same methods. Four of

QUESTIONNAIRE

BUTTERFLIES ON NATIONAL NATURE RESERVES 1990-1992

Please tick or fill in the boxes which apply

NNR name   **Grid reference**

Name of compiler   **Tel. no.**

Address

Is a BMS transect recorded?   Yes in BMS project   □
                                   Yes results not sent to BMS   □
                                   No   □

Has any specific management for butterflies been undertaken since 1990?
                                   No   □
                                   Yes (continue in next section) □

Specify type of management   Ride   □
                                   Coppice   □
                                   Scrub clearance   □
                                   Grassland management   □
                                   Wetland management   □
                                   Other (please summarise below) □

Please complete boxes for butterflies seen for 1990-1992 only, tick in box which applies for those species present on the reserve. For species known to have become extinct since 1980 please give year last observed, for species which have colonised since 1980 please give year first observed in status box.

Figure 5.1   The first two pages of the questionnaire sent out to wardens and site managers seeking information on the status of butterflies on NNRs for the period 1990–92.

Species	Vagrant	Strong	Moderate	Weak	Present unknown
SKIPPERS					
Chequered skipper					
Dingy skipper					
Essex skipper					
Grizzled skipper					
Large skipper					
Lulworth skipper					
Silver-spotted skipper					
Small skipper					
Swallowtail					
WHITES					
Brimstone					
Green-veined white					
Orange tip					
Wood white					
BLUES, COPPERS & HAIRSTREAKS					
Black hairstreak					
Brown hairstreak					
Green hairstreak					
Purple hairstreak					
White-letter hairstreak					
Large copper					

Figure 5.1 Continued.

Table 5.2 Habitat management for butterflies carried out on National Nature Reserves from 1990 to 1992

Type of management	Number of NNRs	Percentage of 81 sites managed
Ride	36	44
Coppice	27	33
Scrub	44	54
Grassland	54	67
Wetland	13	16
Other[a]	12	15

More than one type of management was undertaken on some sites, hence the total for all management activities is greater than 81, and the percentage column totals more than 100%.

[a] The 'other' category includes establishing enclosures for woodland regeneration, removal of conifers, *Pteridium aquilinum* control and selective management, wet heath management, management for *Eurodryas aurinia*, *Coenonympha tullia*, *Satyrium pruni* and *Plebejus argus* and protecting *Ulmus* spp. from river-bank erosion for *Satyrium w-album*.

Management for other species was included under the five major specific categories listed.

the transects outside the BMS were recorded on NNRs with BMS transects. Thus, a total of 77 NNRs (35%) had undertaken systematic butterfly monitoring; this is a substantial proportion of these élite nature conservation sites, and indicates the strong level of support that the BMS approach now receives from those responsible for managing NNRs.

Management for butterflies on NNRs was most frequently directed towards grassland and scrub (Table 5.2), although there was also a substantial amount of ride and coppice management in woodlands. The smaller number of wetlands managed is a reflection of the lower total of wetland NNRs, combined with the fact that there are few wetland butterflies in Britain.

A total of 81 sites received specific management with the objective of conserving their butterfly populations. There was a high degree of overlap between the sites where monitoring was carried out and those where management for butterflies was undertaken; 54 of the 77 sites with a butterfly transect had received management for butterflies. There were 23 NNRs with butterfly transects where no specific management for butterflies was recorded and 27 NNRs which received management for butterflies but where there was no transect to measure the results achieved. This suggests that the availability of accurate information on butterfly abundance from transects increases the likelihood that the needs of butterflies will be addressed by management designed to sustain or enhance their numbers.

Results from the questionnaire survey are presented in Table 5.3, with the number of NNRs given for each category under every species, and the percentage of records in each status category expressed as a proportion of the total number of records for that species.

The results of comparing the NNR questionnaire survey with returns made to the BMS are presented in Table 5.4.

This shows good match between the status assessments derived from these two sources, although there are some inconsistencies. A few species recorded on particular BMS transects were not acknowledged as present in the questionnaire return for that site. Rather more records were included in questionnaire returns of species not noted on the BMS site transect, particularly where these were recorded as being weak populations. This is to be expected where a species is restricted to a part of the site not covered by the transect route. *Thymelicus lineola* (Essex skipper) is not generally separated from *T. sylvestris* (small skipper) on most transects; the former was the most frequent species where a status was given on the questionnaire, but could not be derived from the BMS results. There were some differences in the degree to which assessments coincided according to the recorder/site concerned; also, there were some differences between species, but the significance of these is hard to determine given the relatively small sample size. The three example species shown in Table 5.5 illustrate some of these differences.

Ochlodes venata (large skipper) has populations most frequently assessed as moderate, while *Callophrys rubi* (green hairstreak) tends to occur in small areas and has populations typically assessed as weak. *Pyronia tithonus* (gatekeeper) is often found in high numbers where it occurs, hence it has a higher proportion of sites assessed as having strong populations.

5.4   EXAMPLE CASE HISTORIES WHERE NNRS HAVE PROVIDED STUDY SITES FOR DETAILED RESEARCH ON THREATENED SPECIES

5.4.1   *Mellicta athalia* (heath fritillary) in south-east England

The ecology of this threatened species has been studied in detail by Warren (1987a,b,c). Throughout the period of this study the only woodland site where specific management for the butterfly was sustained was Blean Woods NNR, Kent. The length of the coppice rotation, size of coppice plots and management of rides and glades were all modified to match the needs of the species (Warren, 1987c). The nearby RSPB reserve has also undertaken significant management for this species in recent years, in part with help from British Butterfly Conservation Society (BC) volunteers. In addition to the direct contribution made by these reserves towards maintaining healthy populations of this butterfly, recording its response to special management measures enabled improved management recommendations to be formulated for other sites.

5.4.2   *Carterocephalus palaemon* (chequered skipper) in north-west Scotland

Carterocephalus palaemon was last recorded in England in 1975 after a period of decline and disappearance from NNRs in the East Midlands (Collier, 1986). It was one of several woodland butterflies that were lost from NNRs in this area during a period when there was less woodland management (coppicing and ride

Table 5.3   The representation of resident British butterflies on 217 National Nature Reserves from 1990 to 1992 assessed by a questionnaire survey

Species	Vagrant	Strong	Moderate	Weak	Present	Total	Vagrant (%)	Strong (%)	Moderate (%)	Weak (%)	Present (%)	Sites (%)
Carterocephalus palaemon	0	0	1	0	0	1	0	0	100	0	0	0
Thymelicus sylvestris	0	28	35	15	14	92	0	30	38	16	15	43
Thymelicus lineola	0	4	10	8	6	28	0	14	36	29	21	13
Thymelicus acteon	0	0	0	0	0	0	0	0	0	0	0	0
Hesperia comma	0	1	2	1	1	5	0	20	40	20	20	2
Ochlodes venata	0	14	54	20	17	105	0	13	51	19	16	49
Erynnis tages	1	3	14	12	9	39	3	8	36	31	23	18
Pyrgus malvae	2	1	7	5	6	21	10	5	33	24	29	10
Papilio machaon	1	2	0	0	0	3	33	67	0	0	0	1
Leptidea sinapis	0	0	0	4	1	5	0	0	0	80	20	2
Gonepteryx rhamni	12	16	29	18	12	87	14	18	33	21	14	40
Pieris napi	4	44	56	23	38	165	2	27	34	14	23	76
Anthocharis cardamines	11	14	48	34	21	128	9	11	38	27	16	59
Callophrys rubi	2	5	13	20	22	62	3	8	21	32	35	29
Thecla betulae	0	0	2	2	4	6	0	0	0	33	67	3
Quercusia quercus	1	8	15	7	10	41	2	20	37	17	24	19
Satyrium w-album	1	1	1	6	6	15	7	7	7	40	40	7
Satyrium pruni	0	0	1	1	1	3	0	0	33	33	33	1
Lycaena phlaeas	9	19	37	42	24	131	7	15	28	32	18	61
Lycaena dispar	0	0	0	1	0	1	0	0	0	100	0	0
Cupido minimus	0	3	4	4	5	16	0	19	25	25	31	7
Plebejus argus	0	2	2	1	0	5	0	40	40	20	0	2
Aricia agestis	1	4	7	16	5	33	3	12	21	48	15	15
Aricia artaxerxes	0	2	2	1	1	6	0	33	33	17	17	3
Polyommatus icarus	4	27	52	29	28	140	3	19	37	21	20	65
Lysandra coridon	0	8	3	2	3	16	0	50	19	13	19	7
Lysandra bellargus	0	3	1	2	1	7	0	43	14	29	14	3

Table 5.3 Continued

Species	Vagrant	Strong	Moderate	Weak	Present	Total	Vagrant (%)	Strong (%)	Moderate (%)	Weak (%)	Present (%)	Sites (%)
Celastrina argiolus	7	10	33	27	13	90	8	11	37	30	14	42
Hamearis lucina	0	0	4	3	0	7	0	0	57	43	0	3
Ladoga camilla	3	1	7	8	5	24	13	4	29	33	21	11
Apatura iris	0	0	0	1	3	4	0	0	0	25	75	2
Aglais urticae	5	37	58	25	47	172	3	22	34	15	27	80
Inachis io	15	34	47	16	22	134	11	25	35	12	16	62
Polygonia c-album	12	2	32	33	12	91	13	2	35	36	13	42
Boloria selene	2	7	11	14	17	51	4	14	22	27	33	24
Boloria euphrosyne	1	1	6	8	8	24	4	4	25	33	33	11
Argynnis adippe	1	1	1	3	3	9	11	11	11	33	33	4
Argynnis aglaja	7	8	11	15	16	57	12	14	19	26	28	26
Argynnis paphia	2	6	3	11	7	29	7	21	10	38	24	13
Eurodryas aurinia	2	3	4	3	1	13	15	23	31	23	8	6
Melitaea cinxia	0	0	0	0	0	0	0	0	0	0	0	0
Mellicta athalia	0	1	0	0	0	1	0	100	0	0	0	0
Pararge aegeria	2	28	39	17	22	108	2	26	36	16	20	50
Lasiommata megera	8	15	35	33	16	107	7	14	33	31	15	50
Erebia epiphron	0	0	0	0	1	1	0	0	0	0	100	0
Erebia aethiops	1	3	2	0	6	12	8	25	17	0	50	6
Melanargia galathea	6	13	10	3	3	35	17	37	29	9	9	16
Hipparchia semele	4	14	14	14	9	55	7	25	25	25	16	25
Pyronia tithonus	1	54	22	11	11	99	1	55	22	11	11	46
Maniola jurtina	4	92	28	6	28	158	3	58	18	4	18	73
Aphantopus hyperantus	2	33	24	19	17	95	2	35	25	20	18	44
Coenonympha pamphilus	1	42	35	37	31	146	1	29	24	25	21	68
Coenonympha tullia	0	9	8	0	16	33	0	27	24	0	48	15

Table 5.4 A comparison of NNR questionnaire results for all butterflies from 16 sites with an assessment made from examining BMS data

BMS	NNR Q					
	Absent	*Vagrant*	*Strong*	*Moderate*	*Weak*	*Present*
Absent		4	1	9	12	5
Vagrant	1	**2**	1	2	2	2
Strong	3		**67**	14		
Moderate			36	**61**	15	1
Weak	4	8	7	44	**41**	9
Present				1		**1**

Numbers in bold represent an exact match between the assessments.

BMS indicates assessments made from BMS transect results by TJY; NNR Q indicates assessments made by wardens and site managers for their reserves in questionnaire responses.

Table 5.5 A comparison of NNR questionnaire results with an assessment made from BMS data

BMS	NNRs Q					
	Absent	*Vagrant*	*Strong*	*Moderate*	*Weak*	*Present*
Ochlodes venata[a]						
Absent						
Vagrant						
Strong			**1**	1		
Moderate			2	**5**		
Weak				3		2
Present						
Callophrys rubi[b]						
Absent						2
Vagrant						
Strong						
Moderate				**1**		
Weak				2	**3**	1
Present						
Pyronia tithonus[c]						
Absent						
Vagrant						
Strong			**7**	1		
Moderate			3	**2**		
Weak						
Present						

[a] Data from 14 sites.
[b] Data from 9 sites.
[c] Data from 13 sites.

Numbers in bold represent an exact match between the assessments.

maintenance) than in earlier decades of the twentieth century. It was also lost from Woodwalton Fen NNR, Cambridgeshire, following a summer flood, the last individuals being seen in 1968. After these losses it is more encouraging to report that studies by Ravenscroft of populations in north-west Scotland have made substantial progress in identifying its habitat requirements (for more information see Ravenscroft, 1991, Chapter 12 this volume). His principal study sites were Ariundle Oakwood NNR and Glasdrum NNR, where detailed research was undertaken on each life stage. *Carterocephalus palaemon* is monitored on the BMS transect at Ariundle NNR, the only site where this is possible at present.

5.4.3 *Eurodryas aurinia* (marsh fritillary) in Wales

Eurodryas aurinia has declined considerably during the twentieth century (Heath, Pollard and Thomas, 1984) mainly due to drainage and agricultural improvement of its wet grassland and other habitats. At Rhos Llawr-cwrt NNR, monitoring of adults and larval nests has been undertaken since 1984 (Fowles, 1985). This site supports one of the largest populations in Britain, and conservation of the butterfly is a major aspect of reserve management. Knowledge gained from the successful management of this population will have wider applications for the conservation of *E. aurinia* elsewhere in western Britain. Studies are currently underway to investigate the relationship between grazing pressure, vegetation structure, the condition of the larval food-plant, and fluctuations in the population of the butterfly (Woolley, 1993).

5.5 BUTTERFLIES THAT ARE POORLY REPRESENTED ON NATURE RESERVES

The suites of nature reserves administered by the statutory agencies, the RSPB, RSNC Trusts and BC, each have limitations with respect to the butterflies present in different biotopes. Nevertheless, it is remarkable how well the scarcer, threatened species are represented. It is also encouraging that the National Trust recognizes the significance of its important butterfly sites, and also that it now has a nature conservation adviser with a special remit for butterflies. For some of those butterflies currently found on few nature reserves, there is satisfactory representation on Sites of Special Scientific Interest (SSSIs); however, there are continuing losses of key butterfly species from SSSIs in southern England (Warren, 1993a). These are largely due to difficulties with ensuring that appropriate positive management can be instigated and sustained on these sites. Correctly identifying the management requirements of butterflies and other wildlife on SSSIs, followed by consistent application of the necessary management each year, are essential if these areas are to be better conserved in future.

Butterflies poorly represented on nature reserves in Britain are summarized in Table 5.6 for NNRs and in the discussion on butterflies on RSNC Trusts' reserves and on RSPB reserves below.

Table 5.6 Butterflies reported as having resident populations on less than 10 NNRs from 1990 to 1992

Species	Total sites
Carterocephalus palaemon	1
Thymelicus acteon	0
Hesperia comma	5
Papilio machaon	2
Leptidea sinapis	5
Thecla betulae	6
Satyrium pruni	3
Lycaena dispar	1
Plebejus argus	5
Aricia artaxerxes	6
Lysandra bellargus	7
Hamearis lucina	7
Apatura iris	4
Argynnis adippe	8
Melitaea cinxia	0
Mellicta athalia	1
Erebia epiphron	1

Overall, *C. palaemon*, *Thymelicus acteon* (Lulworth skipper), *Hesperia comma* (silver-spotted skipper), *M. athalia*, *Melitaea cinxia* (Glanville fritillary) and *Erebia epiphron* (mountain ringlet) are the most poorly represented on nature reserves, although they are all present on SSSIs. Several species, including *Argynnis adippe* (high brown fritillary), *M. cinxia* and *M. athalia*, also have important colonies on properties owned by the National Trust and the National Trust for Scotland.

There has been long-standing concern over losses of key butterfly species from nature reserves in Britain (Collier, 1978; Thomas, 1984a; Warren, 1992a). Although the questionnaire concerning butterflies on NNRs from 1990 to 1992 requested information about butterfly extinctions (and colonizations), few responses included this information. This may be because of a lack of clear evidence about extinction and colonization events (which are difficult to document), coupled with staff changes over the period since the previous review. The summary data on changes in representation of butterflies on NNRs, given in Table 5.7, show a general increase in the number of NNRs with populations of the 53 butterflies reviewed here.

This is believed largely to be due to declaration of additional NNRs over the intervening period (the number of reserves with records of butterflies increased from 138 in 1980 to 217 in 1990–92), combined with the influence of improved recording. The apparent decline in NNRs with populations of *E. epiphron* is believed to be due to absence of records rather than showing a genuine decline. While the general increased representation of butterflies on NNRs from 1980 to 1990–92 is a welcome trend, these results should be interpreted with caution.

Table 5.7 Changes in representation of selected butterflies on NNRs between 1980 and 1990–92

Species	Total for NNRs in 1980	Total for NNRs in 1990–92	Change from 1980 to 1990–92
Carterocephalus palaemon	1	1	–
Thymelicus sylvestris	44	92	+48
Thymelicus lineola	12	28	+16
Thymelicus acteon	0	0	–
Hesperia comma	3	5	+2
Ochlodes venata	47	105	+58
Erynnis tages	18	38	+20
Pyrgus malvae	12	19	+7
Papilio machaon	3	2	−1
Leptidea sinapis	2	5	+3
Gonepteryx rhamni	44	75	+31
Pieris napi	91	161	+70
Anthocharis cardamines	51	117	+66
Callophrys rubi	28	60	+32
Thecla betulae	0	6	+6
Quercusia quercus	18	40	+22
Satyrium w-album	5	14	+9
Satyrium pruni	2	3	+1
Lycaena phlaeas	58	122	+64
Lycaena dispar	1	1	–
Cupido minimus	8	16	+8
Plebejus argus	3	5	+2
Aricia agestis	20	32	+12
Aricia artaxerxes	6	6	–
Lysandra coridon	10	16	+6
Lysandra bellargus	3	7	+4
Polyommatus icarus	76	136	+60
Celastrina argiolus	28	83	+55
Hamearis lucina	4	7	+3
Ladoga camilla	6	21	+15
Apatura iris	0	4	+4
Aglais urticae	85	167	+82
Inachis io	66	119	+53
Polygonia c-album	36	79	+43
Boloria selene	26	49	+23
Boloria euphrosyne	14	23	+9
Argynnis adippe	3	8	+5
Argynnis aglaja	35	50	+15
Argynnis paphia	9	27	+18
Eurodryas aurinia	6	11	+5
Melitaea cinxia	0	0	–
Mellicta athalia	1	1	–
Pararge aegeria	40	106	+66
Lasiommata megera	47	99	+52
Erebia epiphron	3	1	−2
Erebia aethiops	11	11	–
Melanargia galathea	16	29	+13
Hipparchia semele	29	51	+22
Pyronia tithonus	51	98	+47
Maniola jurtina	82	154	+72
Aphantopus hyperantus	44	93	+49
Coenonympha pamphilus	88	145	+57
Coenonympha tullia	18	33	+15

A detailed comparison of records from NNRs assessed in 1980 with those also recorded in 1990–92 would be needed to determine the changes in status that have taken place for individual species and butterfly assemblages on NNRs. This is outside the scope of this chapter, although the available data are potentially suitable for such analyses.

5.6 DISCUSSION

This review indicates that British butterflies are generally well represented on NNRs and other nature reserves. There continue to be some gaps in coverage, with some scarce and declining species poorly represented on nature reserves. For *T. acteon* and *M. cinxia*, their absence from NNRs is counterbalanced by the strength and relative security of their populations on Dorset coastal grasslands and the slumping cliffs of the Isle of Wight, respectively. The results from the questionnaire indicate that the presence of a BMS transect on a NNR is linked to the likelihood of specific management being undertaken for butterflies. Comparing status assessments derived from questionnaire returns with an independent subjective ranking of status from BMS transects showed a good match in status allocated; this suggests that a questionnaire survey gives a fair overall picture of the status of butterflies on NNRs, although some inconsistencies were recognized due to differences between observers, coupled with various levels of apparency for different butterflies.

Steel and Parsons (1989) reviewed the representation of 36 scarcer butterflies (termed target species) on 384 RSNC Trusts' reserves for the period 1980–84. Information was obtained from most counties in England and Wales, with nearly 24% of the then total of reserves reporting the presence of target species. There were only four target species not represented on RSNC Trusts' reserves: *C. palaemon*, *N. polychloros*, *M. cinxia* and *E. epiphron*. Seven target species were reported from fewer than 10 reserves: *T. acteon*, *H. comma*, *Papilio machaon* (swallowtail), *Satyrium pruni* (black hairstreak), *M. athalia*, *Coenonympha tullia* (large heath) and *Erebia aethiops* (Scotch argus).

Cadbury (1990) published a review of the occurrence and conservation of butterflies on RSPB reserves. This was based upon records supplied by wardens from 91 reserves, in most cases for the period 1984–89. Of 55 butterflies that breed regularly in Britain, Cadbury stated that 52 species had been recorded, and he regarded 46 of these as breeding regularly on RSPB reserves in the 1980s. The biotope representation of RSPB reserves is broad, with only southern calcareous grassland (which supports few characteristic bird species, but many butterflies) not represented. This lack of southern calcareous grassland has resulted in *Lysandra bellargus* (adonis blue) and *H. comma* not being represented on RSPB reserves. *Lysandra coridon* (chalkhill blue), *Cupido minimus* (small blue) and *Hamearis lucina* (Duke of Burgundy) have been reported only as strays on RSPB reserves from nearby calcareous grassland. Butterfly monitoring transects have been conducted on 36 reserves, so, as with NNRs, many wardens can observe changes taking place in butterfly populations over time and, by comparing site trends with national changes, assess the consequences of local factors (including site management). RSPB reserves support significant populations of several scarce and threatened butterflies. These include *P. machaon* (at Strumpshaw Fen and Surlingham reserves in the Norfolk Broads), *M. athalia* (at Blean Woods, Kent) and *A. adippe* (in its remaining north-west England stronghold at Wharton Crag and Leighton Moss, Lancashire, and also in smaller numbers at Coombes Valley, Staffordshire).

BC has greatly increased its membership and support of butterfly conservation activities in recent years. Butterfly Conservation (1992) summarized its role and aspirations in seeking to reverse the recent declines of British butterflies. With a local structure based around county or regional branches, members undertake monitoring of important butterfly sites and threatened species, some of which are now protected as BC Reserves by purchase or agreement. Although a synopsis of the butterflies represented on these reserves could not be included in this review, it is worth noting that many significant species are now breeding on sites owned or managed by BC, including *P. machaon* at Catfield Fen, Norfolk.

5.7 CONCLUSIONS

The discussion in this chapter indicates that nature reserves in Britain have made a substantial contribution towards conserving our native butterflies. They will continue to do so by maintaining populations of both rare and more widespread species and by acting as refugia, from whence some species may be able to recolonize parts of the surrounding countryside, should more land be managed in ways compatible with sustaining wildlife in future (see Thomas, Chapter 4 this volume). Furthermore, where neighbouring reserves can be linked by corridors managed so as to facilitate dispersal of butterflies, this may increase the chance of successful recolonization after local extinction. Thus, the management of nature reserves should be considered in conjunction with measures designed to restore and enhance the value of surrounding land for butterflies and other wildlife. The challenge over the coming years is to use existing measures (such as designation of Environmentally Sensitive Areas and grants from countryside agencies to deliver positive management for wildlife) until such time as economic use of the country-side is more in line with sustaining a wide range of characteristic species.

Nature reserves can improve their effectiveness in conserving our butterflies by being more responsive to the changes in abundance of species with specialized habitat requirements. This should be brought about by greater use of BMS transects and comparing the results against national and regional trends for each breeding species. The next step is to ensure that, wherever possible, habitat management on nature reserves is compatible with the requirements of breeding butterflies. Furthermore, such management should be 'fine tuned' by assessing whether it is increasing or decreasing butterfly populations in comparison with BMS national and regional trends. The populations of scarcer butterflies, which typically are demanding in their habitat needs, are good targets to assess the success of meeting objectives for nature reserves. They should be used in conjunction with other characteristic wildlife regarded as being of special importance on each site.

Butterfly conservation has the potential to continue to make further improvements to wildlife conservation as a whole. Butterflies are particularly sensitive indicators of some types of environmental changes. They respond more rapidly than plants to changes in the structure and microclimate of their habitats. Thus they can act as sentinel species on behalf of other wildlife, a decline in their abundance (when interpreted against national or regional trends) sending an early warning that improved management is needed on the reserve in question. Nevertheless, it should be borne in mind that butterflies are not associated with all significant habitat features, and so sentinels with other life history requirements should be selected to achieve a balanced approach to reserve management. Other points in favour of butterflies include the practical point that there is a manageable number of species, which are generally conspicuous, with most being easy to identify. Also, the BMS method for assessing the relative abundance of adult butterflies is well established and straightforward to use with relatively little training. The BMS is the only national monitoring scheme currently operational over a substantial number of NNRs and other nature reserves. Thus butterflies can be viewed as having a lead role in evaluating the responses of nature reserves to management and other influences, thereby contributing to wildlife conservation as a whole.

Therefore, monitoring the abundance of butterflies using the BMS method should remain a high priority for many nature reserves, while regular assessments of butterfly populations should be part of the surveillance of most other nature reserves. Monitoring other sites with populations of threatened butterflies is also highly desirable. The survey of NNRs carried out for this chapter indicates that more regular butterfly recording is needed on many of these reserves. There is an important role here for volunteers to assist the statutory conservation agencies, and those other conservation organizations which own or manage nature reserves, with butterfly transects or counts of particularly significant species. It is also vital that different conservation organizations share their skills and experience; BC with its increasing membership and local branches is well placed to help others manage their reserves to favour butterflies.

Nature reserves have a role in providing areas where species can be studied in detail to advance our understanding of their ecology. The examples given in this chapter show that nature reserves, managed primarily to sustain their wildlife interest, have played a significant role in conserving threatened British butterflies through acting as research sites. With a high proportion of our scarcer butterflies needing continuity of habitat management, often with annual inputs to ensure creation or regeneration of early successional stages in biotopes, it is essential to ensure that scarce and valuable research efforts are applied where there is a high probability of continuing appropriate and precisely regulated management. When research is coupled with the presence of a butterfly monitoring transect, which enables the responses of most butterflies to habitat changes to be assessed, there is valuable information available to set research results in context, as well as a better chance of successfully conserving the more specialized and demanding species. It is also important to determine whether specific management activities undertaken for the benefit of butterfly populations are having the desired effects.

Because nature reserves should be managed with the needs of wildlife given precedence over other concerns, there is ample scope for alternative, often experimental, management approaches to be tried, and their effects upon butterflies assessed. In many cases, benefits for butterflies will be mirrored by increases in other wildlife. Cases where there is conflict between the management needs of butterflies and the requirements of other wildlife are, fortunately, very rare in our experience. The greatest challenges in achieving a balance between the needs of different wildlife groups are usually on small reserves, where there may be insufficient scope for sustaining all potential habitat conditions in sufficient abundance over long periods of time.

Finally, nature reserves can provide educational and recreational opportunities for people to see wildlife at first hand. Sadly, in many parts of Britain a significant number of butterflies are largely confined to nature reserves or sites designated as SSSIs. Good access to nature reserves, with interpretive facilities to inform visitors about butterflies and other wildlife, is an essential part of increasing the constituency of support for nature conservation. Butterflies are widely regarded as attractive insects, and they can create the first spark of interest which may lead to observations of other invertebrates, thereby helping to increase our appreciation of the diversity of British wildlife and its conservation needs.

5.8 ACKNOWLEDGEMENTS

We are very grateful to all those wardens of NNRs in England, Scotland and Wales who responded so rapidly and efficiently to our request for data on butterflies on their reserves. James Cadbury (RSPB, Sandy) helpfully supplied a copy of his recent review of butterflies found on RSPB reserves. Caroline Steel (Oxford) and Mark Parsons (JNCC, Peterborough) gave permission for us to use information included in their unpublished review of butterflies occurring on RSNC Wildlife Trusts' reserves. Keith Duff (English Nature) and Mark Parsons (JNCC) commented on a draft of the text. Tony Mitchell-Jones (English Nature) advised on data analysis techniques.

CHAPTER 6

Butterfly conservation on arable farmland

R.E. Feber and H. Smith

6.1 INTRODUCTION

Farmland has undergone some of the most extensive changes of any biotope in Britain over the past 40 years. The post-war drive for agricultural self-sufficiency, augmented by improved mechanization and the introduction of synthetic agrochemicals in the 1950s, has led to drastic intensification of agriculture in lowland Britain (Mellanby, 1981; Sly, 1981). In addition to the well-documented loss of large areas of both farmed and non-agricultural semi-natural habitats (Anon., 1984), the area and quality of smaller, interstitial areas of uncultivated land within the arable landscape has been decimated.

Field boundaries are the primary interstitial habitats which remain common to all lowland British farmland. They are potentially rich habitats, often incorporating woodland, wetland and grassland elements (as hedges, ditches and grass strips) in close proximity. They also form an extensive network across the countryside, linking smaller pockets and larger patches of semi-natural habitats. Despite this potential, they are usually poor habitats for wildlife and can present severe weed-management problems for farmers. Many field boundaries have been removed altogether (Anon., 1985), and those that remain have suffered deleterious changes.

Close ploughing has eroded the width of field boundaries, making the remaining area more vulnerable to agrochemical drift. The nutrient status of field boundary soils has been increased by the use of spinning disc distributors which spread fertilizer across the field margin. As a result, highly competitive species, such as *Urtica dioica* (stinging nettle) and *Cirsium arvense* (creeping thistle) that are able to utilize high nutrient levels, have increased at the expense of floristic diversity. This proliferation of noxious perennial weeds at the crop edge has led farmers to seek control using herbicides. The resulting bare patches of ground are readily colonized by annual species, including pernicious crop weeds such as *Bromus sterilis* (barren brome) and *Galium aparine* (cleavers). A vicious cycle is thus

Ecology and Conservation of Butterflies Edited by Andrew S. Pullin.
Published in 1995 by Chapman & Hall. ISBN 0 412 56970 1

created in which repeated spraying is a necessity and the plant and associated animal communities of the field margins are devastated (Smith and Macdonald, 1989). Deliberate spraying of hedge bottoms with broad-spectrum herbicides was practised by 60% of farmers as recently as 1985 (Marshall and Smith, 1987) and many more field boundaries are affected by accidental spray drift.

Accompanying these changes has been a dramatic loss of wildlife interest, which is particularly well documented for birds (e.g. O'Connor and Shrubb, 1986; Potts, 1986; Marchant *et al.*, 1990). Very little work has been done on butterflies within intensively managed farmland, and there are no published population trends which allow proper comparison of these areas with semi-natural habitats. Nevertheless, arable farmland in lowland Britain is considered to support an impoverished butterfly fauna, consisting mainly of mobile species such as *Pieris brassicae* (large white) and *P. rapae* (small white) (Thomas, 1984a). In other European countries the declining abundance of several species has been linked to agricultural intensification (Kaaber and Nielson, 1988; van Swaay, 1990). In this chapter we first review evidence for the effects of agricultural intensification on butterflies and then examine ways in which simple management techniques can be used to improve the quality of field boundary habitats for butterflies. Finally, we consider the likely impact on farmland butterflies both of current and proposed schemes for reducing agricultural overproduction by fallowing, and of current trends towards low-input agriculture.

6.2 THE EFFECTS OF FARMING PRACTICES ON BUTTERFLY POPULATIONS

In the following sections we review the impacts of changes in the structure of the agricultural ecosystem on the persistence of butterfly populations, and of changes in farming practices on the survival of individual butterflies. All stages of the butterfly life cycle are vulnerable, but larval and adult stages are particularly at risk. They are most vulnerable to loss of resources and may be more exposed to direct mortality factors.

6.2.1 Effects on the persistence of populations

Post-war agriculture has increasingly resulted in reduction in size and fragmentation of suitable patches of butterfly habitat. The small area and frequent isolation of habitat patches increase the probability that populations within them will become extinct, simply as a result of the stochastic nature of demographic processes (MacArthur and Wilson, 1967; Leigh, 1981). Species in field boundary habitats are particularly at risk because the perimeter/area ratio is so high. While the risk of extinction is common to all species in these habitats, those with low mobility, including many butterflies, are relatively unlikely either to recolonize the same patch or to colonize new patches. The number of colonies can be expected to decline progressively. About 85% of British butterfly species are relatively sedentary, needing minimum breeding areas of 0.5–50 ha, depending on the species

(Thomas, 1984a). Thus species with open or migratory populations, such as *Inachis io* (peacock), are frequently recorded on farmland, while *Maniola jurtina* (meadow brown), although not having such exacting requirements for oviposition, or for larval or adult food (Feber, 1993), is less mobile and less common within intensive farmland.

The temporal as well as spatial structure of the agricultural ecosystem militates against the persistence of populations of less mobile species. Major disruptions, such as ploughing, spraying and hedge-cutting, which characterize agricultural habitats, make the supply of resources unpredictable (Macdonald and Smith, 1991). Because of this temporal as well as spatial patchiness, the habitat may be underutilized simply because many butterfly species are insufficiently mobile to respond to the changing distribution of resources.

Potentially, the effects of isolation and fragmentation can be ameliorated where connecting corridors of unfarmed land facilitate the movement of individuals between populations (Merriam, 1984). The data to support the use of field boundaries as wildlife corridors are limited (e.g. Johnson and Adkisson, 1985; van Dorp and Opdam, 1987; Baudry, 1988; Maelfait and De Keer, 1990) and there is no direct evidence for this in butterflies. However, field margins may act as corridors for some butterflies. Many species do not readily cross unsuitable habitats; roads, woodland or agricultural fields may act as barriers to interchange between already fragmented populations, particularly of rarer species (J. Thomas, 1983b; Warren, 1987b; Munguira and Thomas, 1992). As well as facilitating the movement of individual butterflies, field margin corridors which provide continuity of suitable breeding habitat could also facilitate a gradual extension in the distribution of populations.

6.2.2 Effects on larvae

The risks to butterfly populations of local extinctions resulting from the structure of modern farmland are enormously compounded by many aspects of modern farming practice which reduce the probability of survival of individuals.

The quantity and quality of larval host-plants on field margins must have been reduced very substantially by the persistent use of herbicides on nutrient-rich soils. Hedgerows have traditionally been associated with high floral and faunal diversity typical of woodland edge habitats; around a third of the British flora has been recorded from hedgerows (Pollard, Hooper and Moore, 1974). However, herbicide drift can alter the composition of field-edge plant communities, causing both mortalities and sub-lethal effects, including suppression of flowering. Marrs, Frost and Plant (1989) showed that the most severe effects occurred within 2 m of the spray boom, but that some lethal effects occurred over a distance of 5–10 m. Direct application of herbicides to the hedge base has more radical effects. Although such applications reduce populations of some species of pernicious weeds, the simultaneous removal of perennial grass cover also ensures the perpetuation of a species-poor community dominated by annuals (Smith and Macdonald, 1992). For butterfly populations, the result is a decline in abundance or widespread loss of many larval food-plants. Even relatively simple grassy swards required for oviposition by common species, such as *M. jurtina* and *Pyronia tithonus* (gatekeeper), are degraded or destroyed by this practice. Populations of species with more exacting larval requirements are even less likely to be supported.

Where suitable larval food-plants do exist, farming practices may have other deleterious effects on larval populations. There is considerable evidence that insecticide, as well as some herbicide and fungicide, applications within the crop result in changes in abundance of non-target arthropods (e.g. Vickerman and Sunderland, 1977; Burn, 1988; Vickerman, 1991). Butterfly larvae feeding on grass strips or hedges at the field boundary may be vulnerable to the direct effects of insecticide drift, but little research has been undertaken on this since, apart from *P. rapae* and *P. brassicae*, butterflies are of no economic interest. Toxic effects of drift from eight insecticides have been demonstrated on *P. brassicae* larvae, with mortality occurring at a range of doses and in a range of wind conditions (Sinha, Lakhani and Davis, 1990). Although *P. brassicae* is highly sensitive to insecticides, *Pieris napi* (green-veined white), *P. tithonus* and *Polyommatus icarus* (common blue) also showed mortality from one or more insecticides at higher doses (Davis, Lakhani and Yates, 1991).

Major perturbations, such as cutting or direct herbicide application, are typical features of modern field boundary management during the summer. The low mobility of most butterfly larvae makes them particularly susceptible to mortality from these forms of management, which kill or remove the parts of the plant on which they feed. Courtney and Duggan (1983), for example, reported the loss of all surviving third-instar larvae of *Anthocharis cardamines* (orange tip) (which usually feed and complete their development on a single plant) when the flower stalks of over 1000 *Cardamine pratensis* (cuckooflower) plants were cut on their road-verge study site. Other stages of the butterfly's life cycle are also vulnerable to mechanical damage. Thomas (1974) estimated that up to 50% of overwintering eggs of *Thecla betulae* (brown hairstreak) on *Prunus spinosa* (blackthorn) twigs were killed each year in the Weald due to the mechanical effects of hedge-cutting. Although some species may escape the effects of cutting management (for example, the larvae of *Aphantopus hyperantus* (ringlet) rest by day at the base of grassy tussocks), broad-spectrum herbicide use is likely to affect all hatched and feeding larvae. In addition to its longer-term consequences for sward composition (above), direct herbicide application also has more drastic effects than cutting in the short term. Contact herbicides may take only 24 hours to kill the above-ground parts of plants, while systemic herbicides act within 5 days, undoubtedly resulting in the high mortality of butterfly larvae in sprayed areas.

The degradation of field boundaries may have more subtle, sub-lethal effects on larval development. High leaf nutrient status in some plant species, and poor nutritional quality in less competitive species, may have complex effects. Thus, *P. rapae* larvae complete their development more rapidly and spend less time feeding on nitrogen-rich leaves (Slansky and Feeny, 1977; Myers, 1985). Loader and Damman (1991) showed that levels of parasitism were higher on larvae on nitrogen-rich plants, although levels of aerial predation were lower. Increased vulnerability to predation may result from slow development or increased time spent feeding. Dempster (1968) showed that *P. rapae* larval survival was significantly higher after crop spraying with DDT, which eliminated many predators. High soil nutrient status and pesticide use may thus benefit some butterfly species but be deleterious to others.

6.2.3 Effects on adults

Although the nutritional requirements of the larvae of most temperate butterfly species are more specific to particular plant species than those of the adults, the availability of appropriate nectar sources is none the less an essential requirement. Although adults are more mobile than larvae, the nectar requirement of many species must still be met within a restricted area.

Several studies have shown that most butterflies forage on a limited number of the plant species that are available, and of these some are preferred greatly to others (Douwes, 1975; Wiklund, 1977; Wiklund and Ahberg, 1978). Our own studies have shown that the majority of species utilized by butterflies for nectar are perennials (section 6.3.3). Annuals, which are characteristic of early successional communities and dominate the weed flora of arable crops and degraded field margins, have low nectar secretion rates (Parrish and Bazzaz, 1979). Evidence that perennials may have relatively high nectar secretion rates (Pleasants and Chaplin, 1983) led Fussell and Corbet (1991, 1992) to suggest that perennials, which they showed were preferred by foraging bumblebees, provide high nectar rewards. Although butterflies have lower energy requirements than bumblebees, they would nevertheless benefit from this characteristic. Selectivity by butterflies of particular nectar sources suggests that quality, as well as quantity, of nectar supply is important, even to the most generalist species. Nectar quality has direct influences on longevity and female reproductive potential, and indirect influences on the proportion of time butterflies must devote to feeding (Stern and Smith, 1960; Gilbert, 1972; Watt, Hoch and Mills, 1974; Dunlap-Pianka, Boggs and Gilbert, 1977; Murphy, Launer and Ehrlich, 1983; Wiklund and Persson, 1983; Hill, 1989; Hill and Pierce, 1989; Erhardt, 1991, 1992a, Chapter 18 this volume).

The effects of typical field boundary management practices on nectar sources are similar to those on larval food-plants. Management of the field margin by both spraying and cutting reduces flower abundance, and thus nectar availability, at potentially critical times in the butterfly's life cycle. Spraying additionally selects annual species at the expense of perennials, further reducing the nectar supply (section 6.3.3).

6.3 RESTORATION OF BUTTERFLY HABITAT IN FARMLAND

6.3.1 Objectives for farmland butterfly conservation

Over large areas of lowland Britain, field boundaries provide the only ubiquitous habitat in which butterflies can be conserved and enjoyed. However, we have shown above that many aspects of modern farming reduce the probability of survival of all but the most mobile, and least demanding, species. The scale of these problems and the need for conservation measures to be compatible with commercial farming practices means that objectives for conserving butterflies on farmland must be modest. We suggest that the target species for conservation measures should be common grassland and hedgerow species, such as *M. jurtina, P. tithonus* and *Ochlodes venata* (large skipper), which frequently suffer from farmland habitat degradation. The conservation of rarer species is always likely to require protected sites and highly specialized management, but we show in the following two sections that the conservation of many of our most familiar butterflies can be encompassed by farmers within commercially viable systems for field-boundary management. Our evidence comes from two major research projects which have quantified the benefits to butterflies of novel and contrasting field-edge management systems. The first concentrates on the crop headland (the outer 6 m of the crop) and the second on the uncropped field margin.

6.3.2 Conservation headlands

The management system that has become known as conservation headlands was developed by the Game Conservancy Trust's Cereal and Gamebirds Project to increase the abundance of invertebrate food for gamebird chicks reared in arable field margins. The project arose as a result of studies that linked a dramatic decline in *Perdix perdix* (grey partridge) on farmland with a decreasing abundance on cereal field edges of their insect prey (Southwood and Cross, 1969; Potts, 1986). Pesticides appeared to be a major factor reducing their populations (Potts, 1986) and, as a solution to the problem, experiments were designed in which the headlands of cereal fields received reduced and selective pesticide inputs. Reduced herbicide use increased the abundance of broad-leaved species in the conservation headlands compared with the fully sprayed headlands, and the abundance of chick food and chick survival increased substantially (Rands, 1985; Sotherton, Rands and Moreby, 1985; Sotherton, 1991).

The benefits of conservation headlands to butterflies were also quantified. Over 4 years when transects were conducted, more individual butterflies, and more species of butterflies, were recorded over conservation headlands than over fully sprayed headlands (Dover, Sotherton and Gobbett, 1990). Observations of butterfly behaviour in the conservation headlands showed that they were being exploited for nectar and occasionally for oviposition by some species (Dover, 1989).

In 1991, 879 miles of headlands were managed under this system on 118 farms, with considerable benefits for shooting interests (Thompson, 1992). Butterflies may be indirect beneficiaries of this commercial interest. Nectar sources for adult butterflies are frequently in short supply on farmland, and the increased nectar availability arising from the use of conservation headlands may help to maintain field-boundary populations of some of our target species.

However, when conservation headlands are instated next to severely degraded field boundaries, several factors make it likely that their benefits to butterflies are restricted to species outside our target group. First, the nectar resource that they provide is unlikely to support large numbers of adult butterflies. Most of the broad-leaved plant species that increase when headlands are left unsprayed are annuals, and the nectar supply from these may be poorer than from perennials (section 6.2.3). Secondly, headlands cannot in themselves support breeding populations of the majority of butterfly species. Few perennial plants can survive the harvest and cultivation operations in arable crops. Breeding on conservation headlands is therefore restricted to a few bi- or multivoltine species, such as the commoner pierids, whose larvae can utilize annual host-plants and complete their development before harvest. Conservation headlands are thus likely to be most beneficial to butterflies where they augment the resources provided by well-managed permanent field margins.

6.3.3   Extended-width field margins

If farmland is to support breeding populations of butterflies rather than simply attracting aggregations of adults of mobile species, permanent field margins clearly have a vital role. Our own work at the University of Oxford's farm at Wytham has shown that simple changes in the management of field margins can result in radical improvements in the availability of both larval and adult resources for butterflies.

Large-scale experiments designed to test methods for restoring degraded field margins were established by the University of Oxford's Wildlife Conservation Research Unit in 1987 (Smith and Macdonald, 1989, 1992). We postulated that the establishment of diverse, predominantly perennial, grassy swards on field margins could not only restore wildlife interest but also solve agricultural weed-control problems. In semi-natural grasslands it is well established that most annual species are unable to germinate and compete successfully in perennial swards, and that perennial weedy species can be controlled by manipulating the timing and frequency of mowing. We tested the extent to which these principles applied within agricultural systems.

Prerequisites for field-margin restoration were the expansion in width of badly eroded margins and the exclusion of agrochemicals. We expanded the original field margins (usually about 0.5 m wide) on to cultivated land, to give a total width of 2 m. Fertilizer application was excluded from the margins by using a headland deflector disc on the distributor, and spray drift was minimized by turning off the outer one or two jets on the spray boom.

We established swards on the newly fallowed margins either by allowing natural regeneration or by sowing a mixture of wild grasses and forbs (we refer to these as 'unsown' and 'sown' swards, respectively). Plots (50 m long) were established on both of these sward types and managed in the following ways: left uncut, or cut (with cuttings removed) in summer only, spring and summer or spring and autumn. Two further treatments were imposed on unsown plots only:

1.   cut in spring and summer with hay left lying; and
2.   uncut but sprayed with glyphosate herbicide in late June.

The plots were first cut in June 1988, and in subsequent years in the last weeks of

April, June and September ('spring', 'summer' and 'autumn', respectively). Glyphosate was first sprayed in 1989. The 10 treatments were randomized in eight blocks, with each block occupying a single field.

The impact of these treatments on both the species richness and weediness of developing swards, and on selected invertebrate groups, was monitored intensively until 1991. The abundance of adult butterflies on the contrasting experimental treatments was recorded on transects (Pollard *et al.*, 1975) from 1989 to 1991, and in 1991 simultaneous transects were also conducted for 2 months on a nearby, intensively managed, arable farm with conventionally managed field margins (section 6.1). The mobility and behaviour of butterflies on the experimental treatments, and the abundance of larvae of some nymphalid species, was also recorded (Feber, 1993).

Twenty-two species were recorded at the University Farm in 1991, and 16 during the 2 month period of sampling both farms. During this period, *Melanargia galathea* (marbled white) and *P. icarus* were recorded only on the experimental margins at the University Farm. Of the species common to both farms, eight were significantly more abundant on the experimental than the conventional margins, including some more typical of semi-natural grassland such as *Thymelicus sylvestris* (small skipper), *Coenonympha pamphilus* (small heath), *P. tithonus* and *M. jurtina*. No species was more abundant on the farm with conventional field margins. Species commonly associated with farmland, such as *P. brassicae*, *P. rapae* and *I. io*, did not differ significantly in their numbers between the two sites.

Several lines of evidence suggested that species within our target group were able to maintain populations on the field margins. First, the adults of many species were present in large numbers over the 3 years in which they were monitored and have also been recorded in 2 subsequent years. The existence of a breeding population of a species if it is seen in four successive flight periods is an assumption commonly made in relation to data from the Butterfly Monitoring Scheme (Pollard and Yates, 1992). Furthermore, the very rapid and synchronized increases in abundance of *M. jurtina* and *P. tithonus* (Feber, 1993) strongly suggested that they emerged very locally; the field margins were separated by woodland and agricultural fields from the nearest potential sources of immigrants. Oviposition on the field margin swards by *T. sylvestris*, *O. venata*, *P. icarus*, *C. pamphilus* and all three *Pieris* species was frequently recorded, as were the larvae of *A. cardamines*, *M. jurtina*, *P. tithonus*, *I. io*, *Aglais urticae* (small tortoiseshell) and *Lycaena phlaeas* (small copper). Finally, there were no *a priori* reasons to assume that habitat quality on the field margins would result in lower larval survival than in many semi-natural grasslands. The swards were permanent, providing undisturbed overwintering habitat for all developmental stages. The exclusion of agrochemicals encouraged the development of swards rich in larval food-plants (Feber, 1993), as well as nectar sources (below).

Although many more butterflies were recorded on the field margins at the University Farm than on the conventionally managed farm, our different experimental treatments were not equally attractive to butterflies. There were large differences between the experimental treatments in the abundance, species richness and behaviour of butterflies, which enabled us to identify the factors that maintained their populations and which provide the basis for advice on managing field margins for butterflies.

(a) The effects of mowing

In all years mowing had a significant effect on the mean abundance of butterflies (Table 6.1) and on their species richness (Feber, 1993).

Over the season as a whole, butterfly numbers were highest on treatments which were left uncut in the summer. Before the summer cut, other treatments, and particularly sown treatments, were also widely utilized. Mowing in summer affected the majority of species that were on the wing at that time. The response

Table 6.1 The effects of management on the abundance of butterflies on the field margins 1989–91

(a)   Mean numbers of butterflies per 50 m length of field margin

Treatment	Mean number of butterflies per 50 m length of field margin during:		
	1989	*1990*	*1991*
Sown, cut spring + autumn	24.40	82.46	54.21
Sown, uncut	48.48	76.20	47.04
Unsown, cut spring + autumn	36.88	22.41	35.45
Sown, cut spring + summer	25.58	41.62	25.22
Unsown, uncut	29.08	29.06	23.89
Sown, cut summer only	21.22	34.99	18.44
Unsown, cut spring + summer, hay left	14.16	14.86	16.99
Unsown, cut summer only	14.09	14.78	15.40
Unsown, cut spring + summer	26.46	13.86	15.38
Unsown, sprayed	27.92	15.59	10.93

(b)   Significance of selected comparisons between treatments

	d.f.	*1989*	*1990*	*1991*
Main effects[a]				
Block	7,49	**	ns	**
Sowing	1,49	ns	***	***
Cutting	3,49	**	**	***
Sow × cut	3,49	ns	ns	ns
Planned comparisons between means				
Cut v. uncut	1,49	**	**	*
Cut in summer v. not cut in summer	1,49	**	***	***
Cut spring + autumn v. uncut	1,49	ns	ns	ns

Significance levels: ***, $P < 0.001$; **, $P < 0.01$; *, $P < 0.05$; ns, not significant.

[a] Main effects from three-way analysis of variance using the eight treatments that form a 4×2 factorial design (Smith *et al.*, 1993)

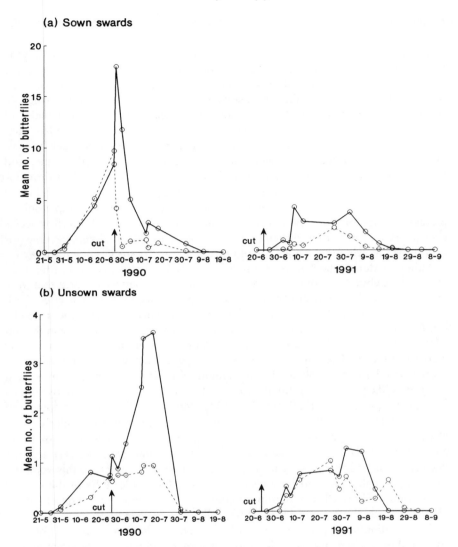

Figure 6.1 The effects of management on the mean number of *Maniola jurtina* per 50 m length of field margin. Plots were either cut − − ○ − − or left uncut —○— in summer. Note the different vertical scales in (a) and (b).

of *M. jurtina*, which moved immediately from cut onto uncut plots (Figure 6.1), was typical.

Munguira and Thomas (1992) noted a similar sharp decline in butterfly numbers following mowing of road verges, and a slow recovery of numbers due to the invasion of mobile species such as *Pieris* spp. and *L. phlaeas*. We also recorded a partial recovery, particularly on sown plots where *Leucanthemum vulgare* (oxeye daisy) reflowered after cutting. However, neither abundance nor species richness regained levels comparable with those on the treatments that were left uncut in the summer.

Treatments that were left uncut, or cut in spring and autumn, had two advantages for butterflies. First, they provided a continuous supply of nectar. Removal of nectar sources by widespread mowing of vegetation during the summer months would make the field margins an unsuitable habitat for all but the most mobile species. Second, they provided an undisturbed area in which adult females of some species could oviposit and the larvae of others could feed or complete their development. Mowing during any of the summer months would be likely to have highly disruptive effects on species that have active larvae at this time, and particularly on those such as *A. cardamines*, whose larvae complete their development on one plant. Furthermore, the variation in the phenology of adult and larval stages, both between and within years, was so great that it would be impossible to predict the best time to mow, even if managing for only one butterfly species (Feber, 1993). It is thus difficult to recommend the least disruptive time for widespread mowing in summer, should this be necessary for weed control.

Although, during the 3 years of this study, we were unable to detect differences in either the abundance or species richness of butterflies between swards that were never cut and those cut in spring and autumn, it is likely that such effects will become apparent in the longer term. By 1990 plant species richness was significantly lower on uncut than on spring and autumn cut swards (Smith and Macdonald, 1992). As these underlying differences in the plant community increase, they are likely to have increasing impacts on the availability of nectar sources and larval food-plants. In the medium term, lack of management of field margins is likely to result in the development of scrub; invasion of woody species was apparent as early as 1989 (Smith and Macdonald, 1989). The development of scrub is likely to result in the loss of butterfly species, and particularly those characteristic of open grassland. Erhardt (1985c) showed that invasion by scrub and decline in plant species richness in abandoned alpine grassland resulted in a decline in butterfly species richness (see also Erhardt, Chapter 18 this volume). Studies on chalk grassland have also attributed undesirable shifts in the species composition of butterfly communities to the cessation of grazing and development of taller swards (Frazer and Hyde, 1965; J. Thomas, 1983a,b).

The creation and maintenance of a heterogeneous habitat, composed of both short and overgrown swards, is likely to maximize the potential of field margins as habitat for a diverse butterfly community, and may provide an acceptable compromise between the needs for weed control and wildlife management. Indeed, the overall success of our experimental field margins in maintaining such a diverse assemblage was probably due in part to their heterogeneous nature, which originated as a consequence of the experimental design.

(b)   *The effects of sowing*

Butterflies were significantly more abundant on sown than on unsown plots in the second and third years of the study (Table 6.1). Our seed mixture contained several important nectar sources, including *L. vulgare*, *Knautia arvensis* (field

Table 6.2 Nectar source utilization by *Maniola jurtina* and *Pyronia tithonus* in 1991

Plant species	Rank abundance of flowers[a]		Percentage of visits [b] by	
	July	*August*	*M. jurtina*	*P. tithonus*
Cirsium/Carduus spp.	13	10	37.0	46.7
Centaurea spp.	45	11	12.0	2.8
Knautia arvensis	19	18	7.7	2.3
Tripleurospermum inodorum	6	2	4.3	2.4
Leucanthemum vulgare	4	5	34.0	10.0
Trifolium spp.	12	44	2.0	0.3
Convolvulus spp.	24	1	1.3	2.3
Pulicaria dysenterica	0	37	0.0	20.5
Ranunculus spp.	25	52.5	0.7	0.0
Rubus spp.	23	39	0.3	7.7
Senecio jacobaea	59	50.5	0.3	2.6
Umbellifer spp.	10	6	0.3	1.7
Hypericum spp.	17	31	0.0	0.3
Achillea millefolium	62	42	0.0	0.3
Dipsacus fullonum	0	40	0.0	0.3

[a] All broad-leaved plant species were ranked in order of the mean abundance of their flowers during summer 1991. Species included in the July list are those on which *M. jurtina* were observed feeding and in the August list those on which *P. tithonus* were feeding.

[b] Data are percentages of total observations of feeding butterflies (300 systematic observations of *M. jurtina* and 351 of *P. tithonus*).

scabious), *Centaurea nigra* (common knapweed) and *Centaurea scabiosa* (greater knapweed), which were heavily utilized for feeding by most butterfly species in the study area (e.g. Table 6.2).

Erhardt (unpublished data) has shown these species to have particularly high nectar secretion rates and sugar levels per flower. In addition to their impact on weed control and amenity on badly degraded field margins (Watt, Smith and Macdonald, 1990; Smith and Macdonald, 1992; Smith *et al.*, 1993), wildflower seed mixtures can thus act as effective supplements to, or replacements for, established plant species on arable farmland. Such mixtures are most beneficial to butterflies where they include early and late flowering species, to provide nectar throughout the season. The criteria used for selecting broad-leaved components of our wildflower seed mixture included their potential as nectar sources, but not their suitability as butterfly larval food-plants (Feber, 1993). The broad-leaved species used for oviposition and larval feeding were present either in naturally regenerated swards or in the vicinity of the experimental margins, for example on ditch banks. Although in our experiment the advantage of sown swards to butter-flies lay primarily in providing resources for adults, broad-leaved larval host-plants can also be included in seed mixtures. In a separate experiment, for example, we found that *L. phlaeas* utilized *Rumex acetosa* (common sorrel) included in a simple grass ley.

(c)   The effects of spraying

The sprayed plots illustrated clearly the devastating effect of this management practice on butterflies (Table 6.1). Butterfly abundance declined over the 3 year period and, by 1991, spraying resulted in significantly lower numbers than any other treatment ($P < 0.001$). By 1990 the numbers of annual plant species on sprayed plots were significantly higher than on other plots, while the numbers of perennials were significantly lower (Smith *et al.*, 1993). Nectar supply and larval food-plant availability declined as a result.

6.4   THE FUTURE

Relatively simple changes in management can radically improve the value of intensively managed farmland for common but declining grassland butterflies. Conservation headlands are already a popular option with farmers interested in game rearing, while the substantial benefits to weed management and amenity of expanded field margins is likely to make their widespread use a commercially viable option. Their instatement is likely to be further assisted by the introduction of incentive payments that recognize the benefits to wildlife and amenity of more sensitive field margin management (Anon., 1993c).

In addition to these possibilities for improving conventionally managed farmland for butterflies, changes in the direction of the European Community's (EC) Common Agricultural Policy since 1988 (Floyd, 1992) and in pressures from consumers (Anon., 1992a), seem likely to lead progressively to the introduction of land-management systems intrinsically less hostile to butterflies.

The UK set-aside scheme, introduced in 1989 in response to an EC directive to reduce grain surpluses, compensated farmers for fallowing arable land (Anon., 1988). Of the options available under this scheme, the most popular was to fallow land for 5 years. In principle the benefits of this option for butterflies were substantial, but in practice they were very limited. Most farmers allowed plant cover to establish on their fallowed land by natural regeneration. Predictably, crop weeds dominated the vegetation in the initial years (Shield and Godwin, 1992; Wilson, 1992). *Cirsium* and *Carduus* species, as well as annual broad-leaved weeds, often provided abundant nectar. Colonizing perennial grasses provided larval food-plants for less mobile species in addition to the annual crucifers and *Urtica* spp., which benefited more mobile pierid and nymphalid species, respectively. However, the voluntary nature of the scheme and the level of compensation payments ensured that the area set aside was small (less than 5% of the total cereal area in the UK: Clarke and Cooper, 1992). Moreover, farmers were required to mow their set-aside fields at least once annually. Although they were advised to avoid the period between April and July to minimize damage to ground-nesting birds, the summer mowing that most farmers employed to control the proliferation of weeds was likely to be very deleterious to both adult and larval butterflies.

In 1992 a second set-aside scheme introduced a compulsory requirement for all larger farmers to fallow 15% of their arable land on a 5 year rotation (Anon.,

1992b). Despite the much larger area involved, the advantage to butterflies of a shifting pattern of annual fallowing continues to be restricted to temporary increase in nectar supply, available only to the most mobile species. Proposals to introduce much longer-term set aside, on which enhancement of wildlife interests is the primary aim (Anon., 1993c), offer radically improved prospects for butterflies in our target group. However, such schemes will require substantial funding if they are to occupy sufficient area to have more than very local impacts on butterfly populations.

An alternative to reducing agricultural commodity surpluses by setting aside agricultural land is to reduce the intensity of agriculture over larger areas by limiting agrochemical inputs. Such 'extensified' agricultural systems have been proposed by the EC as an additional production-control measure. Organic farming systems are the extreme expression of low-input agriculture and are gaining popularity despite the absence of fiscal support, occupying a total area of 25 000 ha in 1991, with a further 24 000 ha under conversion (Redman, 1992). Our review of the problems posed to butterfly populations by modern agriculture suggests that reduced use of synthetic agrochemicals could have many beneficial effects. However, low-input farming systems are unlikely to have radical effects on species in our target group without additional conservation management measures. Crop weed control is as important an objective under these systems as under conventional systems, and so cropped areas themselves are unlikely to support more butterflies. Field boundaries thus remain the key areas for supporting butterfly populations but their floras are likely to improve only very slowly in the absence of positive management (Smith *et al.*, 1993). The use of well-timed mowing and, in some circumstances, of sowing, are as appropriate and useful tools in low-input systems as in conventional systems for producing effective weed control and suitable habitat for our target butterfly species on field margins.

6.5 ACKNOWLEDGEMENTS

We are grateful to Drs Stephen Baillie and David Macdonald for helpful comments on the manuscript. Our own research, as part of the University of Oxford's Wildlife Conservation Research Unit, was supported by English Nature, the People's Trust for Endangered Species and the Ernest Cook Trust.

CHAPTER 7

Butterfly conservation within the management of grassland habitats

M.R. Oates

7.1 INTRODUCTION

This chapter examines the importance of grassland habitats for butterflies, and the range of management options available throughout the spectrum of grassland systems. Examples are mainly taken from calcareous grassland, although the underlying principles apply to grassland habitats generally. The chapter is based on the premise that butterfly conservation should be integrated within wider nature conservation aims.

The majority of British butterflies are insects of grassland habitats, notably of herb-rich sheltered grassland. Approximately half of the British species are strongly or wholly associated with wild or unintensively farmed grassland (Erhardt and Thomas, 1991). The richest sites for butterflies, in terms of species diversity, are all grassland sites, or they incorporate significant grassland habitats. Southern calcareous grassland is particularly rich, with individual sites often supporting over 30 breeding species, several of which occur on the northern limit of their European ranges. Porton Down, on the Hampshire–Wiltshire border is the richest site for butterflies in Britain, boasting 41 species which breed annually, including large populations of seven scarce species. Coastal grasslands, especially dune, soft cliff systems and clifftop grassland, also support rich butterfly faunas. Grassland butterflies also form a significant element of woodland butterfly faunas (see Robertson, Clarke and Warren, Chapter 8 this volume). Thus, of the 37 species of butterfly which were genuine residents in Bernwood Forest during the late 1970s and early 1980s, no less than 24 can be categorized as grassland species, and all bar five are dependent on grassland for nectaring and assembly.

Conversely, upland grasslands usually support only one or two species, some-

Ecology and Conservation of Butterflies Edited by Andrew S. Pullin.
Published in 1995 by Chapman & Hall. ISBN 0 412 56970 1

times none, while improved agricultural grassland generally fails to support any breeding butterflies. The losses of unimproved lowland grassland in England and Wales are quite stunning: Fuller (1987) found that by 1984 unimproved grassland occurred in only 3% of the area it had occupied in 1930. Therefore it hardly needs to be stated that the British grassland butterfly fauna has declined dramatically during the course of the twentieth century.

7.2 A SUMMARY OF RESEARCH INTO GRASSLAND BUTTERFLIES

7.2.1 Historical summary

Scientific study of butterfly ecology began to develop in the early 1970s. With the notable exception of work on *Maculinea arion* (large blue) (Thomas, 1976, 1980) and a status survey of *Melitaea cinxia* (Glanville fritillary) (Simcox and Thomas, 1979), the early studies on butterfly ecology and conservation were all directed towards woodland butterflies (Thomas, 1974; Pollard, 1979; Warren, 1981).

However, since the standard work on grassland ecology and management (Duffey *et al.*, 1974) was published, much work has been conducted on the conservation of butterflies in grassland habitats. During the early 1980s a team of professional and amateur butterfly enthusiasts worked to produce the Nature Conservancy Council book *The Management of Chalk Grassland for Butterflies* (Butterflies Under Threat Team, 1986). Although knowledge on downland butterfly conservation has developed considerably since then, it remains the only published report on the management of any grassland habitat for a range of butterflies, apart from recent publications on woodland-ride management (Carter and Anderson, 1987; Warren and Fuller, 1990).

7.2.2 Present situation

Currently, useful knowledge exists on the ecology and management requirements of some 20 butterfly species in grassland habitats. Effort has concentrated on the scarcer species. Papers and reports have been published, following detailed ecological studies, on *Lysandra bellargus* (adonis blue) (J. Thomas, 1983b), *Carterocephalus palaemon* (chequered skipper) (Young and Ravenscroft, 1991), *M. cinxia* (Simcox and Thomas, 1979), *M. arion* (Thomas, 1980, 1987), *Thymelicus acteon* (Lulworth skipper) (J. Thomas, 1983a), *Hesperia comma* (silver-spotted skipper) (Thomas *et al.*, 1986; Warren and Thomas, 1994), *Cupido minimus* (small blue) (Morton, 1985), *Leptidea sinapis* (wood white) (M.S. Warren, 1984) and *Hamearis lucina* (Duke of Burgundy) (Oates and Emmet, 1990). Table 7.1 shows the current state of knowledge on the scarcer grassland butterflies in Britain, based on published material.

It is surprising that only four of the 17 British grass-feeding species have been studied in any depth, *C. palaemon*, *H. comma*, *T. acteon* and *Pararge aegeria*

Table 7.1   Current conservation knowledge on our scarcer grassland butterflies

Comprehensive knowledge including management details	*Useful ecological knowledge, including management basics*
Lysandra bellargus	*Melitaea cinxia*
Caterocephalus palaemon	*Eurodryas aurinia* (in press)
Mellicta athalia (in grassland)	*Plebejus argus* (in grassland)
Maculinea arion	*Cupido minimus*
Thymelicus acteon	
Hesperia comma	
Leptidea sinapsis	
Hamearis lucina (in preparation)	
Basic knowledge, at least in some habitats or regions, including general understanding of habitat management	*Inadequate knowledge, especially regarding management*
Aricia agestis	*Thymelicus lineola*
Lysandra coridon	*Hipparchia semele*
Argynnis aglaja	*Coenonympha tullia*
Erynnis tages	*Erebia epiphron*
Callophrys rubi	*Erebia aethiops*
Melanargia galathea	
Boloria euphrosyne	
Boloria selene	

(speckled wood) (Shreeve, 1985). In addition, significant aspects of the ecology of many of the other 13 species have been studied in certain regions or in specific habitats such as *Lasiommata megera* (wall brown) (Dennis, 1982–83; Oates, in preparation) and *Erebia epiphron* (mountain ringlet) (Porter, 1989; Thomas and Lewington, 1991).

Moreover, most of the butterflies we currently know least about are grass-feeding species. Rather predictably, these include four scarce species with restricted distributions, *Hipparchia semele* (grayling), *Coenonympha tullia* (large heath), *E. epiphron* and *Erebia aethiops* (Scotch argus). Curiously though, the list also includes some of our more ubiquitous grassland butterflies, notably *Pyronia tithonus* (gatekeeper), *Coenonympha pamphilus* (small heath), *Ochlodes venata* (large skipper) and *Aphantopus hyperantus* (ringlet). Indeed, it is probable that less is known about the ecology of *C. pamphilus*, our most widespread grassland butterfly, than any other resident species. It is also worth noting that our knowledge of *L. megera* is so inadequate that we have no idea as to why it has declined dramatically in many parts of Britain since 1983. Likewise, we do not know why *Thymelicus lineola* (Essex skipper) has expanded its range. These cases of decline and expansion are well documented in the annual *Hampshire & Isle of Wight Butterfly Report*, produced by the Hampshire Branch of Butterfly Conservation (1985–93) (see also Oates, in preparation).

In summary, our knowledge on the ecology of grassland butterflies is locally detailed, generally basic and in many cases merely incipient. There are some alarming gaps in our knowledge, particularly over the habitat-management requirements of species. The situation is complicated further by the ability of species to occur in a diversity of habitats, utilizing a range of food-plants in subtly different ecological niches.

7.3 THE PRINCIPLES OF GRASSLAND MANAGEMENT

7.3.1 General

Grassland management is based on the premise that if left unmanaged, virtually all our grassland would revert to woodland; only high montane grasslands, exposed maritime grasslands on skeletal soils, some thin-soil downland and various dune, floodplain, grassland flushes, gorge and cliff grasslands would remain intact (Erhardt and Thomas, 1991). Consequently, grassland management effort is geared towards hindering or reversing natural succession to woodland. This basic truth has only recently been fully comprehended, and there are examples of early efforts to conserve grassland wildlife by fencing the sites in order to deter public access. For example, the Chipman Valley, a combe near Dizzard on the North Cornwall coast, was acquired by the Committee for the Protection of British Lepidoptera in the late 1920s as a reserve for *M. arion*. The Committee employed a seasonal warden to deter collectors, fenced the site to preclude stock and prohibited the practice of *Ulex* (gorse) burning. The habitat rapidly deteriorated as the sward and scrub grew unchecked and the butterfly died out within a few years (Thomas, Chapter 13 this volume). This example illustrates the importance of maintaining turf at a suitable height for butterflies, as emphasized by Butterflies Under Threat Team (BUTT) (1986).

Such disaster stories emanate from the failure to protect the process (or processes) upon which rare species and valued habitats are dependent. In grassland habitats, grazing is the main management process. There is an inherent problem here in that many grassland sites have been given to conservation organizations, or acquired by them for small sums of money, because they have proved too difficult for economic agriculture and are not viable agricultural grazing land. An ironic, if predictable, consequence of this is that one of the main problems thwarting effective nature conservation in grassland habitats is the dependency of most conservation organizations on farm stock to graze their grasslands.

The conservation movement is seldom able to implement ideal management, even with habitats as uncomplicated as grasslands. Compromising over site management, and accepting second- or third-best solutions, jeopardizes the future of rare specialists which are, by nature, the most exacting in their requirements and the most sensitive to management inadequacies. This problem is particularly acute with poorly mobile invertebrates which need annual continuity of habitat conditions and cannot subsist, like plants, in a dormant state during successive years of habitat unsuitability. The increasing incidence of fragmentation and isolation of habitats further exacerbates this problem.

7.3.2   Objectives

Nature conservation management has to be based on clear objectives. However, stark 'objectives' contained within a few succinct statements in a management plan seldom suffice, as they do not instil the clarity of vision necessary to determine the conditions in which the site in question should appear. To manage a grassland effectively one needs a clear mental image of how the site should look, the knowledge to be capable of realizing the vision, and the practical ability to do so. The scientific conservation movement is still in its infancy, and at present it is seldom that these three requirements are met in full at any one site.

The matter is further complicated by the fact that no two sites are the same. They are all idiosyncratic. Even though two sites may have myriad common factors, there will always be important differences, especially regarding practical management options.

7.3.3   Restoration and maintenance management

There is an important distinction between restoration management and maintenance management. A neglected grassland will often need a period of intensive management to produce conditions which can then be maintained by more simple means. Restoration management can be a short or lengthy phase and can involve dynamic management methods. It is difficult to decide when to change over, but delays can be damaging. Moreover, from a butterfly conservation angle, a short, robust reclamation phase may involve management practices which are too severe for sensitive butterfly populations, in which case a more gentle approach is usually desirable, unless there are strong chances of natural recolonization once management pressure becomes more relaxed.

7.3.4   Mosaic structures and homogeneous grasslands

The richest butterfly sites, in terms of species diversity, are mosaics of grassland, scrub and woodland. The vast majority of mosaic sites are small. One of the main problems with such sites is that they often support a range of species which have conflicting requirements and which are difficult to cater for within a simple management regime.

Homogeneous grassland sites inevitably support fewer breeding species of butterfly, often less than a dozen, although they can support large populations of scarce specialists such as *L. bellargus*. Parsonage Down, a 147 ha chalk grassland National Nature Reserve (NNR) in Wiltshire, supports an exceptional flora but is remarkable for the paucity of its butterfly fauna. The reserve is entirely homogeneous, severely exposed and uniformly grazed; it demonstrates the desirability of mosaic structures to butterfly faunas. The management of homogeneous grassland is simpler than the management of mosaics, although there is always the quandary of whether it is desirable and practical to diversify the structure of the less species-rich types of homogeneous grasslands, some of which may be nationally rare vegetation communities.

7.4 GRAZING

7.4.1 The development of conservation grazing

Since 1980 a great momentum for grazing wildlife sites has developed. Despite the fact that much conservation grazing has been of an experimental nature, there are many examples of successful grazing regimes. Predictably, mistakes have also been made, and there are embarrassing examples of sites where populations of grazing-sensitive butterflies have been radically reduced or even eliminated as a result of 'conservation' grazing regimes. Problems have been especially severe where stock grazing has been reintroduced at a time when rabbit populations were increasing, as illustrated by the recent collapse of the *Argynnis aglaja* (dark-green fritillary) population on the North Downs (Jeffcoate, 1992).

The weakness of communication and co-ordination within the wildlife conservation movement renders it difficult for conservationists to learn from each other's successes and failures. The forthcoming RSNC *Grasslands Management Handbook* should help to solve this. None the less, conservationists' thinking on grassland management is evolving dynamically, impressive success stories exist and grazing is proving to be by far the best form of grassland management (Duffey *et al.*, 1974; Oates, 1992b).

7.4.2 The problems of selective grazing

All grazing animals, both wild and domesticated, graze selectively. Favoured elements of the vegetation are eaten quickly while less desirable plants are left until last, or not grazed at all. There is considerable variation regarding what different species, and indeed different breeds, of animal favour. The same also applies to what they are loathe to eat. Moreover, there are major differences between the grazing abilities of young, mature and old animals, especially in sheep (*Ovis*). Quite simply, different animals find different plants palatable in different situations and at different times of year. The conservation movement only has an incipient knowledge of this complicated and crucial issue, which remains one of the major areas in conservation requiring detailed study.

Selective grazing often produces a juxtaposition of overgrazing and undergrazing, wherein favoured vegetation is effectively overgrazed and less palatable vegetation is untouched, and consequently uncontrolled. This juxtaposition is salient on sheep-grazed downs in north Wiltshire, where the fine turf is heavily and detrimentally grazed, while the coarse *Brachypodium pinnatum* (tor grass) is untouched, and consequently increasing.

The correct choice of animals, and the right timing, intensity and location of grazing are essential in order to overcome the problems of selective grazing. Mixed grazing by more than one type of animal (e.g. cattle (*Bos*) and sheep, or sheep and goats (*Capra*), and rabbits (*Oryctolagus cuniculus*)) can be an effective way of overcoming the problems of selective grazing. All too often conservationists have to make do with whatever grazing local farmers are prepared to conduct. Compromise here jeopardizes the future of both rare species, which are highly sensitive and have critical requirements, and unusual grassland communities. Indeed, the use of inappropriate, or less than ideal, stock is probably the most common failing of conservation grazing at present.

7.4.3 Browsing

Many grasslands do not need much in the way of grazing but require a considerable amount of browsing. All too often conservationists try to use grazing animals to eradicate scrub regrowth, overgrazing the sward in the process and adversely affecting the invertebrate fauna. Grazing animals will tackle scrub regrowth only when the sward no longer offers sufficient 'keep'. The intensity and longevity of 'grazing' necessary to control scrub is seldom justified, mechanical and chemical means of scrub elimination are invariably essential.

7.4.4 Grazing regimes

Conservation grazing regimes involve a complex equation of four factors; timing, intensity (stocking rates), targeting and type of stock. It is imperative to gain the right balance of these factors within the site-management equation.

(a) Timing

This is crucial for the conservation of sensitive species and for the control of problematic vegetation. The key components are time of year, length of each grazing session and length of period between sessions. With butterflies, timing of grazing is important in creating and maintaining the necessary continuity of the juxtaposition of different grassland structures essential for the various stages of butterfly life cycles, including the provision of adequate nectar supplies and coarse grassland pockets necessary for assembly, pairing, night roosting and rough-weather shelter. Badly timed grazing will produce an inappropriate turf height which will cause valued butterfly populations to collapse. For example, colonies of *A. aglaja* and *H. lucina* invariably die out if the summer turf height is lowered too much (BUTT, 1986). Figure 7.1 depicts the standard turf height requirements of the species which can be categorized as being 'grassland butterflies' within the British context.

The autumn–early winter period is the optimum time for grazing for the benefit of butterflies (BUTT, 1986). Certainly, butterfly populations are less likely to be adversely affected by grazing during this period.

(b) Intensity

Stocking rates are particularly important during the growing season, to ensure that sensitive invertebrate populations are not adversely affected and that animals do not graze too selectively. Stocking rates are not so vital during the autumn and winter months, when the vegetation is hardly growing. Similar effects can be achieved by using a small number of animals for a lengthy period or by using a larger number for a short period; thus, 100 sheep for 10 days and 20 sheep for 50 days both total 1000 'sheep grazing days'.

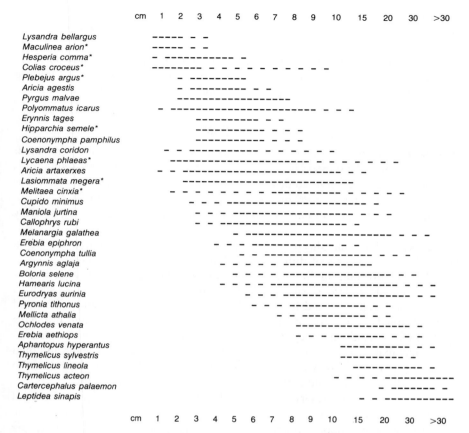

Figure 7.1 Typical turf height preferences of grassland butterflies in Britain (adapted from BUTT, 1986). * = Require sparse turf amongst broken ground.

(c) Targeting

An entire grassland need not be grazed in any one session. Indeed, the concept of compartmentalized grazing is being increasingly applied in conservation management, particularly on sites supporting populations of scarce butterflies. It can be fundamental to the conservation of butterflies within grassland mosaic habitats.

7.4.5 Continuous stocking, cyclical grazing and rotational grazing

Continuous stocking involves grazing stock being present on a site all year round, as is standard practice on common land. If stocking levels are high, butterfly numbers and diversity are likely to be low. Most grasslands are managed by cyclical grazing, wherein stock are moved on and off site. BUTT (1986) gives valuable guidance on this important aspect of conservation. Rotational management is important to the conservation of mosaic sites and can be crucial to

grassland butterfly conservation. In brief, it involves major management operations, notably cutting and grazing, occurring on only part of the site at any one
time.

Many grassland sites do not require annual grazing. Much depends on vegetation type, rainfall, soil depth and fertility. Some grasslands only need grazing
every 3 or 4 years, even less frequently in some cases.

7.4.6 Sheep grazing

Sheep are currently the most regular grazing stock on conservation grassland,
particularly on upland moorland, coastal grassland and limestone grassland. This
preponderance is due more to their availability and relative inexpensiveness, rather
than to any advantages as grazing animals. Indeed, in many situations sheep are
inferior grazers to cattle or ponies (*Equus*), a fact that has yet to be realized by many
conservationists.

The key point about sheep is that there are remarkable differences in grazing
ability between animals of different breed and age. The breeds have evolved to
cope with contrasting conditions, and consequently show different grazing preferences. The older breeds, such as Hebrideans, and mountain breeds, such as
Herdwicks, Swaledales and Rough Fells, graze less selectively than modern and
lowland breeds. They are also hardier and less prone to diseases and problems
such as foot rot, although they are less biddable. The matter is further complicated
by the fact that sheep develop a full set of broad front teeth after 3 years, and then
steadily lose them as they age. Consequently, young and old sheep tend not to
graze as efficiently as middle-aged sheep; they graze more selectively.

At Noar Hill nature reserve, Selborne, Hampshire, the grazing preferences of
15 different breeds and cross-breeds of sheep, of varying age, were studied while
winter grazing in paddocks (personal observations). Major differences in grazing
preference and ability were revealed. Herdwick, Jacob × Herdwick crosses and
Scotch Blackface were the most efficient grazers, eating much *Rubus fruticosus*
(bramble), browsing *Taxus baccata* (yew) and *Thelycrania sanguinea* (dogwood), and
reducing the dominance of coarse grasses. The most inefficient sheep used were
Suffolks and Dorset Horns, which did no browsing and only grazed the finer
grasses and *Dactylis glomerata* (cocksfoot) which all sheep favour in winter. A
thorough scientific study of the grazing abilities and preferences of various types of
sheep in different grassland needs to be conducted as a matter of considerable
importance.

Apart from where tussocky grasses, such as *Molinia caerulea* (purple moor-
grass), are dominant, sheep regimes can detrimentally tighten up the sward, with
the resultant loss of pockets of bare ground necessary for plant propagation and for
many invertebrates. These problems become severe where *Festuca rubra* (creeping
fescue) and various *Agrostis* spp. (bent grasses) are dominant, so much so that these
grasses should be viewed as being as problematic as any 'coarse' grass. However,
tight-sward problems can be overcome by periodic grazing, or rather poaching, by
cattle or ponies.

In 1981, *L. bellargus* was reintroduced at Old Winchester Hill NNR,

Hampshire. The population reached a peak of some 5000 individuals in the second brood of 1984 (Morris and Thomas, 1989) but collapsed during the ensuing poor summer sequence. The last individual was recorded there in June 1989 (Oates and Warren, 1990). The habitat consisted of an exposed south-facing slope, grazed by young Beulahs, a light breed of sheep. In view of the butterfly's oviposition requirement for hot microclimate situations (J. Thomas, 1983b), such as hoof-prints, it is arguable that the colony could have survived had the site received some cattle grazing, or even grazing by a heavier breed of sheep.

7.4.7 Cattle grazing

Cattle differ greatly from sheep in that they eat with a rasping tongue, rather than nibble. They also trample the ground significantly, which, within moderation, is beneficial. Cattle are particularly adept at knocking down and opening up tall coarse vegetation, and browsing scrub, and therefore have a clear role in restoration management (section 7.3.3). However, they can also maintain short turf well. For example, the fine turf of Brook Down, on the Isle of Wight, is kept in pristine condition by Galloway cattle. Again, cattle of different breed and age behave differently. Age is important, as bullocks are active whereas mature milkers and suckler herds are sedate.

It is easy to cause obvious damage to sites by cattle grazing, especially by overstocking during the winter period. Heavy poaching can lead to infestations of *Rumex* spp. (docks), *Senecio jacobaea* (ragwort) and *Cirsium* spp. (thistles), and can also damage archaeological features, although light poaching is desirable. Indeed, many grasslands actually require more in the way of poaching (or opening up) than mere removal of vegetation matter.

Cattle need a copious supply of water and their grazing coverage of a site is greatly influenced by the location of water troughs; localized overgrazing and undergrazing can often be resolved by installing a new trough. Mineral licks can also draw cattle into poorly grazed areas.

Currently, it is difficult to acquire farmstock cattle for conservation grazing. Cattle are worth a lot of money, there is no shortage of quality agricultural grass and, with the exception of dry cows and suckler herds, farmers require cattle to put on weight, which they are unlikely to do on most conservation grasslands. In addition, it is not always practical to graze cattle, due to the occurrence of red water disease or Johne's disease for example. None the less, cattle are far more effective grazing (and browsing) animals than sheep, and their lack of availability is one of the main problems confronting grassland conservation at present.

7.4.8 Horse and pony grazing

Horses and ponies are underrated as grazing animals by nature conservationists, largely on account of the image of 'horsey people', the plethora of examples of overstocked horse paddocks, and the fact that horse latrine areas can be unsightly. Moreover, horses are deemed to graze too selectively. In fact, the patchy nature of horse grazing is often ideal for butterflies, and horse-grazed sites often support unusually diverse invertebrate faunas.

Individual horses behave and graze differently, which adds an element of gambling to horse grazing. Generally, thoroughbreds and halfbreds can be temperamental, difficult to manage and fastidious over what they eat. However, native-breed ponies are radically different.

A survey conducted during 1992 by the National Trust Nature Conservation Section at Cirencester found that horses and ponies can play an invaluable role in grassland management. The survey obtained questionnaire evidence from over 80 horse-grazed sites and found that native-breed ponies are proving to be extremely good grazing animals, adept at tackling coarse grasses, *R. fruticosus* and *Ulex* spp. and especially good at maintaining mosaic structures. Ponies are particularly successful at countering *B. pinnatum* and producing open, herb-rich swards where this grass is co-dominant in a suppressed dwarf form. Indeed, ponies are the only animals to fare well on a diet of this grass.

Clearly, the conservation movement needs to reconsider its thinking on pony grazing, especially as ponies are relatively cheap to buy (unbroken), need minimal handling (worming at least once a year and occasional foot treatment, especially with New Forests), and live a long time. Small herds of hardy ponies, either breeding or gelding herds, could well be the ideal supplement at large sheep-grazed sites where scrub, especially *Ulex*, and coarse grasses, especially *B. pinnatum*, are problematic. Exmoors appear to be the best breed for nature conservation purposes, although Dartmoors, New Forests, Welsh Mountain and even Shetlands are useful.

7.4.9   Goats

Goats are the best livestock for browsing. In scrub-invaded grassland, free-ranging goats can be expected to browse for at least 50% of their time (Bullock, 1991). Scrub, notably *Fraxinus excelsior* (ash), *T. baccata*, *Crataegus monogyna* (hawthorn) and *Hedera helix* (ivy), formed 80% of the diet of feral goats at Cheddar, Somerset (Smith and Bullock, 1993). Indeed, the 'sheep' which played such an important part in creating chalk downland were more like goats in habit and appearance.

Containment and husbandry requirements often make the use of domesticated goats in scrub control a difficult proposition. However, there is much scope for naturalizing feral goats on large scrub-invaded grasslands, especially coastal grasslands (Bullock, 1991). Contrary to conservationist mythology, feral goats can be contained within normal fencing, providing the compartments are large and there is no shortage of browse. The National Trust has recently introduced feral goats to the short turf slopes of Ventnor Downs, Isle of Wight, to counter *Quercus ilex* (holm oak) invasion. The sward there scarcely needs grazing, the essential requirement is for browsing.

7.4.10 Rabbits

It is ironic that rabbit populations are making a dramatic resurgence just as the conservation movement is reinstating stock grazing to neglected grasslands. Rabbit populations have returned to pre-myxomatosis levels in many regions, which means that conservation grazing policy is going to have to be revised. Although localized rabbit activity increases habitat diversity and often enhances butterfly faunas, rabbits are difficult to control and their grazing is hard to direct. High rabbit populations homogenize the turf, altering mosaic structures and suppressing a wide range of plants and invertebrates. The return of the rabbit poses a major threat to the conservation of species-rich varied grasslands, and to invertebrates associated with tall swards. Moreover, rabbit populations tend to fluctuate wildly, which complicates the planning of stock grazing.

Of the 30 or so species of butterfly that occur on calcareous grassland, only six – *Aricia agestis* (brown argus), *Pyrgus malvae* (grizzled skipper), *C. pamphilus*, *H. semele*, *L. bellargus* and *H. comma* – can thrive in rabbit landscapes. However, although the resurgence of the rabbit has enabled *L. bellargus* and *H. comma* to recolonize much former ground, rabbit grazing can be too severe even for them, especially during drought conditions. For example, during the dry summers of 1989 and 1990 many colonies of *Lysandra coridon* (chalkhill blue), *L. bellargus* and *H. comma* were grazed out by rabbits along the scarp slope of the North Downs. One site, Juniper Hill, Reigate, which had supported a sizeable *H. comma* colony, was denuded of vegetation and reduced to a mix of chalk rubble and rabbit droppings. The National Trust responded by erecting rabbit exclosures along the North Downs escarpment.

The problems posed by rabbits are even more acute for butterflies of medium height and tall grassland, including grass-feeding species such as *Thymelicus sylvestris* (small skipper). Indeed, the resurgence of the rabbit is undoubtedly the main threat to the future of *H. lucina* in grasslands (Oates, in preparation) and to *A. aglaja* on downland and sand-dunes. Furthermore, rabbits reduce nectar supplies, to the detriment of all nectar-dependent insects. Although there are sites, such as Broughton Down, Hampshire, which have long been maintained in a species-rich condition by extensive rabbit grazing, the myth that limestone downs were idyllic for butterflies prior to myxomatosis needs to be dissolved.

Rabbit management is becoming an increasingly important issue in nature conservation. With the exception of dune systems and some skeletal soil downs, it is usually easier to produce and maintain desired sward structures by stock grazing rather than through rabbit activity. Situations in which grassland conservation can safely rely on rabbits are rare.

7.5 MOWING, BURNING AND GROUND DISTURBANCE

7.5.1 Mowing

Mowing is seldom a genuine alternative to grazing, as it tends to tighten the sward and reduce herb diversity. Hay meadows are, however, a notable exception. Meadows which are cut late, after mid-July, can support vast populations of two grass-feeding species, *Maniola jurtina* (meadow brown) and *A. hyperantus*, and sizeable colonies of the herb-feeding *Polyommatus icarus* (common blue) and

Lycaena phlaeas (small copper). However, hay meadows that are subjected to aftermath cattle grazing, or worse, sheep grazing, tend only to support low butterfly populations.

When mowing has to be used as a substitute for grazing, it is essential to remove the cuttings, to cut on a rotational mosaic basis and to incorporate deliberate ground disturbance, to simulate poaching. The current preference for early autumn cuts is not universally appropriate. There are many cases where cutting during early or high summer would be more suitable; examples are spring cutting to reduce the dominance of *Deschampsia caespitosa* (tufted hair grass) and *Holcus mollis* (creeping soft grass) and summer cutting to benefit elements of the vernal flora, notably *Viola* spp. (violets), and to generate late summer nectar by delaying the flowering of plants such as *Cirsium* spp. and *Centaurea* spp. (knapweeds).

7.5.2 Burning

Burning the sward can easily do more harm than good, encouraging undesirable vegetation such as B. *pinnatum*, *R. fruticosus*, *Pteridium aquilinum* (bracken) and *Betula* spp. (birch). It often encourages the problem plant it is supposed to incapacitate. Moreover, it adversely affects a wide range of invertebrates, including larvae of *Eurodryas aurinia* (marsh fritillary). However, much depends on the intensity of the burn, with quick, shallow, winter burns causing little damage. Moreover, many butterfly food-plants thrive after burns, especially *Viola* spp., *Primula veris* (cowslip), *P. vulgaris* (primrose), *Succisa pratensis* (Devil's bit scabious) and *Fragaria vesca* (wild strawberry).

There are situations when patchy burning is worthwhile, notably to diversify rank homogeneous grassland which cannot be grazed, or to create palatable regrowth of grasses such as *M. caerulea* and *B. pinnatum* prior to stock grazing. Also, to remove litter and to encourage *Viola* spp. flushes in *P. aquilinum* stands.

Essentially, burning should only be carried out on a mosaic basis on a lengthy cycle. Although the process of burning is quick, the cutting of firebreaks and the policing of the burn can be laborious. The infrequency of suitable weather conditions is another restrictive factor.

7.5.3 Ground disturbance

It is surprising that deliberate mechanical disturbance of the ground is rarely practised in conservation, as continuity of bare or sparsely vegetated ground is necessary for the conservation of significant elements of the grassland flora and fauna, and rich wildlife sites often result from activities such as quarrying. Indeed, quarries tend to make excellent butterfly sites, capable of supporting over 30 species, including *L. bellargus*, *C. minimus* and *H. semele*.

Legislation on Sites of Special Scientific Interest (SSSIs) does not facilitate purposeful disturbance but, more importantly, the conservation movement is failing to realize the potential of activities such as rotovating, turf stripping and topsoil removal. Much effort is being put into trying to invigorate poor-quality grasslands through grazing, when 'earth management' would actually be more worthwhile. It is arguable that what the neglected Cotswold commons need more than anything else is the reintroduction of small-scale surface quarrying, although social values will have to change considerably before this can happen. It is time conservationists started to consider the management of the medium in which vegetation grows.

7.6 CONCLUSION

During the past decade butterflies have come to the forefront of conservationist thinking, especially in grassland habitat management. Gone are the days when grasslands were managed primarily for rare or attractive plants. In many grassland habitats, notably calcareous and coastal grassland, butterflies merit precedence, but there are grassland types, especially upland grasslands, which are of low potential for butterflies and wherein their conservation should only be a low priority.

Far more research into butterfly ecology is required. In particular, butterfly habitat management requirements, such as grazing tolerances, need clarification. Our knowledge of the ecology of most grass-feeding butterflies is merely incipient, and little is known about how to manage for these species. We need to learn more about the effects of different grazing regimes on butterfly populations, and the relationship between butterflies and rabbit activity is a major area for research. We also need to conduct research into the ecology of various plants used by butterflies as food-plants, notably *Anthyllis vulneraria* (kidney vetch), *Hippocrepis comosa* (horseshoe vetch), *Viola* spp. and, especially, various grasses. This learning process will not be a short one.

The management of grassland habitats depends on efficient scrub management and on suitable grazing regimes. Scrub management is a complex issue, especially as there is great variation from site to site regarding the response of different scrub species to different herbicides (Oates, 1993). With grazing, it is now clear that using appropriate grazing animals is as fundamental as timing and intensity. The scarcity of appropriate stock is one of the main problems hindering grassland management. It is clear that conservation grazing can rarely rely on agriculture, as farm stock is only suitable for a limited range of conservation situations and because the ever-changing nature of farming economics hinders conservation management continuity. The conservation movement needs to acquire its own stock. This is best achieved through regional wildlife organization stock-grazing co-operatives, the acquisition of more lay-back land, and by developing share-farming schemes with sympathetic farmers.

However, the main problem within nature conservation is the weakness of communication within the movement and the associated lack of co-ordination of knowledge and effort. There is a need for a Grassland Managers Association, which can co-ordinate research into vital areas, such as the grazing preferences and abilities of different animals, and disseminate knowledge by producing a journal and by holding seminars. Such an organization would help to integrate butterfly conservation within wider nature conservation aims. Until this happens, the flow of knowledge to those implementing practical conservation will remain unnecessarily slow.

CHAPTER 8

Woodland management and butterfly diversity

P.A. Robertson, S.A. Clarke and M.S. Warren

8.1 INTRODUCTION

Woodland provides a diverse range of habitats for British butterflies. Nearly three-quarters of Britain's resident species regularly breed in woodland, and about one-third are found exclusively or primarily in woods through much of their British range (although some are also found in more open habitats in western Britain and in Continental Europe). The true woodland species include *Leptidea sinapis* (wood white), *Quercusia quercus* (purple hairstreak), *Satyrium w-album* (white-letter hairstreak), *S. pruni* (black hairstreak), *Boloria selene* (small pearl-bordered fritillary), *B. euphrosyne* (pearl-bordered fritillary), *Argynnis adippe* (high brown fritillary), *A. paphia* (silver-washed fritillary), *Mellicta athalia* (heath fritillary), *Ladoga camilla* (white admiral), *Apatura iris* (purple emperor), *Nymphalis polychloros* (large tortoise-shell) and *Pararge aegeria* (speckled wood). Three of these are listed as *Red Data Book* species: *N. polychloros* is classified as endangered and probably now occurs only as an occasional migrant to Britain, while *A. adippe* and *M. athalia* are classified as vulnerable (Shirt, 1987) and are both protected under the 1981 Wildlife and Countryside Act. Five others are classed as scarce species in Britain (Nature Conservancy Council, 1989) and all but *P. aegeria* and *L. camilla* have undergone major range contractions in the past 100 years (Heath, Pollard and Thomas, 1984).

Apart from these true woodland species, many typically grassland species will use rides or large glades within a wood. For the purposes of this review we will concentrate on those species found exclusively or primarily in woodland although most of the principles of management will benefit other butterflies as well as a wide variety of other insects.

8.2 HISTORICAL CHANGES IN THE WOODLAND HABITAT

Before human intervention, much of Britain was covered with trees, the prehistoric 'wildwood'. From Neolithic times this was gradually cleared to make way for

Ecology and Conservation of Butterflies Edited by Andrew S. Pullin.
Published in 1995 by Chapman & Hall. ISBN 0 412 56970 1

agriculture, until in 1086 only around 15% remained (Rackham, 1986). Currently only about 2% remains in modified form as ancient semi-natural woodland. In addition, the gradual clearance of the 'wildwood' and subsequent management of the remaining areas has resulted in woodland fragmentation leaving small patches of suitable, or potentially suitable, habitat scattered across the countryside.

The majority of the woods in lowland areas were typically managed as coppice whereby portions were cut on rotation and allowed to regrow naturally from the stools. The extent of this system was first surveyed in 1905 when 30% of woods were coppiced, although this practice was already in decline. Today only 2% of woods are coppiced, the majority of the remaining sites being introduced *Castanea sativa* (sweet chestnut) in Kent and Sussex. While coppicing, and other activities such as firewood collection which can create small woodland glades, have declined, the area of neglected woodland or high forest has increased. Many broad-leaved woods have been replanted with conifers and large areas of open land have been afforested, particularly with conifers in the uplands (Warren and Key, 1991). The decline in coppice management, increased coniferization and high forest has led to British woodlands being 'shadier than they have been for a thousand years or more' (Thomas and Webb, 1984).

Although the area of coppice, with its periodic light, open and sheltered conditions, has declined, the conversion of existing broad-leaved woods to conifers has, in the short term at least, provided another open and sheltered woodland habitat. When young conifers are planted on ancient woodland sites, the conditions prior to canopy closure are quite similar to those found in coppiced areas. However, once the canopy is complete the ground becomes heavily shaded and these dark conditions continue for a number of decades.

8.3 CHANGING WOODLAND MANAGEMENT AND THE STATUS OF INDIVIDUAL SPECIES

Many woodland butterflies depend on open areas, such as rides and glades, while a number are restricted to early successional vegetation in new clearings. Coppicing provided ideal conditions for such species, both in terms of suitable habitat and of an intimate mosaic comprising woodland of different ages. Many woodland species have low dispersive powers and require a continual supply of newly cut areas in close proximity if they are to endure. With the decline in coppicing these sedentary, open woodland species have been particularly hard hit, with severe declines being experienced by *A. adippe*, *B. euphrosyne* and *M. athalia* (Warren, 1987c, Chapter 14 this volume; Thomas, 1991). Their declines have been largely as a result of a plethora of local extinctions as individual woods have become unsuitable following changing management. Clarke and Robertson (1993a) surveyed sites in Hampshire, Wiltshire and Dorset, known to have contained *B. euphrosyne* or *B. selene* in 1970. The number of colonies had declined by 33% or 40%, respectively, and the losses were most pronounced in woods with relatively few remaining areas of coppice, young conifers, rides or glades (Figure 8.1).

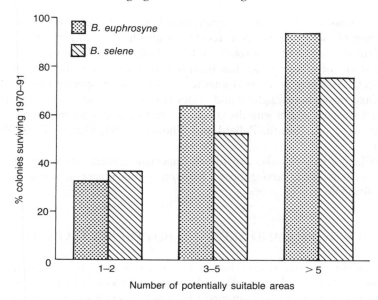

Figure 8.1 Percentage survival (1970–91) of *Boloria euphrosyne* and *B. selene* colonies in relation to the number of remaining patches of potentially suitable habitat in woods in the south of England. In both cases there were significantly higher rates of colony survival in woods with more areas of potentially suitable habitat (chi square tests for linear trends: *B. euphrosyne* χ^2 = 11.9, d.f. = 1, $P < 0.01$; *B. selene* χ^2 = 3.84, d.f.=1, $P < 0.05$).

The extinct colonies were also those in woodlands with the greatest proportion of mature conifers. In a review of prime butterfly sites in central southern Britain, Warren (1993a) found that the highest rates of loss were amongst three predominantly woodland fritillaries, *A. adippe*, *B. euphrosyne* and *B. selene*, with estimated loss rates of 97%, 41% and 38%, respectively, since records began.

The planting of woods with conifers has also affected other species. In the 1970s, over 50% of *L. sinapis* colonies were found in commercial forestry plantations, mainly of young conifers. As these matured and shading increased, many colonies were lost (M. Warren, 1984). The regional effects of these changes can be catastrophic. For example, in Surrey *L. sinapis* declined from 12 large colonies in the 1970s to only one tenuous one in 1988 (Willmott, 1988).

Only two species appear to have benefited from the increasingly shaded conditions within our woods, both having expanded their ranges over the past 50 years, i.e. *P. aegeria* and *L. camilla*. It is notable that they are the only two woodland butterflies that can thrive in shaded woodland conditions (Heath, Pollard and Thomas, 1984; Shreeve, 1986). *Argynnis paphia*, *L. sinapis* and *Aphantopus hyperantus* (ringlet) seem to prefer lightly shaded conditions, despite which all three species have continued to decline (Warren, 1985; Greatorex-Davies, Sparks and Moy, 1993; Thomas, Snazell and Moy, 1994).

Apatura iris and *Q. quercus* are associated with the tree canopy and, due to the inaccessibility of this habitat, their requirements are only poorly understood. Both are characteristic of large broad-leaved woodlands, although the conversion of a proportion of these to conifers in recent decades has almost certainly meant the loss of some colonies of both species (Heath, Pollard and Thomas, 1984).

The status of other species has been linked to more specific changes in our woodlands. *Satyrium w-album* is another elusive canopy species which was well distributed throughout England and Wales after the Second World War. It feeds on *Ulmus* spp. (elms) but with the spread of Dutch elm disease in the 1970s many colonies were lost (Heath, Pollard and Thomas, 1984; Davies, 1992; Warren, 1993a).

Woodland edges can also provide an important habitat, particularly for those species such as *Thecla betulae* (brown hairstreak) and *S. pruni* that require *Prunus spinosa* (blackthorn).

8.4   WOODLAND AREA AND BUTTERFLY DIVERSITY

The woodland habitat is now fragmented compared to the conditions of the 'wildwood'. Does the size of a particular woodland block influence its suitability for butterflies? This can be examined with a dataset compiled by Warren (1993a and unpublished data). A survey of 83 prime butterfly sites, including those that have lost species in the past 10–50 years but which were once considered to be prime, in central southern England can be used to relate the presence of key woodland butterfly species (29 of the 57 resident species) to woodland area (classified into groups to the nearest 10 ha). There was a significant relationship between the area of each wood and the number of key species it contained (Figure 8.2).

However, there was considerable variation due to the management of each individual wood, and some small woods contained high numbers of species. Although a more detailed analysis of individual species occurrence in relation to woodland area is required, it would seem that small woods can contain high butterfly diversities and that large woods may only be better in that they can contain a greater variety of different habitats. This conforms with the findings of Shreeve and Mason (1980). It would seem that woodland area, although related to butterfly diversity, is of less importance than geographical location or site management.

Woodland size may not just influence species abundance, it could also affect the prospects of colony survival. Clarke and Robertson (1993a) examined the survival of 41 *B. euphrosyne* and *B. selene* colonies in Hampshire, Dorset and Wiltshire, known to have been extant in 1970. A re-analysis of these data relating colony extinctions to woodland area found no significant relationships for either species (logistic regression of survival or extinction in each wood against the (log) area of the wood), (*B. euphrosyne* $\chi^2 = 1.62$, d.f. $= 1$, $P > 0.05$; *B. selene* $\chi^2 = 0.93$, d.f. $= 1$, $P > 0.05$). Colonies of these two species were just as likely to have survived from 1970 to 1991 in both large and small woods.

While for many species it appears that small woodlands, if properly managed, can provide adequate conditions for the survival of many species, there are others that do appear to prefer large woodlands or well-wooded areas. *Apatura iris* is most

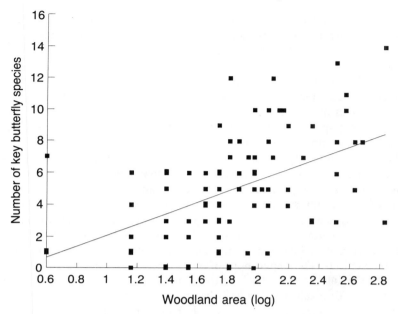

Figure 8.2 Relationship between woodland area and the number of key woodland butterfly species per wood in central southern England ($r = 0.488$, $n=83$, $P < 0.001$). See Warren (1993a) for definition of key species.

commonly found in extensively wooded sites and it appears that it breeds at low densities over a considerable area, congregating only to mate. There are also suggestions that *A. paphia* and *A. adippe* also require large areas of woodland to support a population, but neither of these species is strictly confined to large woods. For some species it seems likely that a cluster of suitable, smaller woods in close proximity may be able to support a population, but further research is needed on this topic.

8.5   OTHER THREATS TO WOODLAND BUTTERFLIES

While changes in woodland management appear to have had a profound effect on many species, in particular those associated with the early stages of woodland succession and coppicing, there are other factors that may have assisted the declines (see Pollard and Eversham, Chapter 2 this volume). Corke (1989) proposed that predation of larvae and pupae by released *Phasianus colchicus* (pheasant) may be important for a number of woodland species (but see Warren, 1989). A more detailed examination of *B. euphrosyne* and *B. selene* colony survival in relation to the presence or absence of pheasant releasing has found no adverse effects. In fact, the rates of colony survival for both species were higher in woods containing pheasant release pens that in other woods not owned or managed by conservation organizations, presumably because pheasant management encouraged the creation of rides, glades and other open areas (Clarke and Robertson, 1993a,b).

Agricultural intensification, in particular the conversion of grasslands to arable production, the increased use of pesticides and hedgerow removal, have certainly affected a great many species (Dover, 1991). One consequence of this has been the increased importance of woodland, in particular rides and glades, as refuges for many species previously found on farmland.

Many woodland butterflies have poor powers of dispersal, but in some regions they can use more open habitats, such as damp meadows, coastal cliffs and moorland (e.g. *B. selene*). The increased intensification of agriculture in many parts of Britain has meant that potentially suitable woods may now be surrounded by relatively inhospitable agriculture as opposed to herb-rich meadows and interconnecting hedgerows, as was once the case. This increasing isolation of small pockets of suitable habitat makes each more vulnerable to extinction and less likely to be recolonized. The implications of these landscape-scale changes in land use have yet to be fully assessed for butterfly conservation, but may be of considerable importance. For instance, *M. athalia* has relatively poor powers of dispersal and often takes 2–3 years to colonize suitable habitat patches just 300–700 m from source populations in south-east England (Thomas, Thomas and Warren, 1992). Verspui and Visser (1992) suggested that *M. athalia* may cover distances of up to 3 km in The Netherlands, but that dense woodland or extensive areas of open ground can act as barriers for their movement. Warren (1991) described how suitable sites for this species in Essex were 40 km from the nearest colonies and were extremely unlikely to be naturally recolonized. Subsequent releases at the Essex sites confirmed their suitability for this species. Perhaps the most sedentary of all woodland butterflies is *S. pruni* which can take many years to colonize a suitable *P. spinosa* stand only a few hundred metres away within the same wood (Thomas, 1975). Following its artificial establishment into unoccupied habitat in Surrey it has spread only 4 km from the release point in 36 years (Thomas, Thomas and Warren, 1992).

Butterfly releases may have a role to play in the conservation of certain species but, to quote Oates (1992a), 'Artificial establishment really amounts to treating the symptoms of butterfly decline rather than the actual causes, and mere propagation is no substitute for true conservation measures.'

Grazing pressure is another factor that certainly influences the availability of many butterfly food-plants and has undergone dramatic changes in the past 40 years. Myxomatosis in the 1950s drastically reduced *Oryctolagus cuniculus* (rabbit) populations, with profound effects on many grassland species; the effects in woodlands where rabbits also graze are less well documented. The current increase in cervid (deer) densities in many parts of Britain is also affecting the ground flora of many woods, although the implications for butterflies are unclear. Cervids have other, indirect effects, as they can greatly increase the cost of re-establishing coppice, making it uneconomic in certain areas (Ratcliffe, 1992). In the New Forest, local decline of *L. camilla* and other species has been attributed to increased grazing pressure by equines (horses and ponies) affecting the abundance of suitable food-plants (Tubbs, 1986).

8.6 CURRENT TRENDS IN WOODLAND MANAGEMENT

Government policy towards woodland has changed rapidly over the past decade. From a position where grant aid was only available for conifer planting, we now have a situation where broad-leaved plants are actively encouraged, together with the provision of open areas within woodland. In addition, management grants are now available to promote the sympathetic management of existing woods. However, the grants for coppicing are still low and there is no special grant for restarting a coppice cycle in neglected woods, an operation that requires investment but provides little financial return for some years, if ever. The consequences of these new initiatives have yet to be seen, but are likely to be generally beneficial.

Potential new habitats may also be created by changing agricultural economics. Short-rotation biomass coppice is being viewed as a potential use of land taken out of agricultural production and may, given time and appropriate management, provide a suitable habitat for certain species, at least in the rides and open spaces it is likely to contain (Robertson and Sotherton, 1992).

8.6.1 Woodland management for butterflies

The consequences of scant management in our broad-leaved woods and an increase in coniferization are likely to be long term. Many woodland butterflies, particularly the Argynninae and Melitaeinae (fritillaries) and Theclinae (hairstreaks) have poor powers of dispersion (Thomas, 1975; Warren, 1987c; Thomas, Snazell and Moy, 1994). Their ability to recolonize sites if suitable conditions are recreated is limited and, without introductions, it is likely that any recovery in their ranges will be slow. Secondly, the long rotations of high forest broad-leafs and conifers may lead to the loss of plant species on sites that were once coppiced. This is certainly true of conifers where the deep shade and changes in soil chemistry result in unsuitable conditions after felling (Mitchell and Kirby 1989). For coppice that has been unmanaged for decades the ability of many plant species, including important larval food-plants such as *Viola* spp. (violets), to survive as dormant seeds is in doubt. Brown and Warr (1992) suggested that many woodland plant species associated with newly cut coppice can survive in the seed bank for at least 50 years. After this there appears to be a rapid decline in their viability. Consequently, the reintroduction of coppicing to many woods may only produce an impoverished ground flora. Furthermore, stool vigour is often reduced (Rackham, 1980) and there may also be a dense growth of *Rubus fruticosus* (bramble) which quickly smothers larval food-plants on the woodland floor (Mason and Long, 1987). The problems and possible solutions of coppice restoration are discussed by Fuller and Warren (1990) and Kirby and Patterson (1991).

This is not to say that coppicing is futile. On the contrary, the reintroduction of

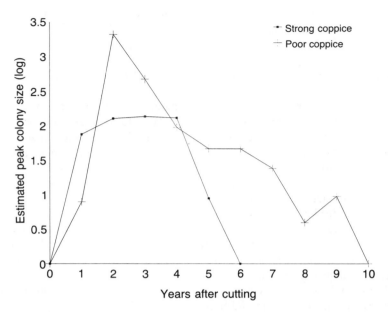

Figure 8.3   *Mellicta athalia* population size (note log scale) in strongly growing (16 sites) and poorly growing (8 sites) *Castanea sativa* coppice of different ages in Kent, 1980–84 (after Warren, 1987c).

coppicing has been shown to be very beneficial to certain rare species, such as *M. athalia.* In the 1980s a concerted effort was made to protect some of the remaining colonies of this species by establishing nature reserves and managing the habitat, in particular by reinstating an active cycle of coppice management. Warren (1987c) described the response of this species to management (Figure 8.3).

Mellicta athalia colonies responded to woodland management, reaching peak population numbers 2–3 years after an area had been cut. In the Blean Wood National Nature Reserve the onset of regular coppice management began in 1979, when the peak population was estimated at less than 20 adults. This rapidly increased to over 1000 adults in 1985. The butterfly has also been successfully re-established on two nature reserves in Essex where coppicing has been restarted (Warren and Key, 1991).

In many woodlands, particularly those used for commercial timber production, the most valuable butterfly habitats are found along rides (e.g. Greatorex-Davies, Sparks and Moy, 1993) as these provide sheltered but sunlit habitats and can contain a variety of food-plants. Moreover, they are a necessary part of forest design for fire protection, access and timber extraction, and they provide public access so that the butterflies can be seen. Large, open and sunny rides provide a habitat for the widest range of butterfly species. The relatively undisturbed vegetation can attract large numbers of grassland species, while the growth of shrubs along ride edges can encourage *T. betulae* and *Callophrys rubi* (green hairstreak). They can also be used by many species more normally associated with coppice or open grassland. As shading increases, rides can become suitable for *A. hyperantus* and *L. sinapis*, while more heavily shaded areas are used by *P. aegeria*. The dappled shade at the edges of the rides can be used by *L. camilla* and *A. paphia* (Warren and Fuller, 1990).

The effects of rides can be dramatic. Robertson, Woodburn and Hill (1988) compared the number of species and individual butterflies seen along fixed routes through different areas of a large woodland in Dorset. In blocks of mature conifer they found only five butterflies per km walked, and these were of only two species, the shade- tolerant *P. aegeria* and *L. camilla*. Overall butterfly density was even lower in unmanaged oak over old hazel coppice regrowth (3 butterflies/km) but the number of species was higher and included *Pieris rapae* (small white), *P. aegeria*, *Maniola jurtina* (meadow brown) and *A. paphia* (NB it is possible that some canopy-living species, such as *Q. quercus*, were missed in these areas). However, in large, sunlit rides cut through areas of mature woodland, the number of individuals seen was considerably higher (89 butterflies/km), comprising 19 species.

There are management problems associated with rides and glades. When first cut they can contain a rich woodland flora, similar to that found in newly cut coppice. However, in subsequent years this can be swamped by invasive grasses. Greatorex-Davies, Hall and Marrs (1992) described the effects of widening rides at their intersections to create glades or 'box-junctions' to promote the growth of *Viola* spp. for *B. euphrosyne*. Although some of these areas were suitable for their first 5–7 years, their value declined thereafter as they were invaded by grasses which outcompeted the *Viola*. Although raking and digging were used to try and maintain the *Viola* populations, neither was successful. The use of selective herbicides or short-rotation *Picea abies* (Norway spruce), for Christmas trees, have both been suggested as possible methods to suppress the grasses, although they remain to be fully tested (e.g. Porter and Nowakowski, 1991).

8.7   CONCLUSIONS

Recent decades have seen dramatic changes in the distributions and abundance of many of our woodland butterflies. Almost every species has declined, particularly those associated with coppicing, while only the two shade-tolerant species have increased. Two trends in woodland management appear to be responsible for these changes; a decline in active coppice management followed by a trend towards coniferization. Although young conifer plantations can provide a suitable habitat for many species, they rapidly deteriorate after canopy closure.

What can be done to alleviate this situation? For the most endangered species, such as *A. adippe* and *M. athalia*, the responsibility lies with nature conservation organizations. However, for most of the other species, more widespread policies are needed to secure their future in state- or privately owned woodlands. Recent changes in the provision of grant aid to forestry should hopefully slow the trend towards coniferization in many areas, but are unlikely to result in a large-scale swing back towards traditional coppice management, probably the best long-term solution for many species. Other, less dramatic changes in woodland management will probably be of greater value in the short term. In particular, the encouragement of ride and glade creation, now helped by grant aid in some woods, should benefit many species. Warren and Fuller (1990) discussed some of the ways in which rides and glades can be created and enhanced to benefit wildlife, while the Forestry Commission is also promoting ride management in many of its holdings (Carter and Anderson, 1987). The creation of these features has benefits for other groups of invertebrates (Warren and Key, 1991), for instance the Heteroptera (Greatorex-Davies and Marrs, 1992), together with migrant songbirds (Fuller, 1992) and *P. colchicus* (Robertson, 1992). An interest in *P. colchicus* shooting can also provide the incentive for many landowners to, incidentally, create features for butterflies by the cutting of rides, glades and small coppices (Robertson, Woodburn and Hill, 1988, Clarke and Robertson, 1993b).

All of these trends may lead to our woods being more attractive to many butterfly species. However, given the fragmented nature of the suitable areas and the limited distributions of many species, together with their poor dispersive powers, recolonization may be slow. As mentioned, reintroductions may have a role to play in this regard but the priority must be the conservation of the habitat. The factors affecting recolonization are also poorly understood; what effects do the land use of intervening farmland have on rates of movement between neighbouring areas of suitable habitat, and do hedgerows really aid dispersal? The factors leading to many of the declines are reasonably well documented and solutions are often available. What is lacking is the financial incentive to implement these solutions on a large scale, and an understanding of how the butterflies will exploit the opportunities for range expansion.

CHAPTER 9

Gardens:
the neglected habitat

M.L. Vickery

9.1 INTRODUCTION

Since the formation of the British Butterfly Conservation Society (BC) in 1968 the attitude of the average gardener to wildlife in the garden has changed considerably. In particular, many gardeners are now actively trying to attract butterflies into their gardens and many books have been published giving advice on ways to do this (e.g. Cribb, 1982; Rothschild and Farrell, 1983; Killingbeck, 1985; Oates, 1985; Payne, 1987), most being based on the authors' own experiences.

Although wild butterfly habitats (e.g. Warren, 1985; Warren and Fuller, 1990) and aspects of butterfly behaviour (Pullin, 1986a, 1987a) have been scientifically studied, the garden as a butterfly habitat has been relatively neglected, with the notable exceptions of research by Owen (1991), over a 15 year period from 1971 in her Leicestershire garden, and by Stephens and Warren (1992), during 1985 in 24 gardens in Dorset.

It was against this background that BC launched the national garden butterfly survey in 1990, with the ambitious objective of discovering the aspects making up the perfect garden for attracting butterflies. This survey also monitors, qualitatively, the year-by-year status of butterflies in typical British gardens.

Although breeding takes place on a small scale in some gardens, I do not consider this aspect to be a prime function of the garden habitat, except for some urban situations. Rural and suburban gardens almost always have areas nearby where butterflies can breed but these do not always contain abundant nectar plants, especially in late summer and autumn when hibernating butterflies need to store lipids (fat reserves) to provide energy during the winter (Pullin, 1987b). It has also been suggested (Porter, 1992) that successful breeding is dependent on the quality and quantity of the nectar butterflies imbibe, thus gardens may help butterflies to breed, even if this does not actually take place within their boundaries.

Ecology and Conservation of Butterflies Edited by Andrew S. Pullin.
Published in 1995 by Chapman & Hall. ISBN 0 412 56970 1

9.2 GARDEN BUTTERFLY SURVEYS

During the past 25 years there have been few published comprehensive surveys of butterflies in gardens. Many individuals have kept records of the butterflies in their own gardens and a few have published these (e.g. Rothschild and Farrell, 1983; Owen, 1991). Although such records are interesting and can be helpful to the management of that particular garden, extrapolation to the garden habitat in general is fraught with inaccuracy. Only surveys involving many gardens can give an accurate picture, and the more gardens involved, the more accurate the picture becomes.

9.2.1 The WATCH butterfly survey

WATCH is the junior wing of the Royal Society for Nature Conservation (RSNC) and in 1983 Dr Jeremy Thomas enlisted the help of members to survey gardens for butterflies (J. Thomas, 1983c). Although the survey involved many gardens, it was carried out only during the period July–September 1981, and so spring butterflies such as *Anthocharis cardamines* (orange tip) were not recorded at all, nor were first broods of the common garden butterflies, such as *Aglais urticae* (small tortoiseshell). Also, butterflies were mostly recorded only in families (blues, browns, etc), with just a few individual species noted.

9.2.2 The Essex Wildlife Trust butterfly survey

The Essex Wildlife Trust survey (Essex survey) carried out in 1991 (Corke and Davis, 1992) involved 460 gardens and butterflies were recorded throughout the season. Recorders were asked to count the butterflies in their gardens on one day each week from April to September. They were also given a list of nectar plants, both wild and cultivated and asked to tick those that flowered in the garden and to note any butterflies that visited the plant. There was also room on the record card for additional plants. A list of caterpillar food-plants was given and recorders were asked to tick those present in the garden and to indicate if caterpillars were seen. The gardens were divided into five categories: normal small, large country, school, churchyard and other.

9.2.3 The RSPB Young Ornithologists Club butterfly survey

The Royal Society for the Protection of Birds (RSPB), Young Ornithologists Club (YOC) survey in July–August 1992 (Harvey, unpublished) covered 288 gardens and species were noted individually. However, again recording took place only during the summer school holidays, so spring butterflies were not included. This survey also noted what the butterflies were doing in the garden, and a useful list of nectar plants was compiled from feeding records.

9.2.4 The Dorset butterfly survey

The Dorset survey by Stephens and Warren (1992) covered the whole 1985 butterfly season and involved 24 gardens within the county. The gardens were divided into rural and urban and the main purpose of the survey was to contrast the garden habitats with wild habitat types that occurred nearby. In this survey a very useful attempt was made to link butterfly density in gardens with garden design.

9.2.5 The BC national garden butterfly survey

The BC survey has been running for 3 years and has involved over 1000 gardens. Butterflies are recorded throughout the flight season and results compared year by year in order to draw up criteria which are important in attracting butterflies to feed in gardens.

During 1991 and 1992 two surveys have been run in parallel: a BC members' survey (BC survey) and a survey involving Womens Institute members (WI survey).

Recorders are given two forms to complete, one form is concerned with the butterflies visiting the garden and the other with plants grown, the latter has only been used for the past 2 years. The common names of 19 butterfly species, established as visiting gardens from the literature and preliminary studies, are printed on one side of the first form, with space for recorders to add further species. On the back of this form are questions about the garden and its environment and space for the name, address and county of the recorder. The second form lists the common names of 32 nectar plants, again deduced from the literature and the 1990 survey, and there is space to add further plants used by butterflies in that garden.

Because of the nature of the survey, which relies heavily on recorders who are enthusiastic but who have no experience of scientific methods and little time to devote to recording, I have had to make some fundamental assumptions. One of these is that a butterfly found to be widespread in gardens is also common in gardens. Recorders are asked to give the date of first sighting, or at the very least to tick that a species has been seen in the garden. There is no attempt at counting numbers, although recorders are asked to estimate the frequency with which the butterfly is seen in the garden and to allocate this to one of three categories: frequent, occasional and once only. It will be shown that there has been a good match between the presence of butterflies in gardens and the frequency with which they are observed, which justifies my original assumption.

In order to determine the criteria making up good garden habitat for butterflies, recorders are asked to tick boxes in answer to a number of questions, including garden size, the direction in which the garden faces, whether or not it is shaded or sheltered, whether it is situated in a rural, suburban or urban environment and whether or not there is wild butterfly habitat within half a mile. I ask recorders to give their counties so that I can divide the records into regions (north, midlands, east, south-east, south, Wales, Scotland and Ireland).

On the 'flowers' form, recorders are asked to tick plants that they grow and also to note any feeding activity they observe on a plant, giving the name of the butterfly. Only about half of recorders trouble to do this for plants other than *Buddleia* spp. and feeding activity is grossly under-recorded. However, enough forms are completed to give much useful information on the value of around 70 nectar plants to butterflies. Particularly valuable is the information being gathered concerning individual butterfly species preferences, although this is not yet complete.

9.3 THE GARDEN HABITAT

9.3.1 Butterfly occurrence

The number of butterfly species recorded in garden surveys varies from 21 (Owen, 1991; single garden) to 51 (Vickery, 1993; 650 gardens). However, many of the latter are transient visitors, unlikely to be recorded more than once in any one garden. By considering only those species recorded in five or more gardens in 1990–91 (around 200 gardens recorded) and 15 or more gardens in 1992 (650 gardens recorded), a more realistic total of 26 species is obtained for the BC survey. This figure compares with the Essex survey which recorded 26 species, and a figure of 24 for Dorset gardens. Of the 26 recorded in the BC survey, *Colias croceus* (clouded yellow) can be discounted as it only visits a sufficient proportion of gardens during the good years of its irregular migrations. *Argynnis paphia* (silver-washed fritillary) and *Quercusia quercus* (purple hairstreak) can also be ignored as these rare garden visitors are only recorded in gardens near woodland where these species occur. Although recorded separately by some, many recorders do not (or can not) distinguish between *Thymelicus lineola* (Essex skipper) and *T. sylvestris* (small skipper), so that *T. lineola* is included with *T. sylvestris*. This leaves 22 species which find gardens useful habitats and these will be considered in more detail.

The most widespread garden butterfly in all 3 years of the BC survey has been *A. urticae*. The WI survey gave *Inachis io* (peacock) first place in 1991 and *A. urticae* in 1992. Owen found *Pieris rapae* (small white) the commonest in her garden, as did Stephens and Warren in their Dorset gardens, while *A. urticae* was the most numerous in the Essex survey. The top five butterflies have remained constant throughout the 3 years of the BC survey and the 2 years of the WI survey. They are *Pieris brassicae* (large white), *P. rapae*, *Vanessa atalanta* (red admiral), *A. urticae* and *I. io*. However, *Pyronia tithonus* (gatekeeper) replaced *V. atalanta* in the Essex survey; and in predominantly rural gardens in Dorset, *P. tithonus* and *Maniola jurtina* (meadow brown) were amongst the top five. A comparison of all 22 species for the BC, WI and Dorset surveys is given in Table 9.1.

The Essex survey cannot be included as the numbers of gardens recording each species are not noted.

It can be argued that presence/absence alone is not a true measure of the value of gardens to butterflies. Recorders for the BC survey were, therefore, asked to estimate the frequency with which they saw the species in their gardens during its flight period(s) for the 19 species listed on the survey form. The comparison of frequency and presence/absence in Table 9.2 shows that there is general agreement between the two measures, and that the latter can be used as an indication of the use of the garden habitat.

If the 19 species are divided into three bands: top five, bottom five and middle

Table 9.1   The 22 species of British butterfly recorded most often in gardens according to three surveys, expressed as a percentage of total gardens recorded

Species	Year:	BC survey			WI survey		Dorset survey[a]
		90	91	92	91	92	
Thymelicus sylvestris		36	38	32	19	19	17
Ochlodes venata		33	25	24	5	10	25
Gonepteryx rhamni		56	60	64	53	56	13
Pieris brassicae		92	94	97	92	92	83
Pieris rapae		93	92	93	84	88	96
Pieris napi		71	69	60	39	40	67
Anthocharis cardamines		79	80	86	77	85	38
Lycaena phlaeas		60	47	42	32	28	0
Polyommatus icarus		46	45	54	46	39	0
Celastrina argiolus		75	73	74	56	62	42
Vanessa atalanta		91	92	98	87	97	75
Cynthia cardui		56	64	72	50	49	38
Aglais urticae		97	98	99	94	99	75
Inachis io		83	88	96	96	95	79
Polygonia c-album		71	81	75	64	62	8
Pararge aegeria		61	68	59	44	55	42
Lasiommata megera		39	30	37	34	41	0
Melanargia galathea		9	8	6	–	6	17
Pyronia tithonus		61	65	70	56	68	88
Maniola jurtina		76	78	80	68	68	83
Coenonympha pamphilus		15	9	11	6	12	0
Aphantopus hyperantus		17	17	21	9	6	8

[a] Compiled from Stephens and Warren (1992).

nine, exactly the same species occur in each of the bands whether judged by presence or by frequency of observation. Those species occurring in the bottom band are also more often reported as occurring occasionally in gardens than as occurring frequently.

A comparison of the three surveys (Table 9.1) shows BC members consistently recording species more often in their gardens than WI members. This is particularly true for species of the middle and bottom bands, but even those of the top band seem to be generally recorded less often in WI gardens than in BC gardens. A similar pattern is shown by the Dorset gardens, except for *P. tithonus* and *M. jurtina* which occurred more widely in Dorset gardens than in BC gardens. However, if only gardens in the southern half of England are analysed, *M. jurtina* occurrence is about the same (BC = 82%; Dorset = 83%), although *P. tithonus* is still more widespread in Dorset (BC = 77%; Dorset = 88%). The differences between the gardens of BC members and non-members could be due to either the garden habitat itself, or to the diligence and accuracy of the recorders. Both could be contributing factors in the WI survey, but the recorders for the Dorset survey were both diligent and experts in butterfly identification, so that the differences can only be due to garden habitat. The fact that BC gardens are on average

Table 9.2 A comparison of the presence of 19 species of British butterfly in gardens with the frequency with which they were observed in 1992, expressed as a percentage of total gardens recorded

Species	Frequently seen	Occasionally seen	% presence
Thymelicus sylvestris	11	13	32
Ochlodes venata	7	12	24
Gonepteryx rhamni	26	27	64
Pieris brassicae	93	2	97
Pieris rapae	82	3	93
Pieris napi	48	19	60
Anthocharis cardamines	51	29	86
Lycaena phlaeas	9	22	42
Polyommatus icarus	14	28	54
Celastrina argiolus	44	22	74
Vanessa atalanta	77	16	98
Cynthia cardui	24	28	72
Aglais urticae	88	5	99
Inachis io	83	11	96
Polygonia c-album	34	34	75
Pararge aegeria	35	18	59
Lasiommata megera	13	19	37
Pyronia tithonus	47	18	70
Maniola jurtina	45	24	80

considerably better butterfly gardens than those of non-members gives much cause for hope for the future, as BC members will have been more likely to have made some effort to attract butterflies to their gardens than non-members.

9.3.2 Garden environment

Very little previous work has been done on the effects of the garden environment on butterfly visitors, but the BC survey examined the garden environment in some detail (Vickery, 1991a, 1992a, 1993). The WATCH survey (J. Thomas, 1983c), Essex survey (Corke and Davis, 1992) and Dorset survey (Stephens and Warren, 1992) showed that rural gardens attracted more species than suburban or urban. This has been corroborated by the BC surveys, where every year there is a steady increase in the number of species recorded from urban through suburban to rural. The Dorset survey recorded 14 species in urban gardens compared with 24 in rural situations, and the BC survey has consistently found a smaller percentage of urban gardens recording over 14 species than suburban or rural. The number of urban gardens surveyed has always been fewer than suburban or rural in both the BC and WI surveys. However, some facts are emerging which indicate that suitably managed urban gardens can be extremely useful butterfly habitats. A number of urban gardens not near any wild habitat have recorded some species more often than those near to wild habitat. Amongst such species are the sedentary and rare garden visitors, *Ochlodes venata* (large skipper) and *Lycaena phlaeas* (small

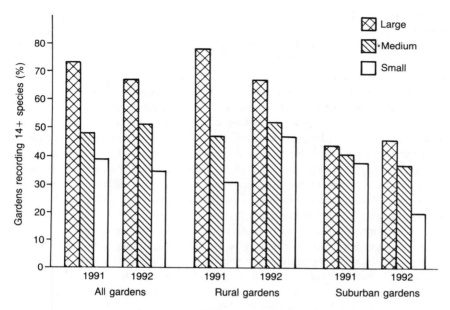

Figure 9.1 The percentage of gardens recording more than 14 species of butterfly according to garden size and situation.

copper). Is it possible that these are the remains of colonies that once existed on the land before it was developed and that they have survived, albeit in small numbers, because the butterflies are breeding in gardens? These questions cannot be answered until evidence of breeding activity is found. Recorders of such urban gardens have been asked to investigate this aspect in 1993. It is interesting that the Essex survey found more *L. phlaeas* in gardens containing the larval food-plants *Rumex acetosa* (common sorrel) or *R. acetosella* (sheep's sorrel) than in gardens without these plants.

The Dorset survey found no connection between garden size and number of species, but the BC survey has shown large gardens always recording more species than small, although the difference is greater for rural than for suburban gardens (Figure 9.1).

Too few records were received to make the analysis of urban gardens by size meaningful.

The direction in which the garden faces does not give such clear-cut answers, but the 1992 survey of 650 gardens showed very clearly that where the garden completely surrounds the house a greater number of species can be expected than for gardens facing in only one or two directions. Fourteen or more species were recorded in 73% of gardens surrounding a house, but in fewer than 50% of gardens facing in only one or two directions. A garden mostly open to the sun has a greater chance of recording 14 or more species than one mostly shaded (53% compared with 10%). While some shelter would seem to be advantageous, this was not reflected in the data, as the percentages of sheltered and unsheltered gardens recording 14 or more species were almost equal, being 52% and 50%, respectively. This is probably the result of the calm conditions during much of the 1992 season. The Dorset survey found shelter and unshaded conditions to be favoured by butterflies.

As would be expected, the region of the British Isles in which the garden is situated has a profound effect on the number of species recorded. Fewer than 10% of gardens in the north of England can expect 18 or more species, whereas around a third of gardens in the south can expect numbers of this magnitude. There were no gardens in Scotland recording more than 12 species. Although the average number of species recorded in the north was less than that in the south, the occurrence of two of the rarer species in gardens, *L. phlaeas* (north = 48%; south = 25%) and *Lasiommata megera* (wall brown) (north = 64%; south = 25%) was significantly greater in the north than in the south in 1992.

A close correspondence was found between the average number of species recorded in gardens and their proximity to wild habitat types in both the BC and WI surveys: the more types the more species. Also, Stephens and Warren (1992) stated that 'position [of garden] with respect to non-garden habitats has an overriding effect on the . . . number of species that can be attracted'. It is interesting that in 1991 farmland was not a good habitat type to be near, but in 1992 it seemed to be improving. Hopefully, surveys in future years will show if this is a permanent benefit of changing farming practices.

9.3.3 Garden design

A factor which was not considered in the BC or WI surveys was that of garden design. However, this was an important feature of the Dorset survey. This found that gardens broken up by internal borders or hedges attracted more butterflies than those containing large areas of hard surface (Warren and Stephens, 1989). This is in agreement with the findings of Wood and Samways (1991) in South Africa, where their investigation of a large garden showed that butterflies prefer to fly along the sides of hedges or over areas of mixed vegetation and that a large area of close-cropped turf acted as a barrier to flight.

9.3.4 Nectar plants

It is generally agreed that butterflies visit gardens primarily in order to feed on nectar plants. In order to create good garden habitat it is therefore of paramount importance to provide an abundance of nectar plants, with a succession of blooms from spring to late autumn. Books on butterfly or wildlife gardening almost always contain lists of butterfly nectar plants derived either from the author's own observations, from previous publications or from hearsay. As butterflies were reported to feed on over 300 plants in the 1992 BC survey alone, but only about 70 of these were recorded in five or more gardens (Vickery, 1993), it is not surprising that many plants that have little overall value appear in these lists. Prior to the 1990s there was little attempt to investigate the usefulness of plants simultaneously in a large number of gardens. A very important fact to emerge from the 3 years of the BC survey is that a plant which is an excellent nectar plant in a single garden may be ignored in most other gardens. Why this should be so is not known and more work needs to be done on this aspect. Extrapolation from observations in a single garden to the garden habitat as a whole has led to many inaccuracies in published lists of supposed good butterfly nectar plants. Conversely, the survey is showing that there are many plants which are consistently used in gardens throughout the British Isles as nectar plants by butterflies (Vickery, 1991b, 1992a, 1993). A list of the most useful, with the number of species reported as feeding on the flowers, the percentage of gardens where feeding occurred and the plant popularity index (PPI) for 1991 and 1992 is given in Table 9.3.

Between them, these plants provide nectar throughout the butterfly season. The PPI is obtained by adding the number of species reported nectaring on the plant to the percentage of gardens recording feeding activity. It effectively discounts plants reported by only a few recorders, as the PPI will be very low. The index remains constant to within ± 10 for most plants between the 2 years, and, despite under-recording of feeding activity, the PPI is a useful measure of the value of a nectar plant to butterflies in general. However, the index can only be calculated when the number of gardens growing the plant is known as well as the number recording feeding activity.

The Essex, YOC and Dorset surveys all contain information on nectar plants, although in the latter two cases spring plants were not generally included. All surveys show that *Buddleia davidii* (buddleia or butterfly bush) is by far the best summer nectar plant. The YOC survey recorded almost 15 times as many feeding instances as for any other plant and the Dorset survey over twice as many. Over the 3 years of the BC survey, feeding activity has been reported in over 90% of gardens growing this shrub. A comparison of the surveys shows that other particularly good summer nectar plants include *Lavendula* × *intermedia* (lavender), *Sedum spectabile* (ice-plant), *Aster* spp. (michaelmas daisy), *Origanum vulgare* (marjoram), *Centranthus ruber* (valerian), *Phlox paniculata* (phlox), *Ligustrum ovali-folium* (privet) and *Rubus* spp. (bramble or blackberry).

In order to attract specific butterfly species to a garden it is necessary to plant those nectar plants used by the species, as each butterfly has its own preferences. For example, even *B. davidii*, which attracts over 29 species, is not a favourite of all. *Coenonympha pamphilus* (small heath) has only been reported feeding once and *Aphantopus hyperantus* (ringlet) only five times. Almost nothing was known about species preferences before the BC surveys, but much is now being learnt. A list of favourite nectar plants is being compiled but is not yet complete. There appears to be a connection between the presence and frequency of a species in gardens and the number of favourite nectar plants; four out of the five top garden butterflies are recorded as feeding on the largest number of nectar plants, with the commonest garden butterfly, *A. urticae*, using the most (Table 9.4).

Butterflies in the bottom band are all recorded as feeding on the least number of plants. It is interesting that common migrants, such as *V. atalanta* and *Cynthia cardui* (painted lady), use fewer nectar plants than common residents of similar percentage presence.

In view of the above, it is not surprising that gardens attracting the most

Table 9.3 Nectar plants used as food sources by butterflies in gardens throughout the British Isles, 1991–92

Plant species or genus	No. butterfly species reported feeding		No. gardens reporting feeding activity as % of no. growing plants		PPI	
	1991	*1992*	*1991*	*1992*	*1991*	*1992*
Alysum spp.	12	17	24	21	36	38
Aster spp. (summer)	14	18	32	28	46	46
Aster spp. (autumn)	14	18	58	46	72	64
Aubretia deltoidea	15	20	45	44	60	64
Buddleia spp.	27	29	97	93	124	122
Centranthus ruber	18	23	56	48	74	71
Erysimum cheiri	12	13	19	18	31	31
Coreopsis spp.	13	15	36	29	49	44
Dianthus barbatus	11	16	18	22	29	38
Erica spp.	13	16	21	16	34	32
Erigeron spp.	14	19	49	38	63	57
Eupatorium cannabinum	16	22	69	62	85	84
Hebe spp.	13	22	44	37	57	59
Hesperis matronalis	9	17	55	53	64	70
Iberis spp.	12	20	35	38	47	58
Lavandula × *intermedia*	21	26	56	49	77	75
Ligustrum spp.	17	18	nr	46	nr	64
Lobelia erinus	18	18	54	34	72	52
Lunaria annua	8	10	49	37	57	47
Mentha spp.	17	19	29	21	46	40
Myosotis alpestris	13	16	20	22	33	38
Nepeta × *faassenii*	13	15	nr	46	nr	71
Origanum vulgare	19	21	69	52	88	73
Phaseolus coccineus	9	13	44	32	53	45
Phlox spp.	9	20	nr	55	nr	75
Primula spp.	6	8	11	12	17	20
Rubus spp.	16	24	nr	55	nr	79
Scabiosa spp.	20	23	54	48	74	71
Sedum spectabile	17	19	63	62	80	81
Solidago canadensis	16	14	32	18	48	32
Tagetes erecta	15	16	21	34	36	50
Tagetes patula	15	17	35	42	50	59
Thymus vulgaris	13	24	22	21	35	45

nr = not recorded.
PPI = number of species + % observed feeding activity in gardens.

butterfly species also usually grow the greatest variety of nectar plants. For example in 1992 it was found from the BC survey that gardens with less than 20 different nectar plants had only a 31% chance of recording 14 or more species, while those with more than 20 plants had a 65% chance of doing so. The density of plants was not recorded in the BC survey but the Dorset survey found that greater butterfly density was associated with sizeable patches of favoured nectar plants.

Table 9.4 The number of nectar plants consistently used by 19 butterfly species, compiled from feeding records

Butterfly species	1991, no. plants with 5+ records	1992, no. plants with 10+ records
Thymelicus sylvestris	5	6
Ochlodes venata	1	0
Gonepteryx rhamni	9	8
Pieris brassicae	24	31
Pieris rapae	33	31
Pieris napi	15	13
Anthocharis cardamines	7	6
Lycaena phlaeas	5	3
Polyommatus icarus	7	5
Celastrina argiolus	14	13
Vanessa atalanta	15	16
Cynthia cardui	6	4
Aglais urticae	37	38
Inachis io	16	23
Polygonia c-album	9	6
Pararge aegeria	5	3
Lasiommata megera	1	1
Pyronia tithonus	20	16
Maniola jurtina	14	14

Total number of gardens recorded was 200 in 1991 and 650 in 1992.

9.4 BREEDING IN GARDENS

Whereas most amateur butterfly recorders are willing and able to record adult butterflies, this is not so for the immature stages. A scattering of breeding records are obtained each year for the BC survey, but these are insufficient to draw any overall conclusions. In general, species are reported breeding mostly in large, rural gardens, although, as mentioned above, there is circumstantial evidence that breeding may be taking place in some urban gardens. The species reported to be breeding are the same as in the Dorset survey: *P. rapae*, *P. brassicae*, *A. cardamines*, *Celastrina argiolus* (holly blue), *A. urticae*, *I. io*, *Polygonia c-album* (comma), *Pararge aegeria* (speckled wood), *P. tithonus* and *M. jurtina*. The Essex survey found *P. rapae* and *P. brassicae* to be breeding, and evidence for the possible breeding of *Gonepteryx rhamni* (brimstone) in gardens growing *Rhamnus cathartica* or *Frangula alnus* (buckthorns) and *L. phlaeas* in gardens growing *R. acetosa* or *R. acetosella*.

A survey of *Urtica dioica* (stinging nettle) patches in gardens was carried out in 1991 (Vickery, 1992b), which found that only about half were used by nettle-feeding nymphalids for egg-laying. The Essex survey also found that garden nettle patches were not generally used. However, work by Pullin (1986a,b, 1987a) has helped to establish the needs of these butterflies and these have been summarized by Corke (1991) and Vickery (1992b). Reports included with the 1992 BC survey give reasons to believe that correctly sited and managed nettle patches will be used, but more data need to be gathered before any firm conclusions can be reached.

Anthocharis cardamines females readily lay eggs on *Lunaria annua* (honesty) and *Hesperis matronalis* (sweet rocket or dame's violet), but few survive to imago. Mitchell (1991, 1992) reported eggs, larvae and pupae on *L. annua* in his garden. However, Courtney (1981) found that survival of eggs and larvae in gardens was lower than in non-garden habitats.

9.5 BUTTERFLY MOBILITY AND DURATION OF STAY

None of the surveys discussed above investigated the mobility or duration of stay of butterflies in gardens. There appears to be a dearth of published work on this aspect, although several unpublished university student projects have been carried out (e.g. mentioned in Owen, 1991). Where results are available they show butterflies to be transient visitors to gardens. Although some Satyrinae stay several days, butterflies from other families seldom stay longer than a few hours.

9.6 CONCLUSIONS

Gardens are not, and never will be, substitutes for wild butterfly habitat. However, they may provide important stepping stones between more permanent habitat patches, and are important habitats in their own right, especially to those species that hibernate as adults and migrants who need to feed as they journey. Gardens can provide large quantities of nectar within a small area, and thus a butterfly does not need to expend as much energy in searching for nectar plants as over large tracts of wild habitat. Often wild nectar plants are in short supply in the spring and autumn, and at these times gardens can be very valuable to butterflies. It is possible that the five most widespread garden butterflies are common simply because they have exploited the garden habitat so well. Could it be that without the support of gardeners *A. urticae* could be as rare in Britain as its larger relative, *Nymphalis polychloros* (large tortoiseshell)?

9.7 ACKNOWLEDGEMENTS

I would like to thank all recorders of butterflies in gardens, especially those who take part in the BC garden butterfly survey. Without their dedication and enthusiasm our knowledge of the garden habitat would be much poorer. Thanks also to Martin Warren for helpful criticism of the first draft of this chapter.

PART THREE

Managing Endangered Species

CHAPTER 10

The ecology and conservation of *Papilio machaon* in Britain

J.P. Dempster

10.1 INTRODUCTION

Papilio machaon (swallowtail) is a widespread and abundant species of butterfly, found throughout the Holarctic, through Europe, temperate Asia to Japan, and in North America. One might then be tempted to ask why we should be concerned with conserving this species in Britain. Anyone who has experienced the excitement of seeing this spectacular butterfly skimming over reed beds and open water in its natural habitat, will consider this a foolish question, but, even so, there are good scientific reasons for wishing to conserve our own swallowtail.

Taxonomically, the species is divided into several recognizable subspecies, so that, in the words of Clarke and Larsen (1986), the species complex 'illustrates evolution in action'. The resident British swallowtail belongs to a separate subspecies, *P. m. britannicus* Seitz, which differs from the continental *P. m. gorganus* Fruhstorfer (= *P. m. bigeneratus* Verity) in colour and in its ecology. Although superficially similar, the two subspecies can usually be distinguished by the dark markings being broader and heavier in *britannicus* with the pale primrose ground colour being much reduced, especially on the hind wings and at the apex of the fore wings. The differences are clear from a series of specimens, but may be difficult to see in single individuals. However, ecologically the two subspecies are quite distinct. *Papilio m. gorganus* is an active migrant, which can be found in a wide range of habitats, e.g. meadows, cliffs and hillsides, which are often very arid, and where its larvae feed on a large range of plants belonging to the Umbelliferae and Rutaceae. In contrast, *P. m. britannicus* is far more sedentary, has become specialized as a fenland insect, and is restricted to the single larval food-plant, *Peucedanum palustre* (milk parsley), which is itself a local plant restricted to fens.

Ecology and Conservation of Butterflies Edited by Andrew S. Pullin.
Published in 1995 by Chapman & Hall. ISBN 0 412 56970 1

10.2   LIFE CYCLE AND ECOLOGY

In Britain, *P. machaon* is confined to the fens and marshes around the Norfolk Broads, mainly in the valleys of the rivers Bure, Ant and Thurne. The adult butterfly is on the wing from late May to mid-July and the females lay most of their eggs on large-sized flowering plants which project above the surrounding vegetation. Eggs are large, spherical, conspicuous and pale straw-coloured at first, but as they develop they darken through brown and plum-coloured to black. They hatch in about 2 weeks to produce a caterpillar which is black with a broad white band across the middle. At this stage, it looks very like a bird's dropping. After the second moult, the caterpillars completely change their appearance, to become bright green, with black and orange rings. When disturbed, the caterpillars erect a bright orange osmeterium from behind the head, which emits a pungent smell, usually described as resembling that of pineapples. As they develop, they tend to move to the top of the plant and feed on the developing flowerheads. They have five larval instars, and by late July they are fully grown and leave their food-plant to pupate, low down on vegetation. Most pupae go through the winter to give adults in the following May, but in some years a few may produce adults in August to give a second generation of caterpillars. Pupal diapause is determined by day length during the larval period, so there tends to be a larger second generation in early years.

   A study of the butterfly in Norfolk over 4 years (Dempster, King and Lakhani, 1976) showed that egg mortality is generally low, and is mainly due to infertility. However, survival of the young caterpillars was always poor, with mortalities of between 30 and 80%. This was due mainly to predation by arthropods, particularly spiders. Birds, especially *Emberiza schoeniclus* (reed bunting), replaced arthropods as the main predators of the larger caterpillars. Large caterpillars frequently run out of food on one plant, and have to move to another to complete their development. In most habitats, other *Peucedanum* plants can be found within a few metres, so this does not usually pose problems. Pupal numbers were always low and they were difficult to find amongst dense vegetation, so that little is known about rates of mortality in that stage. Predation, possibly by small mammals or birds, is thought to be important since many pupae disappear during the winter. While in diapause, pupae can survive submersion under water for long periods. No parasitoids were found during this study, but a specific ichneumonid parasitoid, *Trogus lapidator*, has been reared from pupae of *P. machaon* collected from Norfolk. Obviously, this is a rare species in Britain, and is itself in need of conservation (Shaw, 1978).

10.3   DISTRIBUTION

Although *P. m. britannicus* is now confined to the fens around the Norfolk Broads, at one time it occurred throughout the extensive East Anglian fens south of the Wash, and it probably also occurred in the marshes around the rivers Thames and

Lea, in the Somerset Levels, and possibly as far north as Beverly in Yorkshire (Warren, 1949; Bretherton, 1951; Balfour-Browne, 1958; Wilkinson, 1981).

The past distribution of *britannicus* is made more difficult to unravel, because the European subspecies, *gorganus*, is a regular immigrant to southern England, and early records of *P. machaon* failed to distinguish between the two forms. In the past, *P. m. gorganus* appears to have been resident on the downs in the southern counties of England, during periods with exceptionally warm summers. Thomas and Lewington (1991) list three such periods as 1857–69 near Deal (Kent), 1918–26 near Hythe and during the mid-1940s in Kent, Dorset and South Hampshire, but suggest that odd individual sightings are made in perhaps 1 in 3 years. Most larvae have been found on *Daucus carota* (carrot), *Ruta graveoleus* (rue), or *Foeniculum vulgare* (fennel). As *gorganus* has two generations each year, it seems likely that it dies out in cold summers because it fails to complete its second generation before winter. As already said, *P. m. britannicus* is normally univoltine, although it can produce a partial second generation in some years.

10.4   LARVAL FOOD-PLANT CHOICE

At first sight, it is perhaps surprising that *P. machaon* is confined to damp, cold, fenland habitats in Britain, where it appears to be at the northern edge of its range. One might have expected it to have persisted better in more open, warmer habitats. A possible explanation of the sequence of events which led to *britannicus* evolving is provided by the research of Wiklund (1973, 1974, 1975, 1981) on *P. machaon* in Scandinavia. Two subspecies of the butterfly occur in Fennoscandia (Warren, 1949). In the south it is represented by *P. m. machaon*, but in the north a separate subspecies, *P. m. lapponica*, occurs. The species is found as far north as the Varanger peninsula, i.e. north of the Arctic Circle.

Wiklund showed that, although the butterfly is oligophagous, the female preferentially oviposits on three different host-plants growing in three different habitats. Thus, it lays exclusively on *P. palustre* growing in inland bogs and fens, even when other Umbelliferae are present; similarly, it lays on *Angelica archangelica* (garden angelica) on sea shores, and on river and lake sides in the north, and on *Angelica sylvestris* (angelica) in moist meadows. Thus the butterfly is virtually monophagous in each of the main habitats that it uses. Rearing experiments showed that larval growth is better and pupal weight higher (and presumably fecundity greater) when larvae are fed on *P. palustre*, followed by *A. archangelica* and lastly *A. sylvestris*, and that these three food-plants are superior to other Umbelliferae tested.

Angelica archangelica is not native to Britain, but *A. sylvestris* is widespread. It has been recorded here as a food-plant of *P. machaon*, but usually in the butterfly's second generation, when *P. palustre* is dying back. On the few occasions when I have seen larvae on *A. sylvestris* there has been a badly damaged plant of *P. palustre* close by, from which the larvae may have moved.

Wiklund's studies suggest that the butterfly is monophagous on different plants in different habitats in Scandinavia, so it is likely that this is how the monophagous habit in Britain originated. If so, it seems likely that it was only the fen habitats, which were too wet for early agriculture, that remained extensive enough to sustain the butterfly by the middle of the seventeenth century. All other main habitats were probably lost to the species before that.

10.5 HABITAT DESTRUCTION

Even the fens have now changed dramatically in extent, due to their drainage (Figure 10.1). Rowell (1986) estimates that the Cambridgeshire fens covered 3380 km^2 in the early seventeenth century, but this habitat had been reduced to about 130 ha (1.3 km^2) by 1900. Of course, during that time other fens, which had developed around the old peat diggings that created the Norfolk Broads, became available to the butterfly, but the total available habitat is now a very small fraction of what it once was.

After completion of the drainage of the Cambridgeshire fens, around 1850, the butterfly survived in a small, isolated patch of fenland at Wicken until the early 1950s, when the population went extinct. During that time, Wicken Fen gradually became drier, as the water table was lowered, and oxidation and shrinkage of the peat soils left the land surface on the fen 2–3 m higher than the surrounding farmland. With the drier conditions, the fen was invaded by woody plants, and the area of open fen vegetation was greatly reduced (Godwin, Clowes and Huntley, 1974; Dempster, King and Lakhani, 1976; Rowell and Harvey, 1988). *Peucedanum palustre* is now confined to the open sedge and litter fields, and these can be maintained only by repeated cutting, or burning, to prevent the development of carr. Changes in the management of the vegetation on the fen from a spring to an autumn harvest of the sedge probably further reduced the abundance of *P. palustre* (Harvey and Meredith, 1981). By the time the butterfly became extinct, the area of suitable habitat for it at Wicken had been reduced from about 110 ha to less than 10 ha (Rowell and Harvey, 1988).

10.6 CONSEQUENCES OF ISOLATION

A study of the morphometrics of specimens in museum collections, suggested that differences developed between Wicken and Norfolk butterflies during the time when the habitat was becoming reduced at Wicken, which might be linked to a reduction in the mobility of the butterfly. These changes were later paralleled by similar changes in Norfolk specimens, after 1920, when the butterfly's habitat there also became more restricted (Dempster, King and Lakhani, 1976; Dempster, 1991).

10.7 RE-ESTABLISHMENT ATTEMPTS

Repeated attempts to re-establish *P. machaon* at Wicken have failed. The last attempt was made in 1975 (Dempster and Hall, 1980), but this, too, failed. Initially, the butterfly's numbers increased, but an exceptional drought in 1976 led

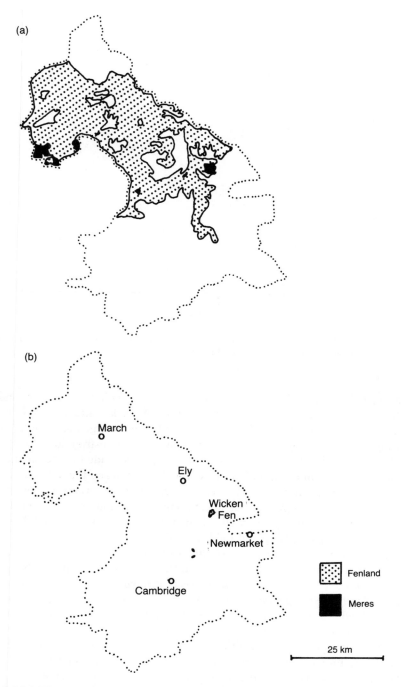

Figure 10.1 The decline in the area of fenland in Cambridgeshire. (a) Early seventeenth century; (b) 1900. (Redrawn from Rowell, 1986.)

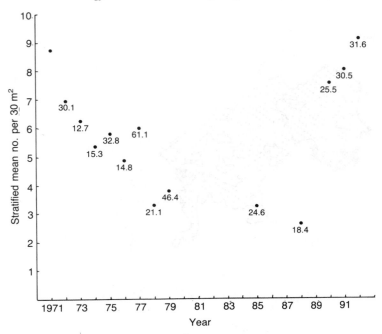

Figure 10.2   Changes in the abundance of *Peucedanum palustre* at Wicken Fen (numbers in the body of the figure are percentage flowering).

to a crash in numbers, from which the population failed to recover, and the butterfly went extinct again in 1979. By the time of this last attempt at reintroduction, the area of suitable habitat on the fen had been enlarged to about 24 ha, by clearing the carr and by repeated cutting. However, the distribution of *P. palustre* on the fen was by then very patchy, and the plants tended to be smaller and shorter-lived than those on the Norfolk Broads. At Wicken the half-life (time to 50% loss) of the plant was about 2 years, compared to 6–7 years in Norfolk, and these differences were entirely due to environmental conditions, since plants from both sites performed equally well when grown in the same conditions in a glasshouse (Dempster, King and Lakhani, 1976).

Since 1971, the numbers of *P. palustre* plants at Wicken have been monitored during the autumn of most years. For this, the number of plants have been counted in a series of 30 m × 1 m transects through each sedge or litter field. Approximately the same path through each area has been taken each year, but the data obtained are not strictly comparable between years, because the rotation of sedge cutting on the fen has led to different areas being cut (and therefore excluded from the sample) in different years. Added to this, new fields created since 1971, by clearance of the carr, have not been included in the survey. Nevertheless, because of a large number of samples (60–80) taken each year, the results do probably give a picture of the broad changes that have occurred over the past 20 years (Figure 10.2).

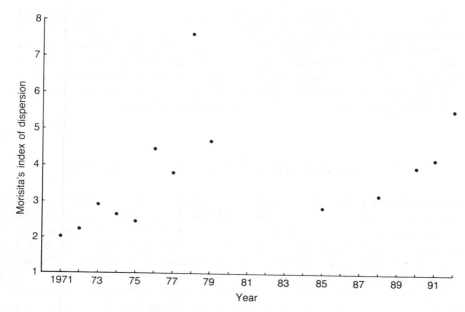

Figure 10.3 Changes in the dispersion of *Peucedanum palustre* at Wicken Fen.

Between 1971 and the mid-1980s, there was a consistent decline in the numbers of *P. palustre* on the fen. Inevitably, the plant was distributed very patchily, as a result of its previous elimination from large areas by the development of carr. However, during its decline after 1971, its distribution became more and more clumped. This can be seen when Morisita's index of dispersion (Morisita, 1962) is calculated from the survey data (Figure 10.3). An index of greater than 1 shows a contagious (e.g. negative binomial) distribution.

Only established plants were counted in these surveys, i.e. seedlings were not counted. Meredith (1979) showed that seed has to go through a winter before it will germinate, so seed set in one year would produce plants which will be included in these surveys 2 years later. Plant numbers were particularly low in 1978, and interestingly, the index of dispersion was highest in that year, 2 years after the 1976 drought. This suggests that the plant was worse affected by the drought on some parts of the fen than others, a view that is supported by the detailed changes in its distribution given by Dempster and Hall (1980), which showed that its numbers had declined most on the higher, drier end of the fen.

The precise effects of the 1976 drought on the butterfly are not known. Large numbers of adults were recorded in that year but they laid very few eggs. Many females must have died without laying, which suggests that either they failed to mate or that their food supply was reduced by the hot, dry conditions. Both of these possibilities are feasible, since difficulties can be experienced in pairing *P. machaon* in dry conditions in a glasshouse (Dempster and Hall, 1980), and it is known that nectar can become too viscous to allow insects to feed in very dry weather (Corbet, 1978; Corbet, Unwin and Prys-Jones, 1979).

Between 1988 and 1990, considerable effort was put into reconstructing and waterproofing the banks that surround Wicken Fen, and as a result it is now far wetter, in spite of the series of very dry years that we have had since then. This has had an immediate and dramatic effect on *P. palustre* numbers, so that the decline that had taken place over the previous 18 years has been reversed (Figure 10.2). As a result, another attempt at re-establishing *P. machaon* at Wicken is being made, with the support of Butterfly Conservation and English Nature.

10.8   CONSERVATION IN THE NORFOLK BROADS

Similar changes to the vegetation that have occurred at Wicken, have also been taking place on the Norfolk Broads. Gradually the Broads are filling and the surrounding land surface is becoming raised with the formation of peat. The resulting drier conditions have led to an invasion of woody plants. Figure 10.4 shows these changes on the Bure Marshes National Nature Reserve (NNR), between 1845 and 1964 (Nature Conservancy, 1965).

The once extensive fenland vegetation has largely been replaced by woodland and scrub. This loss and fragmentation of the butterfly's habitat has been occurring during the past 100 years over much of the Norfolk Broads (Ellis, 1965). Drainage of many of the surrounding marshes has further added to the problem.

On some sites in Norfolk, such as the Bure Marshes NNR, large areas of scrub have been cleared and the area of open fen vegetation has been increased by regular cutting (Figure 10.4), and overall *P. machaon* appears to be holding its own. The main evidence for this comes from the Bure Marshes, the one site covered by the Butterfly Monitoring Scheme (Pollard, Hall and Bibby, 1986). These data (Figure 10.5) show that the butterfly has had a series of good years.

Interestingly, its numbers crashed between 1976 and 1977 in the Bure Marshes, when the introduced Wicken population also crashed.

10.9   CONCLUSIONS AND FUTURE PROSPECTS

Very little research has been done on *P. machaon* during the past 15 years. However, some of the conclusions from our earlier studies (Dempster, King and Lakhani, 1976; Dempster and Hall, 1980) have been put to the test by subsequent events. Our main conclusions were that the butterfly's status depended largely on the performance of its food-plant and that this, in turn, depended on the wetness of the fen habitats that they inhabit.

The waterproofing of the fen at Wicken appears to have had an immediate effect of improving the status of *P. palustre*. It must be stressed, however, that although this is consistent with the view that the increased abundance of the plant has been caused by the improved wetness, it does not prove it. As with any correlation, a causal relationship has to be demonstrated.

It remains to be seen whether the carrying capacity of the fen will now be adequate to support a viable population of the butterfly. The plant is still extremely patchy, and high densities are to be found in very small areas. However, this should improve in the coming years, especially if seed can be spread in the wetter fields.

What, then, of the future? Habitats throughout the range of the butterfly are becoming drier, either by natural processes, such as peat formation, or as a result of human activities. Under the drier conditions, *P. palustre* grows less strongly, its life is shortened, and it is less able to maintain its numbers. Added to this, woody plants rapidly colonize fen habitats, as they dry, and eventually these shade out the herbaceous fen vegetation. Management of these habitats has concentrated on clearance of trees and shrubs, and then introducing a cycle of cutting, or burning, to prevent regrowth. The effects of cutting on *P. palustre* depend on its timing. A spring cut has little effect on the plant (Dempster, King and Lakhani, 1976), whereas cutting in late summer and autumn can cause the death of plants and prevents seed production (Harvey and Meredith, 1981). This difference becomes more important as the habitat gets drier and plants are shorter lived, so requiring more regular recruitment from seed to maintain their numbers. On the other hand, immediately after cutting there is often a surge in germination of seed, as the vegetation is opened. Meredith (1979) also suggested that there is an improved survival after cutting due to reduced seed predation by small mammals.

Cutting is only necessary because the habitats are too dry. Wetter conditions would prevent the establishment of woody plants. Thus a more permanent and satisfactory solution would be to make the habitats wetter, either by raising the water table, as at Wicken, or by lowering the land surface, by peat digging. Although these are relatively expensive operations, they might well prove to be cheap alternatives to repeated cutting over a long time scale. I should like to see some experimental cutting of peat on some of the Norfolk sites, so as to compare the performance of *P. palustre* on those areas with that in cut vegetation.

Although *P. m. britannicus* is clearly far less mobile than its continental form, it is a large insect, capable of covering considerable distances. In Norfolk, adults are regularly seen in gardens, away from their fen habitats, but we know nothing about the extent of natural interchange of individuals between Norfolk populations. In my opinion, it would be very surprising if there were not frequent movements of individuals between populations, at least within one river system, but we have no quantitative information on this. Dispersal is extremely difficult to study, even for a large insect like this.

Similar population crashes of the butterfly appear to have occurred between 1976 and 1977 at both Wicken and in Norfolk. The Bure Marshes population survived but the Wicken one did not, presumably because it was too small. This raises the question of what is the minimum size of a viable population. Of course, there is no real answer to this; it depends on the range of environmental variation that the population has to withstand and the time scale over which viability is judged. Nevertheless, we need to be able to assess the value of sites to the

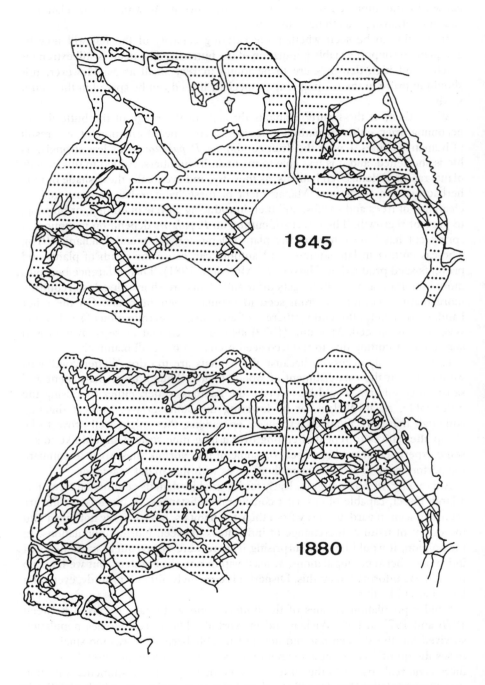

Figure 10.4 Successional changes in the vegetation on the Bure Marshes NNR (redrawn from Nature Conservancy, 1965).

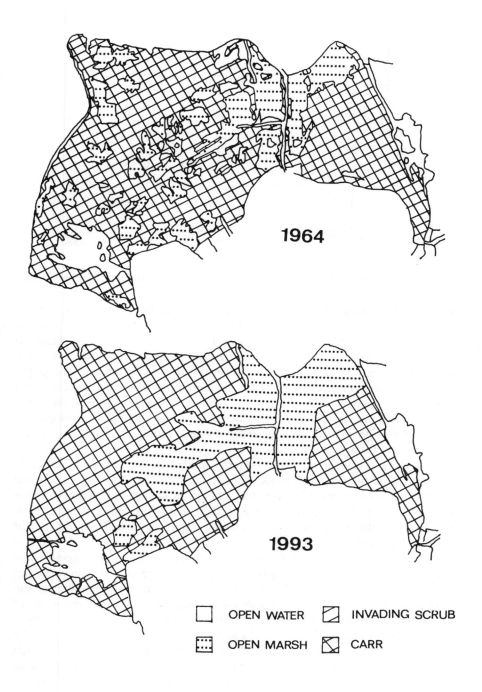

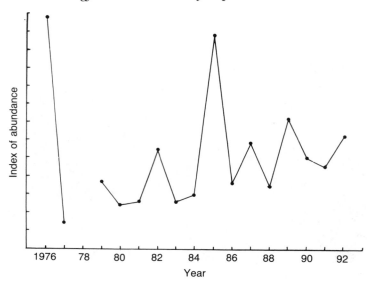

Figure 10.5   Changes in the abundance of *Papilio machaon* on the Bure Marshes NNR.

butterfly. One can do a few back of the envelope calculations, which give a clue to the carrying capacity required of a site. Densities of flowering plants of *P. palustre* on the best sites in Norfolk average about 0.5 per m², i.e. 5000 per hectare. At Wicken, the densities are far lower and egg-laying females would need to search about 5 ha to find the same number of plants. With a maximum of 200 eggs per female and rarely more than one larva supported per plant, these areas might support the progeny of 25 females, assuming that all plants could be found in the time available for egg-laying. Obviously, such calculations are very rough and ready, but they do indicate the large areas of habitat required to support relatively low numbers of this butterfly.

The future conservation of the butterfly in Britain depends upon adequate management of the sites in Norfolk. Although it would be pleasing to return the butterfly to its old haunts at Wicken, this is not going change the need to conserve the species in its Norfolk sites. Nor is it likely that other sites exist away from Norfolk where the status of *P. palustre* is strong enough to warrant consideration of other introductions of the butterfly.

Ecological knowledge is still the main limitation to designing management that will guarantee the future of the species. In particular, we need to know whether existing sites can be managed independently, or whether the mobility of the species means that we are dealing with a single population covering the whole of the Broads. Something in between these two extremes is probably true, but at present we do not know. We also need a better understanding of the impacts of the management options that are open to us. In particular, can we sustain high populations of *P. palustre* with the present cutting regimes, or do we need to base our management on making the fen habitats wetter? More research on these issues is needed.

10.10 ACKNOWLEDGEMENTS

The vegetation map for 1993 (Figure 10.4) was prepared from information provided by R. Southwood, warden of Bure Marshes NNR, to whom I am extremely grateful.

CHAPTER 11

Ecology and conservation of *Lycaena dispar*: British and European perspectives

A.S. Pullin, I.F.G. McLean and M.R. Webb

11.1 INTRODUCTION

Lycaena dispar (large copper) is famous among British lepidopterists as one of the earliest recorded butterfly extinctions and for the early attempts at re-establishment. In the rest of Europe it is a rapidly declining butterfly with a complex set of ecological requirements, which urgently needs further study before it suffers the same fate as in Britain. This chapter introduces a programme of research which aims to combine the conservation of the species in The Netherlands with its re-establishment in England through a study of its habitat requirements using the restoration ecology approach. The priority of re-establishment for us in Britain may seem minor in the European context, but as we hope to show, the challenges of conservation in Europe and re-establishment in Britain (and possibly other countries) can benefit from an integrated approach.

11.2 THE BIOGEOGRAPHY OF *L. DISPAR*

The species is distributed over a large area of Europe and into Asia. In textbooks of British butterflies it is commonly divided into three subspecies: *L. dispar dispar* Haworth, confined to England; *L. d. batavus* Oberthür confined to The Netherlands and *L. d. rutilus* Werneburg., which is widespread in Europe. More detailed treatments describe other European subspecies. Bretherton (1966) lists ssp. *gronieri* Bernardi, which was last seen in 1890 and formerly occurred in marshes round St. Quentin in north France; ssp. *carueli* le Moult, which occurs in Belgium, Luxembourg and north-east France possibly south to Loiret, Cote d'Or,

Ecology and Conservation of Butterflies Edited by Andrew S. Pullin.
Published in 1995 by Chapman & Hall. ISBN 0 412 56970 1

Isere and Bosses Alpes; ssp. *burdigalensis* Lucas, which is isolated in south-west France from Vendee to Lot-et-Garonne; ssp. *centralitaliae* Vty in the mouth of Arno, Tuscany and the Pontine marshes near Rome (although it may be extinct from here); ssp. *rutilus* under this treatment is scattered from Denmark through Germany, Austria, northern Italy, into Serbia, Croatia, Bosnia, Macedonia, Albania, Greece and Turkey. The above nomenclature has generally been followed by later authors and will be used here.

Unfortunately, there is very little information on the species taxonomy in eastern Europe and Asia. It is possible that the formation of subspecies increases towards the western edge of its range as the distribution fragments, but it is more likely to be the result of more intensive study of the species in western Europe.

11.2.1 Historical biogeography

Looking back as far as the recent ice ages it is likely that *L. dispar* found refuge from glacial incursions in the eastern Mediterranean as it is still locally common in this area and it does not occur in Spain, the alternative western refuge. Following the most recent ice age it probably recolonized Britain around 10 000 years BP, during the pre-Boreal period (Dennis, 1992). At this time the area now under the North Sea was above sea-level and connected eastern England with The Netherlands (Figure 11.1).

At this time the species may have been abundant in this low-lying and presumably largely wetland area. As the sea-level rose, populations would have been pushed farther east to The Netherlands or west to England and they probably became isolated as recently as 7000 years ago.

The morphological differences between *L. d. dispar* and *L. d. batavus* were studied by Bink (1970) using sizes of spots and banding on the hindwing. Although he found variation in pattern, there were no significant differences between the two subspecies. The small but seemingly inconsistent differences may be typical of that recorded for populations of species which have become isolated at the edge of their range (Dennis, 1992). The similarities between these subspecies are more striking, both having only one generation per year and feeding solely on *Rumex hydrolapathum* (great water dock). Bretherton (1966) also includes the extinct *L. d. gronieri* in this group.

Lycaena d. rutilus and *L. d. carueli* are more easily distinguished morphologically from the above, they have a bivoltine life cycle and feed on a number of species in the genus *Rumex*. These observations lead to the general view that the north-westerly subspecies *dispar*, *batavus* and possibly *gronieri* diverged from each other more recently than either diverged from the more widespread European subspecies.

11.3 LIFE HISTORY

Virtually nothing is known of the life history of the extinct English subspecies except what is inferred from *L. d. batavus*. The latter lays its eggs on the leaves of

Figure 11.1   Possible changes in the distribution of the univoltine subspecies of *Lycaena dispar* during the past 8000 years. When sea-levels were much lower (former coastline shown as – – –) univoltine populations would have occurred in the low-lying river valleys now covered by the North Sea (hatched areas). As sea-levels rose, some populations would have been forced westward to found the recent *L. d. dispar* populations and some eastward to found the recent *L. d. batavus* colonies.

R. hydrolapathum in open fenland areas, usually during July and August. On hatching the larvae begin feeding on the underside of the leaves, creating a characteristic 'window' since the upper epidermis is not eaten. This is easily seen by the experienced observer, making detection and counting of larvae relatively easy. Larvae reach the second instar and begin diapause (hibernation) in mid-September, characterized by cessation of feeding and migration down the plant and on to the senescing leaves around the base. They do not appear to move on to surrounding vegetation. The diapause state is not obligatory as it can be averted by exposing the first two larval instars to long photoperiods and high temperatures (Pullin, unpubublished), but under normal field conditions the shortening photo-periods of early September induce diapause. The larvae stay in their overwintering site until warm spring weather (around April) encourages them to crawl back up the plant on to young growing shoots. They then continue feeding until pupation in June, which takes place either on the plant or on nearby vegetation. The adults are on the wing in July and August. Little is known about adult behaviour and particularly dispersal, partly due to the difficulty of following the butterfly in fen habitats and the low density at which populations seem to exist. Some evidence from population monitoring in The Netherlands suggests that the adults may be able to colonize habitats as far as 20 km away from existing populations, presumably by flying along water courses (C. van Swaay, personal communication).

The subspecies *carueli* and *rutilus* are both bivoltine and eat a range of dock species. Ebert and Rennwald (1991) record *L. d. rutilus* feeding on *R. hydrolapathum*, *R. obtusifolius* (broad-leaved dock), *R. crispus* (curled dock) and *R. acetosa* (common sorrel) in Germany, although *R. hydrolapathum* appears to be the main food-plant. The flight periods are May–June for the first generation and August for the second. These differences aside, the ecology of the bivoltines seems to be similar to that of the univoltines, but with the former being less restricted in habitat use, small populations occurring at low density in marshland and river valleys.

11.4 RECENT HISTORY OF *L. DISPAR*

This species was never common in England and was only first recorded in 1749 from Dozen's Bank near Spalding, Lincolnshire (Heath, 1983). It was at the time supposed to be *Palaeochrysophanus hippothoe* (purple-edged copper) but Haworth (1803) recorded it as a new species, giving it the name *dispar*. There is good evidence that its range in the early nineteenth century included Lincolnshire, Huntingdonshire and Cambridgeshire across to the Norfolk Broads. Less reliable evidence points to more westerly populations on the Somerset Levels (but see Sutton, 1993) and the Wye marshes in Monmouthshire.

The species was probably already in decline when it was first recorded, as it seems to have disappeared rapidly from Lincolnshire and was last recorded in Huntingdonshire at Holme Fen around 1847 or 1848 and in Cambridgeshire at Bottisham Fen in 1851. The last British record may well be from the Norfolk Broads as two specimens labelled 'Ranworth 1860' and 'Woodbastwick 1864' were examined by Irwin (1984), who regarded them as authentic.

There is a consensus of opinion that the decline was an inevitable consequence of habitat loss due to widespread drainage of the fenlands (see Figure 10.1), the history of which is comprehensively reviewed by Godwin (1956). The draining of the complex of meres adjacent to Holme and Woodwalton fens during the 1840s, culminating in the drainage of Whittlesey Mere in 1851 (at the time the second largest lake in England, famous for sailing and skating), gives some indication of the extensive drainage activity at the time of the butterfly's disappearance (Poore, 1956). Once confined to small, isolated populations its extinction may have been accelerated due to overcollection and/or random fluctuation of small populations, but these can only be seen as secondary causes.

The Dutch subspecies was not discovered until 1915 at Nyetrine near Wolvega in Friesland, and was probably already declining at this time because of similar widespread drainage of habitat. Since then it has been recorded at just 13 sites (excluding introductions) and by 1970 it was reduced to small populations within a marshland area in north-west Overijsel and south-east Friesland (Bink, 1970). In 1992 only four populations were known to exist, and only one of these was substantial and within an extensive wetland area (van Swaay, unpublished). The species was not recorded in the Wieden in 1992 and may have been lost there (J. van der Made, personal communication). This is a large nature reserve where considerable effort has been made to manage areas for the species (Evers, Maaren and Made, 1987). In 1993, it was only recorded in the Weerribben National Park and this subspecies is now considered endangered.

In central Europe, the subspecies *rutilus* is also declining and has disappeared from many areas. Kudrna (1986) documented its disappearance from south Bohemia following large-scale drainage of marshes and fens beginning in the late sixteenth century. In the River Po region of northern Italy the butterfly actually increased in numbers as a result of rice cultivation. This is thought to have been because the digging of new drainage ditches increased availability of *R. hydrola-pathum*. Unfortunately, the subsequent increased intensification of cultivation and associated use of selective herbicides has led to a rapid decline in numbers of food-plant and butterfly, to a point where the latter's status is threatened (Balletto and Kudrna, 1985). In south-western Germany some areas still hold good populations, such as around Mannheim (Settele, 1990), but it is local, and endangered in the Rhine valley south of Mannheim (Ebert and Rennwald, 1991) and no longer occurs in Switzerland (Geiger and Gonseth, 1992). In eastern Germany decline has been even more rapid, and it is now considered very rare in the Brandenburg region, with only a few sitings in 1992 (M. Kuhling, personal communication). It is also considered endangered in Poland as populations have become increasingly fragmented (Dabrowski and Krzywicki, 1982).

The other widespread subspecies, *carueli*, is still present in southern Belgium and Luxembourg (Wynhoff, Made and Swaay, 1990) and in the Alsace region of France, where it has declined substantially in recent years (A. Erhardt, personal communication). Fortunately, it is apparently still abundant in the Burgundy region of central France (Dutreix, 1992).

A single population of unknown status was found in Finland in 1983 and has been given the benefit of protective legislation (Mikkola, 1991).

The extent of decline of this species has merited its inclusion in the Bern Convention list of threatened invertebrates (see Fernández-Galiano, 1992). An approximation of the species' current distribution in western and central Europe is shown in Figure 11.2.

11.4.1 The history of re-establishment attempts

Since the extinction of *L. d. dispar*, a number of attempts have been made to re-establish populations in England by introducing one of the European subspecies. The first recorded attempt was made by G.H. Verrall, who released 'a few' larvae of *L. d. rutilus* on Wicken Fen, Cambridgeshire, in 1909 (Verrall, 1909). This was

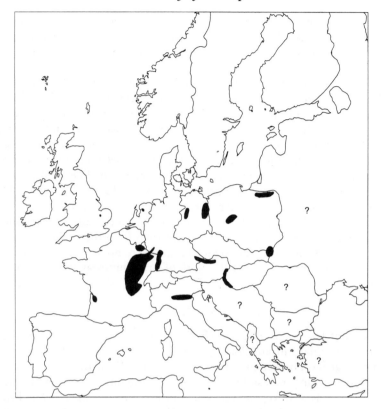

Figure 11.2 Estimation of the present distribution of *Lycaena dispar* in western and central Europe.

unsuccessful, possibly because the food-plant was not in sufficient abundance (Duffey, 1968). A number of introductions were then started by W.B. Purefoy in Eire which cannot be regarded as true re-establishment attempts since there is no evidence of the species ever having occurred in this country. However, in 1926 stock of *L. d. rutilus* from Eire was introduced to Woodbastwick marshes in the Norfolk Broads. This colony survived for only 2 years and its failure was attributed to lack of food-plants in the open marsh (Ellis, 1965). However, there is no scientific evidence for this, and the fact that it is a bivoltine race may have been a more fundamental problem.

In 1926, when a re-establishment was planned for Woodwalton Fen, Cambridgeshire, by the newly formed Committee for the Protection of British Lepidoptera, the recently discovered subspecies *L. d. batavus* was considered more appropriate because of its greater similarity to *L. d. dispar*. In 1927, 38 butterflies were released in a specially prepared area of the fen. The colony did well initially but subsequently has only been maintained by protective caging of larvae in spring and through reinforcement by stock from a greenhouse-reared population. Discussion of the Woodwalton population is continued in more detail below.

A second attempt was made to re-establish the species at Wicken in 1930; this time on Adventurer's Fen, where the food-plant was abundant, and using *L. d. batavus*. This colony survived unaided until that area was reclaimed for agriculture in 1942.

Another attempt was made to re-establish in the Norfolk Broads, this time in the Yare Valley near Surlingham in 1949. This colony survived until high tides inundated the area in April 1951 (Ellis, 1965). Interestingly, Ellis believed that the introduction of *Myocastor coypus* (coypu) to the area made it unsuitable for further re-establishment attempts because of the widespread destruction of the food-plant, a situation which has now changed since *M. coypus* has been exterminated.

Re-establishment attempts have also been made in The Netherlands. Bink (1970) listed four attempts made between 1928 and 1940, all of which failed.

The value of careful recording of re-establishments can be seen clearly from experience with *L. dispar*. Although we have gained some information from previous attempts, few lessons have been learned from their failure. The mistake made in the past was that many such attempts were not carried out on a scientific basis. The following account of the work of Eric Duffey at Woodwalton Fen is a notable exception.

(a)   The Woodwalton population

This account has been taken largely from Duffey (1968, 1977). In the period between the initial introduction to Woodwalton in 1927 and 1955 there is no evidence that the population survived for any length of time without protection from predators and parasitoids by caging larvae in the spring. However, no extensive records were kept and no scientific assessment can be made. From 1955 the warden kept records of the numbers of larvae and pupae and timing of the life cycle, and in 1960 an experiment was set up to compare the survival of a wild and a caged colony (Duffey, 1968). The wild population never reached high numbers from the 24 pupae introduced, and the population went extinct in 1966 and may have done so sooner if some immigration of adults from the caged population had not occurred.

An abnormal flood in July of 1968 caused a dramatic decline in the number of eggs in the remaining population. By 1969 only five adults remained and these failed to reproduce. The Woodwalton colony was completely wiped out (Duffey and Mason, 1970). However, several captive populations of the Woodwalton stock remained, and over 1000 adults were used to re-establish the colony in 1970 (Duffey, 1977). These were added to, from the same sources, from 1971 to 1973. Subsequently, large annual reductions in size of the population made it necessary to constantly replenish the wild stock with individuals reared under greenhouse conditions.

11.5   ECOLOGY OF *L. D. BATAVUS*

Pioneering work by Fritz Bink in The Netherlands and Eric Duffey on the Woodwalton population has given us some insight into the ecology of the subspecies and into the factors contributing to the failure of re-establishment attempts. Duffey (1968) studied mortality in the field by counting eggs, larvae at emergence from spring hibernation and pupae, during the period 1960–66. He suggested that the key factor regulating the population was mortality during the period from oviposition to spring emergence. The mortality over this period averaged 95.6% and protection from vertebrate predators did not significantly reduce it. However, vertebrate predation was significant on those remaining alive in the spring, with unprotected individuals suffering 84.9% mortality, compared with 24.8% of protected larvae. Parasitism by *Phryxe vulgaris* (Diptera: Tachinidae) was also recorded during spring feeding, but was not such a significant contributor to mortality.

Bink (1972) also found high mortality during early larval stages and towards the end of hibernation, the latter possibly caused by fungal attack. However, tachinid flies do not appear important in The Netherlands. Other parasitoids have been recorded, including *Trichogramma* sp. (Hym. Chalcididae, egg parasitoids), *Anisobas hostilis* and *Hyposoter placidus* (Hym. Ichnuemonidae, larval parasitoids). Some predators of early-stage larvae are also listed: *Forficula* sp. (Dermaptera: Forficulidae), Gymnocerata, Coccinelidae and Chrysopidae (Bink, 1962, 1972).

Since flooding is common during winter at Woodwalton Fen, Duffey set up an experiment to look at the effect of submergence on larval survival. Bases of food-plants on which larvae were being fed were submerged in water during the autumn and kept flooded until late winter. This had the effect of preventing some larvae from migrating to the base of the plants to overwinter but mortality was lower than in control groups (81.4 and 86.5%, respectively). In contrast, a second experiment submerging larvae already in hibernation for 140 days from November to March caused significantly higher mortality than in controls (91.8 and 75.6%, respectively) (Duffey, 1977).

Research on female egg-laying preferences during this same period also yielded some partially conflicting results. Initial experiments suggested that females laid eggs on smaller *R. hydrolapathum* plants in the open fen rather than large plants on dyke sides (Duffey, 1968). However, there were a number of uncontrolled variables which may have influenced the results, such as shading, surrounding vegetation and nutritional/chemical status of the food-plants. Indeed, in one group Duffey (1968) notes 'The dyke-side plants are all large with many leaves while about half the fen-margin plants are of poor growth and smothered by surrounding vegetation. This is probably why more eggs per plant were recorded on the . . . dyke-side docks.' Such variation in preference highlights the requirement for more studies in this area.

The number of eggs laid was related to temperature and sunshine over the flight period at Woodwalton during 1960–66. This is in agreement with work by Bink (1972) indicating that 28 °C is the lower threshold for flight activity, leading him to suggest that weather is a major factor in population regulation and that population crashes may be caused by poor summers.

Nectar sources are another possible limiting factor. In The Netherlands common nectar sources are *Lythrum salicaria* (purple loosestrife), *Valeriana officinalis* (common valerian), *Eupatorium cannabinum* (hemp agrimony) and *Cirsium arvense* (creeping thistle) (Bink, 1962). All of these species occur at Woodwalton in abundance and appropriate management should ensure that these are not limiting. In addition, *Cirsium palustre* (marsh thistle) has been recorded as a nectar source at Woodwalton (M. Webb, personal observation). Ebert and Rennwald (1991) also lists *L. salicaria* as a nectar source for *L. d. rutilus*, along with *Valeriana procurrens*, *Pulicaria dysenterica* (common fleabane), *Mentha longifolia* (horse mint), *Cotoneaster salicifolius* (narrow-leaved cotoneaster) and *Rudbeckia hirta* (black-eyed Susan).

11.6   CURRENT STUDIES

11.6.1   Population monitoring at Woodwalton Fen NNR

In 1987 a joint Butterfly Conservation (BC) and English Nature (EN, then the Nature Conservancy Council) project started recording changes in the wild population established at Woodwalton Fen. Michael Chinery (BC) and Ian McLean (EN) co-ordinated annual counts carried out by volunteers from the Cambridgeshire and Essex Branch of BC.

A large release of adults was made from greenhouse stock and was followed by annual counts of adults, eggs and larvae on the fen. Even with the large numbers of adults present that year and the fine weather during the flight period, transect counts along pathways around the principal areas where the butterfly was known to breed resulted in only a few sightings. Subsequently counts were confined to eggs and larvae.

In the summer of 1987, 250 *R. hydrolapathum* plants were marked in three compartments (fen areas surrounded by water-filled drainage ditches commonly referred to as dykes) using labelled canes. Plants were chosen to include examples over the range of conditions present; 70 in compartment 37, 110 in compartment 39 and 70 in compartment 53. Over the period 1987–93 some plants perished and some were not relocated when marker canes were lost. Consequently the estimates of abundance given in Table 11.1 are in the form of mean numbers per plant for each compartment.

Counts of eggs and young larvae were carried out in August or September and counts of post-hibernation larvae were undertaken in May. The number of leaves on each marked plant, plant height, presence or absence of inflorescence, height of surrounding vegetation, egg numbers (unhatched, hatched or parasitized) and larval numbers were recorded. Additionally, the spring counts recorded damage to marked plants caused by browsing *Hydropotes inermis* (Chinese water deer) and when plants were found in standing water.

The results give a good indication of recent changes in numbers for early stages, as well as some information about egg-laying and larval survival on plants growing in different conditions. A rapid decline was recorded from 1987 to 1989 for both eggs and larvae. From 1989 onwards numbers have been more stable at a low level of absolute abundance per plant. The butterfly may have been able to persist at these low numbers at least partly due to a sequence of years with fine weather during the flight period.

Table 11.1 Mean number of *Lycaena dispar batavus* eggs or larvae per *Rumex hydrolapathum* plant in three compartments at Woodwalton Fen NNR, Cambridgeshire from 1987 to 1993

	87/88	88/89	89/90	90/91	91/92	92/93	93/94
Compartment 37							
eggs	31.5	9.3	0.5	0.4	0.5	1.1	0
autumn larvae	4.7	1.0	0.2	0.3	0.2	0	0
spring larvae	0.36	0.13	0	0.02	0.02	0	
Compartment 39							
eggs	28.0	5.8	1.2	1.1	1.8	0.2	0
autumn larvae	4.3	1.0	0.3	0.2	0.6	0	0
spring larvae	0.25	0.06	0	0.04	0.04	0.05	
Compartment 53							
eggs	11.1	1.6	0.1	0.02	0.7	0.01	0.07
autumn larvae	2.2	0.2	0.03	0	0.3	0	0
spring larvae	0.17	0.03	0	0	0	0.04	

Note: in 1992 the autumn larvae count was undertaken later than usual, when larvae may have entered hibernation sites and were therefore not recorded.

The initial rapid decline suggested that the population might disappear about 3–4 years after release. That this did not happen is particularly interesting and it may be that some factors causing density-dependent mortality may be reduced when numbers are low, and/or there has been selection in favour of genotypes which can survive better in the wild, in contrast to the survival of genotypes which may have been selected for at the high densities and warmer climate experienced by the captive greenhouse stock. Nevertheless, at such low abundance the species must be considered very vulnerable at Woodwalton Fen to any additional mortality brought about by adverse weather conditions, or other natural or human-induced factors, and, as suggested in Table 11.1, the population may have gone extinct in 1993.

11.6.2 Larval mortality

A new project was started by two of us (ASP and IFGM) in 1989, with the aim of studying the ecology of the Woodwalton population, leading to management proposals which would allow a self-sustaining population on the fen. Work is concentrating on larval mortality, following the results of Duffey (1968) showing high mortality during this stage, and on the effect of management on the population. Additionally, the project aims to identify sites in eastern England (within the species' former range) which are suitable for trial re-establishments.

In the autumn of 1991, MRW and ASP set up a field experiment to assess the contribution of different factors to larval mortality from the beginning of hibernation to pupation in newly released greenhouse stock. Fifteen pot-grown *R. hydrolapathum* plants were spaced 2 m apart in an area of compartment 39, and 10 hibernating larvae were placed on each plant on 18 September. A system of

Table 11.2 Survival of *Lycaena dispar batavus* larvae at Woodwalton Fen NNR during the pre-diapause stage (August–October), given different levels of protection from natural enemies (see text for explanation)

	Open			Mesh			Muslin			
	N	S	%	N	S	%	N	S	%	
(a) Open fen vegetation										
1992	45	10	22	40	10	25	52	20	39	ns
1993	64	10	16	64	12	19	64	32	50	$P<0.001$
(b) Dyke-side plants										
1992	28	0	0	–	–	–	–	–	–	–
1993	40	3	8	40	4	10	40	17	43	$P<0.001$

N, Number of larvae at start; S, number surviving the stage.
Significance levels are shown for a G-test on treatments versus survival.

cages was used to protect larvae from different sources of mortality. Five plants were surrounded by muslin cages, which excluded most invertebrates and all vertebrates, and all other vegetation was removed from the cage to minimize the numbers of invertebrates inside. A second group of plants was caged in mesh, which allowed invertebrates free access but excluded all vertebrates. A third group was left uncaged as a control. The difference in mortality between control and mesh-caged larvae will therefore indicate the level of vertebrate predation, and the difference between mesh- and muslin-caged will indicate the level of invertebrate-induced mortality. Lastly, the mortality in the muslin cage is a direct measure of other mortality factors combined. Subsequent counts were made of larval numbers on emergence from hibernation and just before pupation.

The following year the experiment was repeated and extended to include more replicates (eight instead of five), two control groups (open fen and dyke-side plants) and lengthened to include the period from hatching to hibernation. Currently this experiment is being repeated for a third year and is being replicated in the Weerribben NP.

Preliminary results are shown in Tables 11.2–11.4. Survival of larvae from hatching in August to hibernation in October is higher in muslin cages in both years investigated, but this was only significant in 1993 (Table 11.2a).

This survival difference is also significant on dyke-side plants, where overall survival is lower (Table 11.2b). This suggests that invertebrate predation is an important mortality factor during this stage. However, even the muslin cages do not prevent over 50% mortality occurring. This may be partly due to our manipulation of early-instar larvae and requires further investigation, although observations on untouched larvae suggest that this is not the case.

The results given in Table 11.3 show no significant difference in survival between treatments during hibernation. The implication is that the mortality that does occur is due to factors other than predation (by vertebrates or invertebrates)

Table 11.3 Survival of *Lycaena dispar batavus* larvae at Woodwalton Fen NNR during winter diapause (October–April), given different levels of protection from natural enemies (see text for explanation)

	Open			Mesh			Muslin			
	N	S	%	N	S	%	N	S	%	
91/92	50	15	30	50	16	32	50	19	38	ns
92/93	64	7	11	64	10	16	64	5	8	ns

N, Number of larvae at start; S, number of larvae surviving the stage.
Significance levels are shown for a G-test on treatments versus survival.

Table 11.4 Survival of *Lycaena dispar batavus* larvae at Woodwalton Fen NNR during the post-diapause stage (April–June), given different levels of protection from natural enemies (see text for explanation)

	Open			Mesh			Muslin			
	N	S	%	N	S	%	N	S	%	
1992	15	3	20	16	16	100	19	19	100	$P<0.001$
1993	19	13	68	22	21	96	18	17	95	$P<0.05$

N, Number of larvae at start; S, number of larvae surviving the stage.
Significance levels are shown for a G-test on treatments versus survival.

and parasitism. Such factors may be abiotic, pathogenic or factors inherent in the population rendering them maladapted to the conditions experienced. This is obviously an important target for further study since mortality was 70–90% over this period.

The difference in mortality between years, shown in both Tables 11.2 and 11.3, may be due to the extensive flooding which occurred during the 1992–93 winter. On 23 September 1992 the fen was flooded due to heavy rainfall. Dyke-side plants were completely submerged for a time and all plants were partially submerged. When the flood receded sufficiently to count larvae, on 15 October, no larvae were recovered from dyke-side plants (Table 11.2b). Overall mortality increased from 72.3% to 91.8% over that three-week period.

The flood continued up to April 1993 and significantly increased (G-test $P < 0.01$) mortality during the hibernation stage, from 70% in 1991–92 to near 90% in 1992–93 (Table 11.3). It is evident that flooding during active larval feeding in autumn or spring is more detrimental than winter flooding when larvae are in diapause. The ability of larvae to survive submergence is the subject of laboratory experiments currently under way at Keele.

During the spring feeding period (April–June) there is a significant difference between the two caged treatments and the control, the latter suffering significantly higher mortality in both years (Table 11.4).

This suggests that vertebrate predation is important during this period. Some parasitoids (tachinid flies) have been recovered from larvae, but they have not caused sufficient mortality to produce a significant difference between the mesh- and muslin-caged populations.

Our conclusions from these preliminary results can be summarized as follows.

1. Floods before the hibernation period cause high mortality, particularly on dyke-side and low-lying plants, and prolonged floods probably cause additional mortality during hibernation.
2. Invertebrate predators may be a key biotic mortality factor during the pre-hibernation period.
3. A factor causing high mortality during overwintering is still unknown.
4. Vertebrate predators cause significant mortality during spring feeding in some years.

11.6.3   Female egg-laying preference

The distribution of the species is clearly not limited by its food-plant, since *R. hydrolapathum* has a far wider distribution and is widespread in Britain. Clearly, the species is confined to particular habitats but food-plant choice may also depend on microhabitat conditions and the nutritional status of the plant.

The food-plant is found in a number of different fen microhabitats, in the open fen, along dyke sides and along drove sides (pathways cut through the fen). Additionally, management practices provide plants in different stages of growth and in different vegetation. It is clearly of importance to conservation strategy to find out if females show a preference for any of these states and situations, and to follow this up by looking at survival of eggs and larvae on different plants. A comparison of egg-laying in different microhabitats carried out in 1992 involved searching for eggs, larvae and larval feeding damage during the autumn on plants in different situations. The results, shown in Table 11.5, suggest that most types of plant are used, but in this particular year young plants in an open situation caused by cattle grazing were favoured.

Table 11.5   Percentage of *Rumex hydrolapathum* plants on which *Lycaena dispar batavus* larvae, or their characteristic feeding damage, were found in four habitat situations at Woodwalton Fen NNR during September 1992

Management type	% Plants used
Mixed fen, 1 year since mowing	15
Mixed fen, 2 years since mowing	10
Mixed fen, cattle grazed in June	40
Uncut dyke side	25

Twenty plants from each area were selected at random for inspection.

Interestingly, dyke-side plants were used more often than plants in 1- or 2-year-old reed fen, which is not entirely consistent with the view of Duffey (1968). More research is planned in this area.

11.7 REQUIREMENTS FOR CONSERVATION

11.7.1 Europe

In those areas where the species is still locally abundant the priority must be the conservation of traditional habitats. Ebert and Rennwald (1991) list factors that are endangering these habitats in the Rhine valley: as drainage, intensive fertilization, large-scale mowing and mowing during the flight period. These factors must also apply to many other areas of Europe.

Appropriate management of areas which have been secured, such as nature reserves, is equally important. From our experience this includes maintaining an open fen vegetation with a high density of *R. hydrolapathum*. Evers, Maaren and Made (1987) suggest that this may also be important for male territoriality and list the vegetation types associated with male territories in the Wieden, The Netherlands. Keeping the open vegetation structure presents the greatest uncertainty. Spring, summer and autumn mowing will all inevitably cause mortality, yet they are often the only effective way of maintaining open fen vegetation. Evers, Maaren and Made (1987) recommend delaying mowing from June–July until August–September, but are uncertain how this would help. Cattle grazing has been used at Woodwalton and the Wieden during summer months with some success, because light grazing leaves the food-plant relatively untouched and in a more open position for egg-laying by females.

11.7.2 England

The attempts to re-establish *L. dispar* in England, after its extinction in the middle of the nineteenth century, have spanned the entire period of organized efforts to conserve any butterfly or any other insect. Sadly, success in the shape of a self-sustaining population still eludes us. This may be partly because very few resources have been put into the project at any time, including today, and sufficient research has simply not been done up until now to enable the true ecological requirements of the species to be determined. This is a situation which we hope will change in the near future as a result of a new and more scientific approach which is being taken to the problem. Of course, this does not guarantee success, but at least we will learn from experimentation so that future efforts may have a better chance of achieving the goal.

The current priorities are to continue to study the ecology of the Woodwalton population to see if a self-sustaining population can be supported there by appropriate management and to investigate the feasibility of re-establishment at other sites. Our current view from preliminary results is that management should aim for the following conditions.

1. Food-plants in relatively open, sunny positions using cattle grazing and/or biannual mowing. Females preferentially lay eggs on these plants and the open position may reduce invertebrate predation during the pre-hibernation stage.
2. Plants should be actively growing at the time of egg-laying to provide nutritious food for larvae, enabling them to accumulate sufficient energy reserves to survive the winter.
3. Plants should not be in low-lying positions which will be severely affected by flooding, and flooding should be avoided before and after hibernation.
4. Large areas of open fen should be maintained for male territories with adequate nectar sources during the flight season.
5. A network of sites meeting these requirements, and containing patches with high densities of *R. hydrolapathum*, is needed to compliment the dispersive, opportunistic strategy of the butterfly. Populations may frequently go extinct and this needs to be balanced by the availability of sites for colonization.

However successful the above management is, it is clear that for a true re-establishment one population is not enough and will never represent security. The goal must be to re-establish a number of populations, preferably in sufficiently close proximity that migration and gene flow may be possible between them. The only feasible area today is the Norfolk Broads, and this possibility will be intensively studied over the next few years. The Woodwalton population has been captive-bred under greenhouse conditions for over 20 years and undoubtedly the genetic bottlenecks and selection pressures experienced, together with inbreeding, have changed the gene pool; probably substantially. This has almost certainly made it unsuitable for further re-establishment attempts and we must consider using new stock of *L. d. batavus*. Our problem here is that this subspecies is currently endangered, as mentioned above, and we therefore have a clear link between our wish to re-establish the species in England and the conservation of the species in The Netherlands. Efforts to re-establish *L. dispar* can be justified in the same way as for any other butterfly; they are valued species in our national fauna and significantly enhance the beauty of our countryside. But there is an equally important reason. Although our subspecies is extinct, the establishment of another subspecies (arguably the same subspecies) which is itself endangered, provides an important reserve population which will help ensure its survival. Additionally, research on its re-establishment will hopefully prove useful to other conservation attempts in The Netherlands and across Europe. Only close collaboration between us and our European colleagues will ensure that *L. dispar* is truly a member of the British butterfly fauna once again.

11.8 ACKNOWLEDGEMENTS

We thank English Nature, Butterfly Conservation and the Dutch Butterfly Foundation for their support; and Ron Harold, Alan Bowley, Michael Chinery and many Butterfly Conservation members for their help.

CHAPTER 12

The conservation of *Carterocephalus palaemon* in Scotland

N.O.M. Ravenscroft

12.1   INTRODUCTION

To date, most butterflies that have received detailed study of their conservation needs have been those confined to the southern regions of England where they are at the northern limits of their range, surviving in habitats that produce artificially increased temperatures, such as heathland or calcareous grassland, e.g. *Hesperia comma* (silver-spotted skipper) (Thomas *et al.*, 1986), *Lysandra bellargus* (adonis blue) (J. Thomas, 1983b) and *Plebejus argus* (silver-studded blue) (Thomas, 1985). Often, the factors limiting the presence of these species are found to be the abundance of features that maintain this microclimate, such as unusually short food-plants or high proportions of bare ground.

There are, however, butterflies that are confined in the British Isles mainly to Scotland, for example *Erebia epiphron* (mountain ringlet), which is found on high ground, and *Erebia aethiops* (Scotch argus), which frequents boggy areas. The most famous northern recluse is one that used to be regarded as peculiar to the heart of England: *Carterocephalus palaemon* (chequered skipper). This butterfly is unusual in Britain as it has undergone a decline from south to north, in opposition to the majority of butterflies that have declined over a similar period. Currently, it has one of the most curious distributions and is confined in the British Isles to the west coast of Scotland (Figure 12.1).

It was regarded as common in England until the 1960s when it underwent a rapid decline that culminated in its extinction by the mid-1970s. Little was known of the butterfly at this point and it remained somewhat of an enigma until studies on its ecology were commissioned by the former Nature Conservancy Council. This chapter reviews the history of the species in the British Isles and research on its ecology, and presents the current understanding of its ecology in Scotland.

Ecology and Conservation of Butterflies Edited by Andrew S. Pullin.
Published in 1995 by Chapman & Hall. ISBN 0 412 56970 1

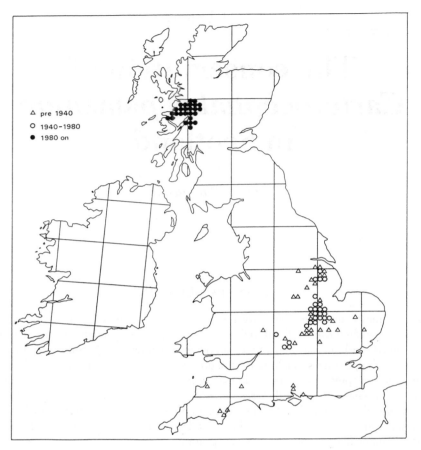

Figure 12.1 The past and present distribution of *Carterocephalus palaemon* in the British Isles.

12.2 THE DISTRIBUTION OF *C. PALAEMON*

The butterfly is found across the northern hemisphere: from Britain across Europe and Asia to Japan and in North America. It has a distinctly boreal distribution throughout. In North America (where it is called the Arctic skipperling) it is most common in Canada and extends north into Alaska where its distribution closely follows that of the tree line (Scott, 1986). In Europe, it is found from Scandinavia south to Greece (Higgins and Riley, 1970), but the species is local throughout and it is absent from large areas, mainly to the south, including much of Italy, Spain and France as well as Denmark, England, Portugal and Ireland. Declines have been noted in most European countries, especially Britain (Heath, Pollard and Thomas, 1984) and The Netherlands (Tax, 1989), as well as Czechoslovakia, Denmark, France, Poland and Switzerland (Heath, 1981a). Only in Scandinavia, Germany, Austria, Hungary and Romania was the butterfly still regarded as widespread in 1981 (Heath, 1981a).

In Britain, *C. palaemon* was one of four butterfly species afforded full legal protection by the Wildlife and Countryside Act of 1981. This protection has since been removed, although the species remains on schedule 5 with respect to trade; a licence is required to sell or exchange livestock or specimens. It is listed in the *British Red Data Book* of insects, one of only seven British butterflies (Shirt, 1987). Furthermore, it is one of three British butterflies on the European red list of globally threatened animals and plants published by the United Nations in 1991. The species was classified as vulnerable on a European basis by Heath (1981a).

12.3 THE HISTORY OF *C. PALAEMON* IN BRITAIN

Carterocephalus palaemon was first recorded in Britain by a Dr Abbott at Clapham Park Wood, Bedfordshire in 1798. It was subsequently found to be fairly common and widespread in woods on the chalky boulder clay of Bedfordshire, Rutland, Huntingdonshire, Cambridgeshire, Northamptonshire, extending north to Lincolnshire and south to Oxfordshire. A number of records from outlying areas suggest that in the nineteenth century the species had a wider range, including Hampshire, Suffolk, Merioneth, Dartmoor and Devon (Coleman, 1860; Meyrick, 1895; Newman and Leeds, 1913). Museum specimens also exist that are labelled from Kent (Collier, 1986) and it was recorded there late in the 1890s (Sassoon, 1938). But perhaps the most surprising record is one from the Lake District, near Coniston, made by Salter (1880). An extract from his diary (J. Mitchell, personal communication) reads: '16th July 1880. Made a butterfly hunting excursion in the neighbourhood of Coniston – in a marshy spot we found one of the [chequered] skippers'.

Whatever the original range of the species, it was the East Midlands of England that supported the species into the twentieth century, and here it was always regarded as common (Pilcher, 1961). There was a partial decline in the middle of the century when the species became more confined, particularly to Lincolnshire, Rutland and Northamptonshire, but this was followed by a partial expansion in range in the 1950s when old sites in Cambridgeshire were colonized, and the species was never thought to be in any decline or in danger of extinction. At one site in 1957 it was 'incomparably more numerous than it was then [30 years ago], and abounds in the rough grass . . . of the wood' (Lane and Rothschild, 1957). However, late in the 1960s a severe decline occurred. By 1971 the species was seen in only 6 of 80 former sites (there were over 20 throughout the 1960s) (Farrell, 1973) and in 1972 and 1973 the Biological Records Centre received no records for the species. Surveys were organized in 1974 and reports were received of skippers at two sites, both in Lincolnshire (Farrell, 1974). Lamb (1974) drew a blank from an exhaustive search of former sites. The species was recorded from one site in 1975 (Stark, 1975) but the last known sighting was at the same site in 1976 (Archer-Lock, 1982), thus completing a rapid extinction in England. Further surveys have been organized, such as one in 1980 (Heath, 1981b), but to no avail.

The history of *C. palaemon* in Scotland is equally remarkable. It was not known from here until 1942 when it was found in 'Western Inverness-shire' (Mackworth-Praed, 1942) and seen again in 1943 (Mackworth-Praed, 1945). It was subsequently revealed that a specimen had been taken near Loch Lochy in 1939 and this is the earliest confirmed record (Evans, 1949). There is a much earlier reference to a 'skipper', seen at Glenshian within the modern range of the species, but the identity of this was not revealed (Joicey and Noakes, 1907). No other skipper species are known to occur in this part of Scotland and it seems likely that this was the first record in Scotland.

Up to 1971, the butterfly was still only known from five sites in Inverness-shire, all in a very small area north of Fort William. Efforts were spurred on by the plight of the butterfly in England, and Scottish Wildlife Trust surveys revealed a number of sites in Argyll (Moffat, 1975; Shaw, 1975). By 1984, 39 sites were known (Collier, 1986) and recent surveys have extended the range of the species (Ravenscroft, 1991). Currently, there are about 50 localities where *C. palaemon* has been seen in the past 10 years (Young and Ravenscroft, 1991), and a survey by Butterfly Conservation in 1993 has added a further dozen or so sites to this list (Ravenscroft, 1994e).

The range of the species extends to the north to Loch Arkaig (although there is an unconfirmed record of the species at Drumnadrochit, Loch Ness, 40 km north of Arkaig) and along Glen Spean to where the ground starts to rise towards Loch Laggan and the inland mountains around Creag Meagaidh. In the east, the Nevis range of mountains, and those of Glen Coe, restrict the butterfly to the east shore of Loch Linnhe, and Loch Etive and Loch Creran to the south. To the west, the species extends as far as Moidart but does not venture far on to the Ardnamurchan peninsula (although a specimen was seen near Kilchoan; R.V. Collier, personal communication) nor into Morvern, where it is only known from a few sites.

12.4   ECOLOGY

12.4.1   Food-plant and habitat

The two other butterflies peculiar to Scotland, *E. epiphron* and *E. aethiops*, are both found in habitats that are more or less confined to Scotland (in Britain), mountains and bogs, but outwardly there is nothing particularly unusual about the habitats of *C. palaemon* in Scotland. They are found around open woodland which is dominated by *Quercus petraea* (sessile oak) or *Betula pubescens* (downy birch) in herb-rich wet grassland, usually on south-facing hillsides, often by the sides of lochs. Adult butterflies emerge in late May and fly until late June. They can be difficult to locate, even at known sites, but are particularly fond of taking nectar from the flowers of *Ajuga reptans* (bugle), found in sheltered spots in woodland or in more open meadow vegetation.

In England, the species frequented the rides of damp deciduous woodland growing on boulder clays and associated areas of limestone grassland (Collier, 1986), and also occurred in scrub areas of fenland on peats (Bretherton, 1981).

Here the flight period was slightly earlier, and the mean of the earliest dates for each available year on specimens kept in the British Museum of Natural History was 23 May between 1845 and 1967, compared to 31 May for Scottish specimens (1942–79) (Ravenscroft, 1992). On the continent, *C. palaemon* appears to have a similar flight period to the old English populations (Tax, 1989; Warren, 1990b), but it has been reported from a wider variety of habitats. In The Netherlands, 51% of sites are associated with woodland, 29% with grassland and 10% with heaths (Tax, 1989). Of 45 sites surveyed by Warren (1990b) in France, Germany, The Netherlands and Belgium, 33% occurred among glades and rides of damp woodland on heavy clay soils and 33% in tall fen or carr vegetation on peats (the main habitats in England), with the remainder in acidic marshy grassland (14%) (similar to the main habitat in Scotland) or among calcareous grassland and scrub (20%). In North America, the species inhabits bogs and wet grassland of clearings along the edges of the boreal forests of Canada into Alaska (Scott, 1986). Here, the flight period is much later than in Europe, with a mean earliest date on Canadian specimens in the British Museum of Natural History of 1 July, but there is a large variation from late May to the end of July (Ravenscroft, unpublished).

Detailed study in Scotland has shown that the primary habitat used by adults is wet acidic grassland and bogs on peaty soils, often dominated by *Juncus* spp. (rushes) or *Myrica gale* (bog myrtle), but with patches of abundant nectar sources (Ravenscroft, 1994a). The habitats of larvae also occur in wet situations where flushes of the food-plant occur under *M. gale*, *Pteridium aquilinum* (bracken) or saplings of *B. pubescens*. A common feature of all habitats chosen throughout the distribution of *C. palaemon* is their dampness (with perhaps the exception of calcareous grassland), especially in the north of their range in Scotland, Scandinavia and North America, and also the occurrence of saplings of *Betula* spp., *Alnus glutinosus* (alder) or *Salix* sp. (sallows). These areas are often associated with mature woodland, either along the edges or in large clearings, but adults can be found in suitable areas away from woodland, i.e. fenland-type vegetation. There also seems to be a pattern of increased association of *C. palaemon* with woodland in the north of its range, which may be a reflection of the increased requirement for shelter in the cooler temperate climates.

There was a wide variety of reported food-plants in the extinct English sites, including *Brachypodium pinnatum* (tor grass) (Collier, 1966), *Bromus ramosus* (wood brome) (Frohawk, 1892), *Cynosurus cristatus* (crested dog's tail) and *Plantago major* (broad-leaved plantain) (Coleman, 1860), but most often suggested was *Brachypodium sylvaticum* (wood false brome) (e.g. Rollason, 1908; Wood, 1908; Frohawk, 1934; Collier, 1966) and this is now regarded as its main former food-plant. On the continent, *C. palaemon* has a wider range of food-plants, being reported from *Molinia caerulea* (purple moor-grass) and *Calamagrostis canescens* (purple small-reed) in The Netherlands (Warren, 1990b), and *Dactylis glomerata* (cocksfoot), *B.pinnatum*, *M.caerulea* and *Festuca* spp. (fescues) in Germany (Weidemann, 1988). In France, egg-laying has been seen on *Milium effusum* (wood millet), *Calamagrostis epigejos* (wood small-reed) (Warren, 1990b) and *Poa* spp. (meadow grass) (Ravenscroft and Davis, 1992) and larval signs have been found on *C. epigejos* and *B. sylvaticum* (Ravenscroft and Warren, 1992). In Scotland, however, *B. sylvaticum* is thought to be used only occasionally (Collier, 1972; Sommerville, 1977), and *M. caerulea* appears to be used almost exclusively (Houston, 1976; Hockey, 1978; Ravenscroft, 1992).

12.4.2 The life cycle in Scotland

Adults are on the wing in Scotland from the middle of May until late June and, rarely, early July. Females emerge up to a week later than males and both sexes can be difficult to locate. Mating pairs are rarely seen. Eggs are laid singly, are white, spherical, around 1 mm in diameter, and larvae hatch after about 20 days. They begin, almost immediately, to construct a shelter from the grass blade within a few centimetres of the eggshell. They attach strands of silk to the edges of the blade, gradually drawing it together until tubular. The larva retreats into the shelter and emerges only to feed on the blade and this is eventually stripped for a distance above and below the shelter, leaving this suspended by the midrib. Often the blade snaps off above the shelter (Figure 12.2).

The larva then constructs a second shelter, often below the first. A series of shelters of increasing size is constructed on other blades as the larva develops. Shelters probably have an anti-predator role as they do in other butterflies (Ruehlmann, Matthews and Matthews, 1988), but may also be important in maintenance of the larval microclimate (Knapp and Casey, 1986). This stage of the life cycle lasts about 2 months into early September when the sixth or seventh shelter is constructed by sewing two or three blades together. The larva retires into this for 10–14 days to undergo a protracted moult. On emergence, the green final instar becomes free-living amongst *M. caerulea*, feeding extensively, and leaving characteristic feeding damage of v-shaped notches to the mid-rib of grass blades (Figure 12.2). The final shelter is a large structure formed from two or three blades and sealed at either end and is used for winter hibernation. Free-living larvae seem relatively immune from attack by other insects although they remain fully exposed on the grass blades. Mortality occurs in the early shelter stages, and overall 25% of hatched larvae survive to hibernation (Ravenscroft, 1992).

Although the hibernaculum is initially constructed well above the ground, *M. caerulea* soon dies back and the grass blades fall as litter. Larvae may spend the winter in pools of water, become exposed to the winter weather or blown away with strong winds. They emerge, straw-coloured, from hibernation in late March or early April and take up a position underneath a dead blade of *M. caerulea* where they begin to pupate, held to the blade by a girdle of silk. This stage lasts 4–6 weeks before the adults emerge.

12.4.3 Habitat preferences in Scotland

Adults occur in distinct aggregations for much of the flight period (Ravenscroft, 1994a). Females spend much of their time feeding in herb-rich areas, usually damp meadow vegetation dominated by rushes but rich in nectar sources such as *A. reptans* and *Cirsium palustre* (marsh thistle) or basic flushes in woodland

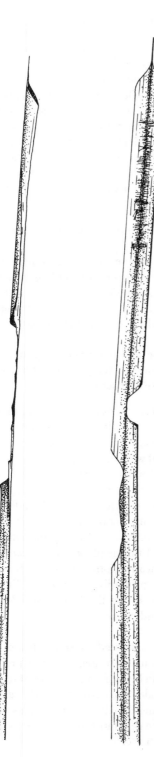

Figure 12.2 The characteristic feeding damage caused by larvae of *Carterocephalus palaemon*. Left: early instars, which form tubes in blades of *Molinia caerulea*, feeding on the blade above and below (the blade has broken off above the shelter); and right, damage due to final instars, which are free-living and make notches to the mid-rib below a silk pad from which they feed on the blade above.

clearings. These areas are often found in bays along the woodland edge, or may be completely exposed with only *Salix* saplings providing shelter.

After males emerge they may spend a day or so around the eclosion site, perhaps to locate emerging females, but then they disperse. They are territorial and occupy areas that satisfy the requirements of mate location, which provide good visibility over the surrounding vegetation and elevated temperatures in which they can maintain high rates of activity (Ravenscroft, 1994c). These requirements are met by scallops in the woodland edge, where *P. aquilinum* grows over exposed ground, or arcs of saplings on more open ground, or even a sheltered bay along the edge of a *P. aquilinum* front. The patches selected for territories have certain features in common (Davis, 1990; Ravenscroft, 1992):

1. they are relatively open compared to the surrounding vegetation, with a few well-spaced perches;
2. the heights of perches chosen vary between patches, but specific heights are selected according to the visibility they provide over surrounding vegetation and the temperature of the environment; taller perches are cooler (Ravens-croft, 1994c); favoured perches are generally *B. pubescens* saplings or *P. aquilinum* fronds, although any platform of the correct height may be used (usually 0.4–0.5 m); and
3. they are sheltered, very hot, and often close to female nectaring areas; the aggregations may occur because males are following temperature gradients, or male areas may be characteristic to females (Ravenscroft, 1992).

There is one feature that both male and female habitats have in common: they are both extremely humid. In a survey of the populations in 1993, Ravenscroft (1994e) noted that apparently suitable areas for male territories, such as sparse fronds of *P. aquilinum* with *B. pubescens* saplings in sheltered wooded sites, were vacant if the sites were particularly dry. Similarly, both males and females will feed from *Hyacinthoides non-scriptus* (bluebell) (Hockey, 1978; Kelly, 1983; Collier, 1986; Ravenscroft, 1992), yet large areas of this plant are often avoided by adults. In 1993, it became apparent from visiting many sites that adults will take nectar from this plant where it grows in unusually damp conditions, often by the side of a stream (Ravenscroft, 1994e). The other favoured nectar plants of *C. palaemon* – *A. reptans*, *C. palustre* and *Dactylorhiza maculata* (heath-spotted orchid) in Scotland (Kelly, 1983; Collier, 1986; Ravenscroft, 1992); primarily *A. reptans* in England (Collier, 1986); and *Lychnis flos-cuculi* (ragged robin) on the Continent (Warren, 1990b) – are all characteristic of humid areas, and other plants growing in these situations are occasionally used.

Suitable areas for males and females can be separated by some distance, or occur side by side. Little interference of feeding females by males occurs, probably because the confusion of flying insects at these patches makes detection of females unlikely (Ravenscroft, 1994c), although some males do set up territories around the edges of nectar patches and intercept incoming females. Even in areas where males are successful in detecting approaching females, most of their time is spent investigating other insects, usually bees or other butterflies.

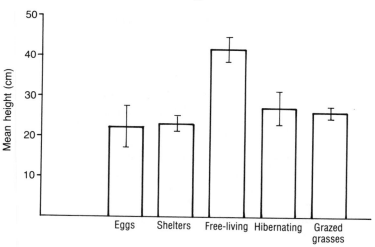

Figure 12.3 The mean heights (±95% confidence limits) of the various stages of *Carterocephalus palaemon* compared to the mean height of grazed grasses (after Ravenscroft, 1992).

Larval habitats are quite distinct from those of adults. They are most frequently found where the food-plant *M. caerulea*, which is found almost everywhere in this part of Scotland, grows in flushes under either *P. aquilinum*, *B. pubescens* saplings or *M. gale* and sometimes under *A. glutinosus*, usually along the woodland edge but often quite distant from woodland (Ravenscroft, 1994a). Again, all habitats selected are extremely damp. The vegetation is generally tall and dense scrub is characteristic. Larvae may also occur along the edges of clearings under partial canopy in quite shaded environments, and some degree of shading is a feature of all sites. Throughout development, larvae are found high on the food-plant (Figure 12.3) and the accumulated feeding damage of larvae over the season becomes quite easy to find in the autumn.

Large tracts of *M. caerulea* are avoided by *C. palaemon* and larvae only survive where the food-plant grows under certain conditions (Ravenscroft, 1994b). These preferred food-plants have greater leaf concentrations of nitrogen and stay green well into the autumn. The latter may well be the most important. *Carterocephalus palaemon* has one of the longest larval development periods of British butterflies (up to 140 days; Ravenscroft, 1992) and requires food-plant until late October in most years, November in some. The areas avoided by the butterfly were found to die back much too early in the year, often in mid-September (Ravenscroft, 1994b). *Molinia caerulea* is a nutrient-poor plant, and it seems probable that partially shaded flushes of the plant are preferred, perhaps because the favourable water regime prolongs summer growth or delays winter senescence, or because the nutrient status of these plants is raised under flushed conditions (Ravenscroft, 1994b). Even the feeding behaviour of the final-instar caterpillars, which notch the blade to the mid-rib before feeding, is thought to reduce nutrient loss

(Ravenscroft, 1994d) as the plants withdraw nutrients for winter storage as they approach autumn senescence (Morton, 1977).

12.4.4 Dispersal

Carterocephalus palaemon inhabits an expansive landscape in Scotland with almost unbroken semi-natural habitat with ill-defined boundaries. This differs from its haunts in England where its former habitats now exist as islands amongst developed land. Here, *C. palaemon* was found to be sedentary (Collier, 1986). Butterflies that occupy such habitats are usually found to be highly sedentary (e.g. *P. argus*; Thomas, 1985) and the majority of British butterflies form colonies with closed populations from which there is little or no emigration (Thomas, 1984a). The mobility of some species is thought to have been restricted in response to increasing isolation, as the individuals selected are those that stay in the habitat patch (Dempster, 1991).

Studies on *C. palaemon* in Scotland, however, suggest that adults, particularly females, are highly mobile (Thomson, 1980; Ravenscroft, 1992). Even at the best colonies, adults of both sexes occur at low densities and can be difficult to find, which has also been noted for continental populations (Warren, 1990b). Recapture rates of females during studies in Scotland were very low, and estimates suggested that they may spend only a few days in the flight area before emigrating (Ravenscroft, 1992). Males were less mobile, especially after they had selected an area in which to wait for females, to which they exhibited a high degree of fidelity.

There is a large abundance of the larval habitat in the region and it seems that females would not have to travel far from the flight area before suitable areas are encountered. Despite this, colonies are not closely grouped: the most concentrated group of 'sites' are still separated by over 2 km each. Even so, it is probable that there is a high degree of interchange of individuals between 'sites' and few are genetically isolated. During studies, single females were often encountered in areas that were unlikely to support a population and long distances away from potential habitat, even high up mountains (Ravenscroft, 1992).

Carterocephalus palaemon is therefore a species that moves about freely and is one that occurs at low density throughout its range, with concentrations of adults in favoured areas. There are inferences for populations of butterflies that exhibit poor mobility and are isolated on fragments of habitat. *Carterocephalus palaemon* indicates that such behaviour may have been altered and that freedom to disperse and interchange between populations is more normal.

12.5 CONSERVATION

12.5.1 Distribution and status

The dispersal of female *C. palaemon* has a number of implications for conservation. Many records of the species are single butterflies, at sites which are unsuitable breeding areas. The species is not as abundant as might first appear after omitting these records (Figure 12.4).

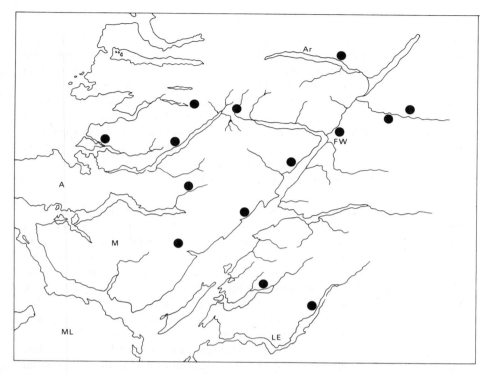

Figure 12.4   The approximate distribution of the main populations of *Carterocephalus palaemon* in Scotland. Ar, Arkaig; FW, Fort William; A, Ardnamurchan; M, Morvern; ML, Isle of Mull; and LE, Loch Etive. Scale, 1 cm: 7.5 km.

However, most suitable areas of habitat should be occupied within its range. Also, the large distances that often separate adult and larval resources, and the low densities of adults and their dispersal, means that large areas of habitat are required to support populations (Ravenscroft, 1992). Only one of the nature reserves on which it occurs (the largest) is likely to encompass enough habitat for the population to be self-supporting in the event of surrounding areas becoming unsuitable or being lost altogether. Similar patterns have been found in continental populations which also occur at low densities (Warren, 1990b; Ravenscroft and Warren, 1992). Here, there is still an abundance of marginal habitats, such as meadows, adjoining the population foci.

Four of the populations in Scotland occur on nature reserves (three National Nature Reserves (NNRs) and one Scottish Wildlife Trust) and several are on Sites of Special Scientific Interest (SSSIs). However, of the 14 main populations identified in Figure 12.4, only six are protected and most of the remaining 40 or so sites where *C. palaemon* has been recorded are unprotected. Many of these occur on Forestry Commission ground, or amongst private forestry. Presently, the Highland Branch of BC is investigating the possibilities of designating a reserve specifically for *C. palaemon*.

12.5.2 Management

Although usually seen in association with woodland, *C. palaemon* is primarily a butterfly of bog, scrub and meadow habitats. Many old English records refer to the species in fens and grasslands (e.g. Lane and Rothschild, 1957; Bretherton, 1981) where, no doubt, nectar sources would have been abundant. Similarly, *C. palaemon* is known from fen habitats on the continent and some of the highest densities of adults occur in these habitats (Warren, 1990b). In Scotland, woodland is the most frequently used habitat, but here the main function of this relationship appears to be the shelter it affords, the association of scrub and meadow habitats with its edges and therefore the provision of nectar sources. Poor soils cover most of the species' range in Scotland and favoured areas tend to occur around the richer soils of woodland.

Adults require the most open habitats, and are not tolerant of dense regeneration. Experiments at one reserve, where saplings of *Salix* were cleared to expose concentrations of *A. reptans*, were highly successful, as adults gained access to previously shaded plants. This led to an apparent increase in adult numbers (this may have been just an effect of concentrating them, but it is difficult to imagine where this population was feeding before). After 6 years, the same areas have become overgrown and numbers of butterflies have fallen (Ravenscroft, 1994e). Regular clearance of wayleaves through woodlands for the maintenance of power lines by Hydro Electric has produced suitable habitat for adults at a number of sites, especially where this occurs on damper ground. These rides are cut when regenerating trees start to interfere with the overhead cables, so there is no strict interval between clearance, but this usually occurs every 7–20 years. They are cut to widths varying between 15 and 30 m, but the larger rides are those most suitable for *C. palaemon* which tends to be absent from those less than 20 m wide (Ravenscroft, 1994e). These areas are not immediately suitable; only after several years' growth when *B. pubescens* and *Salix* saplings are abundant and around 2 m tall, with bays and damp hollows supporting *A. reptans* and *C. palustre,* does the habitat reach peak suitability.

Periodic clearance of woodland and control of scrub invasion are therefore important factors in the presence of *C. palaemon.* Before modern mechanical clearance, many of the woods supporting the species were coppiced on roughly 20 year cycles as a source of charcoal for iron furnaces in the region. This started in the middle of the eighteenth century and continued until late in the nineteenth (Lindsay, 1975). The oak woods were also important sources of bark, used in tanning processes, and trees 18–24 years of age were favoured. Before this, the only apparent pressures on the woods would have been as sources of domestic fuel. Nowadays, there is little management of the woodlands. Measures that are necessary for the conservation of the butterfly usually concern clearance of mature trees and dense scrub shading nectar patches, or removal of dense scrub over larval resources, although the latter are usually less limiting than adult resources. The sheltered structure of the wood needs to be retained.

The other main factor affecting the woodland structure is grazing, a particularly important and limiting factor in the regeneration of Scottish deciduous woodlands.

Overgrazing by large numbers of *Cervus elaphus* (red deer) is held responsible for poor woodland regeneration in most areas, although red deer graze woodlands mainly in the winter. *Capreolus capreolus* (roe deer) are primarily browsers and only take grasses in the spring (Rudge, 1983), but these are present at *C. palaemon* sites all year round. The presence of roe deer is important to the species as they prevent succession in larval areas, and keep adult areas open where scrub invasion is detrimental. Culling of roe deer on one reserve has had the desired beneficial effect on regeneration, but it has necessitated the use of labour to maintain areas for *C. palaemon* by clearing encroaching scrub. Browsing and limited grazing pressure is therefore beneficial to the butterfly. However, any livestock that alter the structure of the food-plant during the summer, when larvae are feeding high on the leaves, are detrimental, and *C. palaemon* can only tolerate low densities of these, particularly sheep (*Ovis*). Some former sites are heavily grazed, as are some marginal sites and most superficially suitable sites (Ravenscroft, 1994e) and a reduction in grazing levels on these would be beneficial to the butterfly.

In Scotland, *C. palaemon* also occurs in and around recently planted coniferous plantations. This is mainly because these have been planted up to the edges of deciduous woodland, or over areas of bog, both of which may have already supported the species, or because they have provided shelter for adults. These sites have limited life spans, however, and many of them are approaching the limits of suitability. A programme of clearance of trees around the edges of plantations and widening of rides will be necessary to maintain sites.

12.5.3 Monitoring

The population size of *C. palaemon* at one NNR has been monitored each year since 1986. The transect has recently been incorporated into the Butterfly Monitoring Scheme, but it remains the only site where the species has been monitored regularly. It is a difficult species to monitor, as adults occur at low density and also because they are generally difficult to see, but nevertheless numbers recorded on transects, albeit small, do correlate strongly with actual population estimates generated from recapture studies (Young and Ravenscroft, 1991), suggesting that this is an effective method for *C. palaemon*. Recently, two more transects have been organized, one on another NNR and the other on a Scottish Wildlife Trust reserve. Future strategy should promote the continuation of these and the involvement of the major landowners and managers, particularly the various forestry organizations, in the scheme. Increased involvement by these powerful bodies in Scotland is essential as so many sites of the species occur on land under their control.

12.6 THE EXTINCTION IN ENGLAND

A number of factors have been discussed that are important to *C. palaemon* in Scotland, and it seems appropriate to finish this chapter by considering the mystery of the extinction of the butterfly in England around 20 years ago.

Similarly, there are a number of potential explanations of the extinction, arising from the studies in Scotland and on the continent.

We have seen that large areas of habitat are required to support populations, and adult and larval resources are often widely separated. The species occurs at low density and shows a tendency to disperse. Rarely do sites occur that could support a healthy population in the event of all other areas becoming unsuitable. In Scotland there are few barriers to such movement and suitable breeding areas are scattered throughout the range, joined by semi-natural habitat. On the continent also, there is still an abundance of marginal habitats linking population foci. The situation in England was quite different at the time of the species' extinction. Here, the loss of marginal habitats such as scrub and grassland around woodlands (which may actually have been the primary habitat of *C. palaemon*) and, perhaps even more important, hedgerows connecting populations, may have been the most influential factors in its extinction.

The demise of traditional coppice management and increased shading in woodlands have been implicated in the declines of many woodland butterflies (Warren and Key, 1991) including *C. palaemon* (Farrell, 1973; Collier, 1986; Warren, 1990b). This is likely to have had a pronounced effect on populations that were becoming increasingly isolated in woodlands where the species depended on large, open rides and clearings (Collier, 1986).

Many woodlands were also underplanted with conifers after the Second World War. At the same time, myxomatosis was decimating the rabbit population of England, accelerating succession in grassland areas, a factor responsible for the declines of many grassland butterflies (Thomas, 1984a). Initially, these factors would have produced favourable conditions for *C. palaemon*, as they do in Scotland where early-stage plantation, scrub and ungrazed grassland combine. Indeed, there was a brief expansion in the number of sites occupied in the 1950s. Eventually, however, increased shading levels would have had a similar effect on woodland populations as lack of coppicing, and because of the short period over which planting took place, sites would have become rapidly and simultaneously unsuitable, a pattern that was reflected in the pattern of the species, decline.

The butterfly depends on damp habitats on peaty soils throughout its life cycle in Scotland. On the continent, most sites are similarly wet, either damp woodlands or fens, although some occur on calcareous grassland. The English sites occurred on the peats of fenland in East Anglia or in the wet woodlands growing on the band of heavy boulder clay of the East Midlands. A further factor worth considering, given the predilection of *C. palaemon* for wet ground, is long-term changes that may have occurred to the water table as a result of demands on the water supply, leading to a drying of habitats. Similarly, on a smaller scale, woodlands may have been drained as a precursor to planting with conifers.

At present, research is being undertaken by BC into the possibility of re-establishing *C. palaemon* in England. Investigations of the ecology of the species on the continent have been made (Warren, 1990b; Ravenscroft and Warren, 1992) and great care is being taken over the selection of potential sites and stock, by comparing feeding preferences and development of larvae from Scotland and the continent. Re-commencement of management activities in many of the woodlands that used to support *C. palaemon* suggests that some of these may be suitable for the species again. Assessment of conditions in these, and comparison with those on the continent and in Scotland, will confirm the feasibility of bringing the butterfly back to English woodlands.

12.7 ACKNOWLEDGEMENTS

Many of the studies in Scotland were financed by the Nature Conservancy Council. Thanks are due to Dr Ian McLean (English Nature) and Dr Mark Young (University of Aberdeen) for their roles in this. Many thanks to Ray Collier (Scottish Natural Heritage) for information and comments on the text, and especially Butterfly Conservation for financing studies on the continent and additional work in Scotland.

CHAPTER 13

The ecology and conservation of *Maculinea arion* and other European species of large blue butterfly

J.A. Thomas

13.1 INTRODUCTION

Five of the world's six species of *Maculinea* butterfly live in Europe. They are among the few insects for which specific conservation measures have been taken, and are regarded as 'flagship' species by many Western conservationists (Anon., 1993a). Thus *Maculinea* butterflies are regularly used for logos and on stamps (Anon., 1981, 1993b), have been the subject of numerous radio and television broadcasts and have frequently featured in magazines as diverse as the *National Geographic* and *The Economist* and in newspapers ranging from the *European* to the *Sun*.

The unusual appeal of this genus stems from a combination of three factors: their beauty as adult butterflies; a specialized life style that involves living underground in red (*Myrmica* spp.) ant nests for 11 months of the year; and the severe declines that are being experienced by all species over most of their ranges. There have already been several national extinctions of *Maculinea* species in western Europe (Tax, 1989; Thomas and Elmes, 1992), and the world status of all five species is considered to be 'endangered' or 'vulnerable' by the International Union for the Conservation of Nature (IUCN) (Anon., 1990).

13.2 THE CONSERVATION PROBLEM

13.2.1 Biotope loss

One obvious factor causing the extinction of many *Maculinea* populations has been

Ecology and Conservation of Butterflies Edited by Andrew S. Pullin.
Published in 1995 by Chapman & Hall. ISBN 0 412 56970 1

the fundamental destruction of their biotopes, for example by drainage, ploughing, the intensification of agriculture and afforestation. This has harmed every species, and has been the principal cause of decline of the three inhabitants of wet heath, moist grassland and fen: *M. alcon* (alcon blue), *M. teleius* (scarce large blue) and *M. nausithous* (dusky large blue) (Table 13.1; Thomas, 1984b).

At present, biotope destruction is a major threat to *Maculinea* populations in Mediterranean Europe, especially Spain (Munguira, 1987), southern France and northern Italy (Balletto and Casale, 1991), and in all eastern European countries (L. Peregivits and M. Woyciechowski, personal communications). This partly reflects the fact that most wetland breeding areas have already been destroyed in other parts of Europe or, in a few instances, are protected as nature reserves.

13.2.2 Extinctions in unchanged biotopes

An equally serious problem has been the disappearance of *Maculinea* populations from sites that have not changed in any obvious way, and where the early larval food-plants and *Myrmica* ants remain abundant. For example, by 1978, 48% of the UK's 91 recorded *M. arion* (large blue) populations had disappeared from land that still looked suitable to experienced entomologists (Thomas, 1980); by the early 1980s, all the European nature reserves known to me that once supported *M. arion, M. teleius* and *M. nausithous* had lost their populations of these butterflies.

The aim of this chapter is to describe the factors causing these more puzzling disappearances, and to suggest how populations can be maintained on isolated nature reserves, given sufficient knowledge of the population biology of each species and the resources to manage land with this specific aim. The British experience of re-establishing *M. arion* on former sites, and of creating a new breeding area, is used as an example.

13.2.3 Background to recent research on *Maculinea*

The first reports of *Maculinea* populations disappearing from suitable-looking sites were from nineteenth-century Britain, and concerned *M. arion* (e.g. Marsden, 1884). Since then, many hypotheses have been advanced to explain these local extinctions, all taking the premise that its habitat was indeed unchanged. Fuller details are given by Spooner (1963) and Thomas (1980, 1984a, 1989, 1991), but the most widely accepted hypotheses blamed collectors, insecticides, fragmentation (leading either to increased homozygosity in surviving biotopes or to the destruction of obligate metapopulations), a cooling of the British climate or air pollution.

For 50 years, these and other hypotheses formed the rationale for many different projects that were undertaken annually with the specific aim of conserving *M. arion* in Britain. Unfortunately, every attempt failed, and populations fell from at least 100 000 adults in the mid-1950s to one colony of about 250 adults in the early 1970s (Thomas, 1980, 1989). Clearly, some unrecognized factor was responsible and, equally clearly, more information was required about the biology of the butterfly before conservation measures were likely to succeed. For although

Table 13.1 The food-plants, host ants, biotopes and IUCN status of European species of *Maculinea*

Species	Initial food-plant	Myrmica host	Biotope	IUCN status
M. arion	*Thymus* spp., *Origanum vulgare*	*M. sabuleti*	Warm, dry grassland	Vulnerable
M. teleius	*Sanguisorba officinalis*	*M. scabrinodis*	Moist grassland; fen	Endangered
M. nausithous	*Sanguisorba officinalis*	*M. rubra*	Tall, moist grassland; fen	Endangered
M. rebeli	*Gentiana cruciata* and other *Gentiana* spp.	*M. schencki*	Very dry grassland; dunes	Vulnerable
M. alcon	*Gentiana pneumonanthe* and other *Gentiana* spp.	*M. scabrinodis* (southern Europe)	Moist, acid grassland	Vulnerable
		M. ruginodis (Netherlands)	Humid heath and tall, acid grassland	
		M. rubra (Sweden)	Tall, acid grassland	

the basic life cycle and dependency on *Myrmica* had been described nearly 60 years earlier (Chapman, 1916; Frohawk, 1916, 1924; Purefoy, 1953), there were many gaps concerning the behaviour and ecology of *M. arion*, and nothing was known about its population dynamics.

Research duly revealed that *M. arion* was distinctly more specialized than had previously been suspected, and offered a new explanation for its decline, while demonstrating that every hypothesis on which the conservation measures of the previous 50 years had been based was untenable (Thomas, 1976, 1977, 1980, 1984a, 1989, 1991; Elmes and Thomas, 1992; Thomas and Elmes, 1992). Meanwhile, growing concern across continental Europe about declines among the other (rarer) species of *Maculinea* was crystallized in a Council of Europe review of the continent's butterfly populations, which called for immediate ecological research on the genus, on which to base urgently needed conservation measures (Heath, 1981a). This was echoed by the IUCN and by the Societas Europaea Lepidopterologica (Thomas, 1984b).

13.3 THE BIOLOGY OF *MACULINEA* BUTTERFLIES

13.3.1 Population structure and life cycle

All five European species of *Maculinea* live in predominantly closed populations, with negligible interchange between populations that are separated by more than 2–10 km of unsuitable land (Thomas, 1976, 1977, 1984a,b and unpublished; Munguira *et al.*, in preparation). Typical populations consist of a few tens or hundreds of adults, supported by 0.5–5 ha of land, although much larger colonies have occasionally been recorded (Thomas, 1984a,b; Munguira, 1987; Munguira, Martin and Balletto, 1993).

Adult females oviposit among the flower buds of one or two specific food-plants in summer (Table 13.1), and the young larvae feed in July and August on the flowers and seeds. They develop quickly, entering their fourth and final instar after 2–3 weeks. However, larvae acquire little weight from the food-plant, and at the start of their final instar are no more than 1–2% of their ultimate biomass (Elmes, Thomas and Wardlaw, 1991; Thomas and Wardlaw, 1992).

The final instar possesses the full array of secretary organs that has evolved among phytophagous Lycaenidae to attract, excite and appease ants (Malicky, 1969; J. Thomas, 1992a); it also has the ability to mimic the stridulations of *Myrmica* workers (DeVries, Cocroft and Thomas, 1993). On the evening after its final moult, the small larva falls off the seed head and waits to be found by foraging *Myrmica* workers (Purefoy, 1953; Thomas, 1977). Contrary to most accounts, which are based on observations made in unnatural situations or in the laboratory (Frohawk, 1924; Schroth and Maschwitz, 1984), wild *Maculinea* larvae do not actively seek ant nests (Thomas, 1977, 1992b; Fiedler, 1990), and they die (or are killed) after 1 or 2 days if they do not occur within the foraging range of a *Myrmica* colony.

The behaviour of young fourth-instar larvae differs between species once they are discovered beneath their food-plants by a *Myrmica*. *Maculinea arion* (Frohawk,

1924) and *M. teleius* (Thomas, 1984b) start the complex sequence of behaviour first observed by Frohawk (1916) and Purefoy (1953). *Maculinea rebeli* and *M. alcon*, which I have suggested are the most evolutionary advanced members of the genus (Thomas *et al.*, 1991; Thomas and Wardlaw, 1992), are immediately picked up by *Myrmica* workers (Elmes, Thomas and Wardlaw, 1991), as is *M. nausithous*, which shares some attributes of 'primitive' and 'advanced' species of *Maculinea* (Thomas, 1984b; Elmes and Thomas, 1987; Fiedler, 1990). But whatever the mechanism, the effect is the same: the butterfly larva (hereafter called caterpillar) mimics an escaped *Myrmica* larva, and tricks the worker into carrying it into the ant nest, where it is placed among the larger ant brood (J. Thomas, 1992b).

The behaviour of different *Maculinea* species again differs in the *Myrmica* nest. The 'primitive' species, and *M. nausithous*, are predators of ant brood, feeding infrequently in bouts and avoiding the worker ants for most of the day (Figure 13.1a; Thomas and Wardlaw, 1990, 1992).

Maculinea alcon and *M. rebeli* continue to mimic ant larvae, and are cossetted by the nurse ants and fed directly with trophic eggs, regurgitations and prey, although they also occasionally attack ant larvae (Figure 13.1b; Chapman, 1919; Elfferich, 1963; Elmes, Thomas and Wardlaw, 1991). Both these types of caterpillar, which can be termed predacious and 'cuckoo', respectively, hibernate deep in their ant nests and resume feeding in spring, which is generally the main period of growth (Elmes, Wardlaw and Thomas, 1991; Thomas and Wardlaw, 1992). All species pupate in early summer in the upper cells of the ant nest, to emerge as adults 2–3 weeks later.

13.3.2 Population ecology

Population studies were made of a predacious species, *M. arion*, in Britain during 1972–79 and 1983–93 (Thomas, 1976, 1977, 1980, 1989, 1991; Thomas *et al.*, 1989; Thomas and Wardlaw, 1990, 1992), and of a cuckoo species, *M. rebeli*, in 1983–93 (Elmes, Thomas and Wardlaw, 1991; Elmes, Wardlaw and Thomas, 1991; Hochberg, Thomas and Elmes, 1992; Thomas and Elmes, 1993; Thomas, Elmes and Wardlaw, 1993; Hochberg *et al.*, 1994). Nineteen parameters that affect dispersal, natality or mortality were identified and measured in the first study, whereas 18 parameters were measured for *M. rebeli*.

Models, and all other analyses of the population data, suggested that variation in one factor, the survival of larvae in *Myrmica* nests, determines both the intrinsic rate of growth (r) and the carrying capacity of a site (K) for both types of *Maculinea*, although in other respects the population dynamics of the predacious and 'cuckoo' species are very different. The abundance of the initial food-plant is unimportant *per se*, but its distribution can be critical, for only those ant nests that are within 2–3 m of a host-plant can be parasitized by the caterpillar.

Four factors are responsible for the key mortalities of caterpillars in ant nests:

(a) The species of Myrmica

Up to eight species of *Myrmica* may forage under the initial food-plants on a *Maculinea* site, although it is unusual for more than four species to be common. All discover and adopt the young *Maculinea* larvae with equal alacrity (Thomas, 1977, 1984b, 1992b; Elmes, Thomas and Wardlaw, 1991). However, once underground, the caterpillars gradually die or are killed in the nests of all but one of those species of *Myrmica*. By the following June, the process is usually complete, and all but an insignificant proportion of the adult population emerges from the single host ant species (Thomas *et al.*, 1989).

(a)

(b)

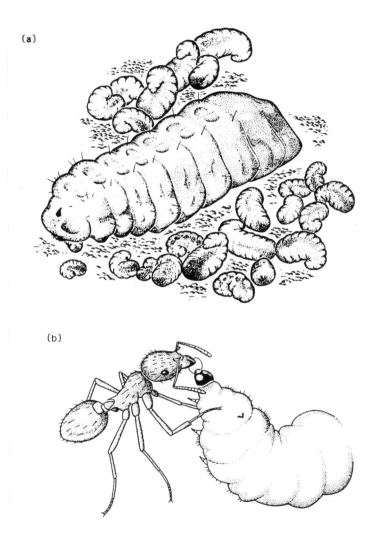

Figure 13.1 Two methods of feeding that have evolved in the genus *Maculinea*. (a) A predacious species (*M. arion*). The well-armoured caterpillars periodically enter the brood chamber to engulf the largest available ant larvae (from Thomas and Wardlaw, 1990). (b) A 'cuckoo' species (*M. rebeli*). The caterpillar is permanently tended by workers in the brood chamber. It begs for food like an ant larva, and is fed directly by the ants (from Elmes, Thomas and Wardlaw, 1991).

In our early studies, confined to the UK, Sweden, The Netherlands, France and Poland, we found a very clear pattern of host specificity, with each species of *Maculinea* emerging from the nests of a different species of *Myrmica* (Table 13.1); it appeared that the genus *Maculinea* had speciated by evolving as the parasites of each of the four commonest *Myrmica* species in the western Palaearctic or, in the case of *M. rebeli*, of an ant adapted to the arid habitats of its main food-plant, *Gentiana cruciata* (cross-leaved gentian) (Thomas *et al.*, 1989).

More recent studies of the five species in other parts of their range, including Spain, have confirmed this specificity for *M. arion*, *M. teleius*, *M. nausithous* and *M. rebeli*. But *M. alcon* is different: individual populations still depend upon a single host ant, but the species varies across the continent. Southern populations, at least up to Normandy, use *M. scabrinodis*, some in The Netherlands use only *M. ruginodis*, whereas those in central Sweden appear to use only *M. rubra* nests (Elmes *et al.*, 1994). Further research is required to establish whether these three types of population represent local races of *M. alcon* subspecies, or even undescribed cryptic species of *Maculinea*, comparable to the question earlier this century as to whether *M. rebeli* was a true species or mere subspecies of *M. alcon* (Berger, 1946; Bernardi, 1947).

The fact that a *Maculinea* population requires a single, rather than any, species of *Myrmica* was the key discovery for the conservation of the genus, for it means that each species inhabits a much narrower niche than had previously been supposed.

(b)   *The effect of the two feeding strategies of* Maculinea *on the carrying capacities of ant nests*

The 'cuckoo' method of feeding bestows three advantages on *M. rebeli* and *M. alcon* over their 'primitive' predacious congeners.

1. About six times more 'cuckoo' butterflies can be reared by a standard-sized *Myrmica* colony, because it is more efficient to intercept food directly from nurse ants than to let it be fed to ant larvae that are then eaten by caterpillars (Elmes, Thomas and Wardlaw, 1991; Thomas and Wardlaw, 1992).
2. Ant colonies frequently adopt more caterpillars than they can support (Elmes, Thomas and Wardlaw, 1991; Thomas and Wardlaw, 1992; Thomas, Elmes and Wardlaw, 1993). When this occurs, *M. rebeli* and *M. alcon* compete for the attention of nurse ants, which select as many caterpillars as they can feed, leaving the surplus to starve. The result of this contest competition is that a fixed number of caterpillars survive in crowded nests, determined by the size and physiological state of the ant colony (Thomas, Elmes and Wardlaw, 1993). By contrast, predacious caterpillars experience strong scramble competition in crowded nests, and every caterpillar dies of starvation once all the ant brood is eaten. The only caveat is that caterpillars sometimes have an opportunity to prey on a second *Myrmica* colony if they do not exhaust the brood of their original colony before spring. This occurs because workers from the first colony often desert when their brood has disappeared, leaving

the *Maculinea* caterpillar(s) in the vacated nest site. The caterpillar has an unusual ability to fast, and, before it starves, the nest is often colonized by an offshoot of an adjacent *Myrmica* colony, bringing in a fresh supply of ant brood (Thomas and Wardlaw, 1992). But even with this adaptation, it is unusual for any *M. arion* adults to emerge from nests that adopted more than four individuals.

3. Predacious species of *Maculinea* usually destroy the host-ant colonies that adopt them, leaving vacant nest sites that may be colonized by different, unsuitable species of *Myrmica* (Thomas and Wardlaw, 1992). This reduces the carrying capacity of the site for the next generation of butterflies. 'Cuckoo' species weaken their hosts sufficiently to disturb the balance in favour of non-host species of *Myrmica* in some parts of sites, but the damage is less and they seldom cause their hosts to desert (Hochberg *et al.*, 1994).

(c) The physiological state of the ant nest

There is much variation in mortality even when caterpillars are adopted at low densities into nests of their host *Myrmica* species. This reflects the physiological state of the ant colony, which may range from being benign to aggressive (Elmes, Thomas and Wardlaw, 1991). We have yet to identify every factor involved, but one variable is whether the nest contains a queen ant (queenright) or is queenless. In queenright colonies, the queens induce nurse ants to rear workers rather than rival queens, which is achieved by nipping and later attacking large queen-potential ant larvae (gynes) (Brian, 1979). By selecting large ant larvae as prey (Thomas and Wardlaw, 1992), *M. arion* (and presumably all predacious *Maculinea*) becomes contaminated by the pheromones of any gyne larvae in the nest. This renders the caterpillar liable to attack after about 2 weeks in the nest, when it reaches the size of a large gyne larva. The result is that the mean mortality of *M. arion* larvae in queenright nests is about three times greater than when caterpillars are adopted into queenless ones (Thomas and Wardlaw, 1990).

'Cuckoo' species of *Maculinea* do not experience the 'queen effect', possibly because they mimic some attributes of adult queen ants (Elmes, Thomas and Wardlaw, 1991 and unpublished). However, laboratory experiments suggest that although the biomass of *M. rebeli* reared by queenless and queenright ant colonies is the same, the former rear fewer, but heavier individual butterflies (Thomas, Elmes and Wardlaw, 1993).

(d) Parasitoids

Despite being among the rarest and most specialized insects known to science, *M. rebeli*, *M. alcon*, *M. nausithous* and *M. arion* each has a specific parasitoid that may kill up to 20–25% of a population. There is, again, a great difference in the life style of the parasitoids of 'cuckoo' and carnivorous species of *Maculinea*, which are hosts, respectively, to *Ichneumon* and *Neotypus* species of ichneumonid. For reasons described by Thomas and Elmes (1993), *Ichneumon* parasitoids have evolved to sting their host caterpillars inside the ant nest. Their adaptations include an ability to determine, from above ground, which *Myrmica* nests contain the specific host ant and which of these also contain *Maculinea* caterpillars; they are also adapted to enter the ant nest, sting the host and escape unscathed. *Neotypus* parasitoids sting their hosts on the initial food-plant, but nevertheless emerge as adults in the ant nest. *Neotypus* parasitoids have considerably narrower niches than their host *Maculinea*, and exist only in a small number of host populations (Thomas and Elmes, in preparation).

13.4   DEFINING THE HABITATS OF *MACULINEA* SPECIES

13.4.1   General

The data in section 13.3, together with measurements of fifteen additional parameters affecting survival, dispersal or natality, were combined with studies of the behaviour of each species and with analyses of occupied and former *Maculinea* sites to define the habitats of each species more precisely than hitherto. This has included constructing models to estimate the area of habitat needed for a population to survive in isolation, and how this area varies with the quality of the habitat.

13.4.2   Predacious *Maculinea* (*M. arion*)

A minimum density of host ant must be present for a population to survive at all: in the case of the predacious species, *M. arion*, the estimated intrinsic growth rate (r) of a population falls below 1 (i.e. the population declines) if fewer than about 50% of the initial food-plants occur within the foraging range of its host ant (Thomas, 1991). In other words, the poorest habitat in which *M. arion* can survive in Britain has half the *Thymus* plants within about 2 m of an *M. sabuleti* nest. The quality of habitat increases at higher densities of host ant, to reach an estimated maximum of $r = 2$ when every *Thymus* plant is near a *M. sabuleti* colony (Figure 13.2).

Under these conditions, a population of *M. arion* should, on average, double each year over a period of time, although there is much annual stochastic variation, with greater increases after periods of favourable weather punctuated by occasional crashes, for example following a drought (Thomas, 1980).

Values of r calculated for different assemblages of *Thymus* and *M. sabuleti* represent the inherent growth rate of a population of *M. arion* at low densities in those habitats. As numbers rise, an increasing proportion of individuals experience extra density-dependent mortalities, the most important being the high mortality when an ant colony adopts more than one caterpillar. Eventually, the population reaches a density at which these additional mortalities inhibit further growth. This ceiling represents the carrying capacity of a site (K). For *M. arion*, the value of K is crudely determined by the total number of *M. sabuleti* nests that are within 2 m of a flowering *Thymus* plant. This is a function of the size of the site, the density of *M. sabuleti* and the distribution (but not the absolute abundance) of *Thymus*.

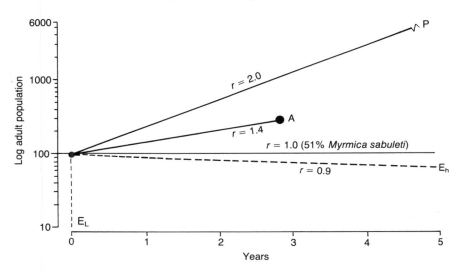

Figure 13.2 Estimated values of the intrinsic rate of increase (*r*) on *Maculinea arion* sites supporting different densities of the host ant (from Thomas, 1991). *P* = highest potential value when 100% of food-plants coincide with *M. sabuleti*. A = value on the last British site for *M. arion* (spot marks its carrying capacity, *K*). E_h and E_l, respectively, represent the highest and lowest values of *r* on 40 other former sites that superficially looked suitable for *M. arion* in the 1970s.

This simplified account of the science behind recent conservation proposals for *M. arion* was sufficient to set targets for site managers that were easy to perceive and to monitor (section 13.5). At its simplest, it involved changing the management of a potential site so that the density of *M. sabuleti* is as high as possible over as large an area as is available. At present, we aim to maintain sites at *r* values of >1.8 (= >85% of *Thymus* in foraging range of a *M. sabuleti* nest), and reject sites that have a minimum potential carrying capacity of fewer than 1000 adults (equivalent to about 5000 available *M. sabuleti* ant nests). The high value of *r* increases the probability that a population will recover quickly after an occasional catastrophe, such as drought, while that of *K* should (in theory) ensure that the population never falls to such low numbers as to be liable to chance extinction under the most adverse sequence of mortalities that is likely to occur every 1000 years under the current British climate.

These criteria may be stricter than is necessary, but probably not by much. For example, the last British population of *M. arion* became extinct on a site with a carrying capacity of 250–300 adults and a value of *r* = 1.4 (Figure 13.2). Extinction occurred after a sequence of 4 years when mortalities were exceptionally high, due to severe overcrowding, followed by an unusually cold, wet summer, followed by the two severest droughts recorded in south-west England for more than 250 years (Thomas, 1980, 1991).

As knowledge and management techniques improve, it may prove possible to support populations using lower densities of *M. sabuleti* and smaller areas of land.

13.4.3 'Cuckoo' *Maculinea* (*M. rebeli*)

The principles described in section 13.4.2 apply also to 'cuckoo' species of *Maculinea*. However, Hochberg *et al.* (1994) predict that they should be much easier to conserve than the predacious species because of their efficient exploitation of *Myrmica* colonies and because caterpillars experience contest rather than scramble competition in crowded ant nests. Thus their populations appear to inhabit broader niches, allowing greater latitude in site management; they are intrinsically more stable, and when perturbations do occur the higher value of *r* (for sites with comparable host ant densities) results in a quicker recovery than is experienced by predacious *Maculinea* populations; and 'cuckoo' populations can apparently be supported by much smaller areas of land (Hochberg, Thomas and Elmes, 1992; Hochberg *et al.*, 1994). For example, we estimated that a population could just survive (i.e. *r* = 1) on sites where as few as 5% of the initial food-plants grow within the foraging range of the host ant (Hochberg, Thomas and Elmes, 1992). These predictions are largely confirmed by site analyses, although to date we have yet to find an occupied site where fewer than 12% of gentians are within range of *M. schencki* (Thomas and Elmes, unpublished).

13.5 CONSERVATION OF *M. ARION* IN BRITAIN

13.5.1 Explaining previously puzzling local extinctions

The 42 former British sites of *M. arion* that still looked suitable for the butterfly were surveyed and reappraised using the new definition of its habitat (section 13.4.2). Forty sites still had a sufficient distribution of *Thymus* to support more than 1000 adult butterflies. However, although non-host species of *Myrmica* were generally common, *M. sabuleti* was either absent or occurred at such low densities that on only one site was the estimated value of *r* greater than 1 (Figure 13.2; Thomas, 1991). This exception supported the last British population of *M. arion*, but it had a very small *Thymus* population and, as described in section 13.4.2, its carrying capacity was too low to withstand a series of unusually adverse events.

Analyses of occupied *M. arion* sites elsewhere in Europe confirmed this narrow definition of *M. arion*'s habitat (Thomas *et al.*, 1989 and unpublished), supporting the conclusion that a major decline in *M. sabuleti* had occurred unnoticed on British *M. arion* sites, and was responsible for the puzzling extinctions of the butterfly.

I therefore analysed the habitat of *M. sabuleti*, and found that on British sites, the ant was abundant enough to support *M. arion* only on southern south-facing slopes where the turf was grazed to less than 3 cm tall (Figure 13.3; Thomas, 1976, 1980, 1993), in contrast to *Thymus* populations, which were most dense in turf 3–5 cm high.

By varying the intensity of grazing, it was possible to produce rapid changes in the abundance of *M. sabuleti*, with the host ant being replaced by other unsuitable species of *Myrmica* when the site became overgrown (Figure 13.4).

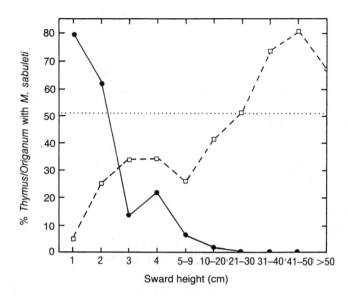

Figure 13.3 Variation in the abundance of *Myrmica sabuleti* in swards of different heights in Britain (solid line, *n*=347) and in Dordogne, where summer temperatures are warmer (broken line, *n*=227) (from J. Thomas, 1993). About 50% of early larval food-plants must be within the foraging range of this ant for a site to support *M. arion*.

Myrmica sabuleti is abundant only on heavily grazed, south-facing, southern sites in Britain because it requires a warmer microclimate than most other *Myrmica* species. On British *M. arion* sites, the ground temperature beneath *Thymus* in spring is about 7 °C warmer in swards that are 1 cm tall compared with those only 5 cm taller (and denser). But in Dordogne, where summer temperatures are about 3 °C hotter, the butterfly and ant occur only in swards more than 20 cm tall, and on all aspects of land except south-facing slopes (Figure 13.3; Thomas, 1993). This is a much commoner type of grassland and explains why *M. arion* has survived much better in central/southern Europe.

This explained the enigma of why *M. arion* became extinct on many sites, for the poor pasture on steep unfertilized slopes had progressively been abandoned by farmers over the previous century in Britain. However, extinction was postponed on many sites, because large rabbit populations provided adequate grazing. The introduction of myxomatosis removed this grazing force in the mid-1950s, and *M. sabuleti* declined, along with its parasitic butterfly, over the next decade (Thomas, 1980). Since conservationists were unaware of the specific need for high *M. sabuleti* densities, they too failed to provide the heavy grazing needed to maintain this ant on nature reserves. Instead they maintained slightly later seral stages, in which *Thymus* and non-host species of *Myrmica* were abundant. Unfortunately, by the time this information was available, and site management could be altered, the last British population of *M. arion* had become extinct.

13.5.2 Practical conservation

(a) Manipulating former sites and creating a new one

By the early 1980s, evidence that *M. arion* had disappeared from about half its former sites simply because they were less heavily grazed than before, was convincing enough for the new conservation programme to be started. The goal was to re-establish the species on at least six sites where the management objective was to encourage high densities of *M. sabuleti* and *Thymus*. This had the additional attraction of testing my conclusions.

The first site had been the last to support *M. arion*. It was unusual in that it was already more closely grazed, and had much higher densities of *M. sabuleti*, than any other in Britain (Thomas, 1980). Nevertheless, it was too small to support a population over a period of overcrowding and adverse weather (section 13.4.2). The plan was to improve the quality of habitat in the traditional breeding areas, by encouraging higher densities of host ant, and to increase the carrying capacity of the whole site further by doubling the size of grassland available for breeding.

This site (which was traditionally called X), consists of south-facing shales and requires regular management of its *Ulex* (gorse) scrub as well as intensive grazing. Our first objective was achieved by reducing the rotation time for scrub clearance and by slightly increasing the stocking density in late winter. This resulted in a slightly shorter sward, and although *Thymus* populations remained roughly stable, the density of *M. sabuleti* gradually increased, until, in 1993, 97% of food-plants

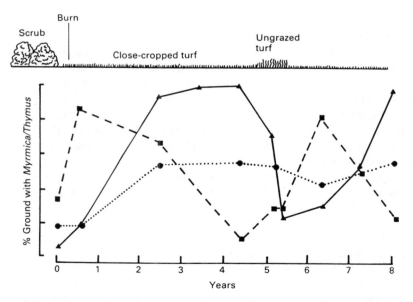

Figure 13.4 Changes in the abundance of *Maculinea arion*'s host, *Myrmica sabuleti* (solid line), initial food-plant (dots) and an unsuitable ant, *M. scabrinodis*, (dashes) following different forms of management. *M. sabuleti* can largely be replaced by its similar-looking congener after just one season of undergrazing (from Thomas, 1984a)

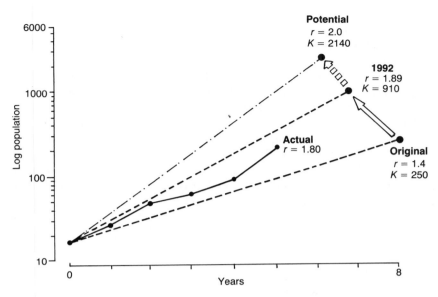

Figure 13.5 Changes in the estimated values of *r* and *K* for *Maculinea arion* on site X following conservation management (dashed lines). The potential for the site is also shown (dotted dashes), as is the actual annual increase of the introduced *M. arion* population in 1983–89.

were within the foraging range of this ant, compared with 66–75% when it had supported *M. arion* in the 1970s. The second objective was achieved by burning and grazing a similar-sized (4 ha) adjoining area of land that had been under a closed canopy of dense *Ulex* for many years. This was rapidly colonized by *M. sabuleti* (the data shown in Figure 13.4 were obtained from this area), but not by *Thymus*. This was as expected: *Thymus* is notoriously slow to reach new patches, although the first colonists then spread rapidly. I estimated that it might take decades, or centuries, for natural colonization to occur. The process was therefore accelerated by planting cuttings taken from the old grassland on site X. By 1993, the new area had a similar density of *M. sabuleti* to the improved, former breeding area, and a more widely distributed population of *Thymus*.

In formal ecological terms, we have manipulated site X so that its estimated values of *r* and *K* for *M. arion* have been considerably increased (Figure 13.5). In theory, it should be possible to continue this improvement (for *M. arion*) until the site is capable of supporting about 2000 adult butterflies.

The second site (Y) to which *M. arion* was introduced involved a more ambitious project. It is the neighbouring hillside to site X, and has the same aspect and geology. However, there was no history of *M. arion* on this site, which for several decades had consisted of about 10 ha of solid 1–2 m tall *Ulex* scrub. We began an attempt to create grassland on the hillside in 1974, when much scrub was cut. The regrowth was grazed heavily by cattle during winter, and more lightly by ponies in spring and summer, following the successful regime on site X. For 2 years, when the mean height of the 'sward' was about 20 cm in summer, I was unable to detect any colony of *M. sabuleti*. Gradually, a discernable turf, dominated by *Agrostis curtisii*, developed over about 5 ha of the site, and after 15 years we achieved a sward structure that met my definition of *M. sabuleti*'s habitat. The ant indeed increased to reach densities approaching those on site X by 1993 (Figure 13.6).

Thymus was also absent and, as predicted, has failed to colonize any control area in 20 years. However, cuttings were introduced to other areas, where they established and spread. As a result, site Y has fulfilled my definition of the habitat of *M. arion* since 1991, having originally been totally unsuitable for the butterfly (initial *r* and $K = 0$, Table 13.2).

A large-scale programme to introduce *Thymus* to unoccupied areas began in 1993, with the aim of raising the estimated carrying capacity of site Y to more than 1000 adult *M. arion* within 5 years. But even if this fails, this site provides an example of the dramatic changes in host ant population that can occur on British sites that are abandoned or reclaimed for heavy grazing (Figure 13.6).

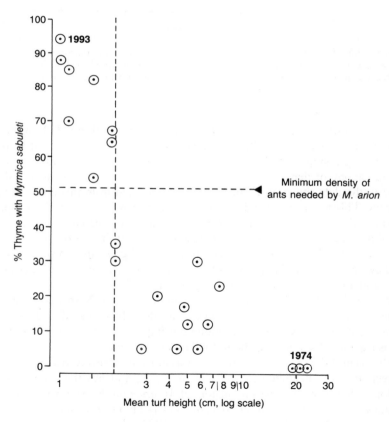

Figure 13.6 The relationship between vegetation height and the percentage of thyme plants within foraging range of *Myrmica sabuleti* over a 20 year period on a site (Y) created out of dense scrub in the early 1970s.

Table 13.2 Estimates of original and current intrinsic rates of increase (*r*) and carrying capacities (*K*) of *Maculinea arion* populations; the actual change in size of the butterfly population on each site in 1992–93 is given in the final column

Site	r/K	Original habitat	Habitat in 1992–93	Population change in 1992–93
X	*r*	1.40	1.89	1.74
	K	250	910	
Y	*r*	0	1.81	1.99
	K	0	270	
C	*r*	0.26	1.87	1.88
	K	0	5500	
S	*r*	0.90	1.86	3.33
	K	0	1250	

Sites X and Y are unusual among British *M. arion* sites in that *Thymus* grows at very low densities. Even when the planting programme is complete, it is envisaged that 50–75% of *M. sabuleti* colonies will not have *Thymus* growing within their foraging ranges. It is possible that this will result in more stable populations of *M. arion* than those on sites where *Thymus* coincides with every *M. sabuleti* nest, because when individual ant nests are destroyed by the caterpillars, there will usually be unparasitized nests in the adjoining grassland which can quickly recolonize any vacant nest sites (section 13.4.2; Hochberg *et al.*, 1994).

The other two introduction sites (C and S) are limestone hillsides that once supported the butterfly. *Thymus* was abundant throughout the sward on both, and almost every *Myrmica* nest could potentially be parasitized by *M. arion*. However, in the 1970s to mid-1980s, both sites were too overgrown to support adequate densities of *M. sabuleti*. Ant populations increased with heavier grazing during the late 1980s, until the estimated carrying capacities of both sites exceeded those of X and Y (Table 13.2).

(b)   Introductions of Maculinea arion

The ability of conservationists to manipulate populations of *Myrmica* and *Thymus* has been established. The next step was to test whether the requirements of *M. arion* had also been correctly defined.

Introductions of the butterfly have recently been made to the four sites described above. It is too early to be certain of success, but the results are broadly promising. The oldest population was introduced to site X in 1983. Small numbers were used, and the population increased for 6 years, roughly following the mean estimated annual rate of increase for a low-density population in the improved habitat (Figure 13.5). The population then fell sharply in 1989 and 1990, during 2 years of extreme drought. This also matched predictions, based on the decline of the last English population during the droughts of 1975 and 1976. In the next 2 years the population again increased at roughly the predicted rate for a non-drought year.

The population on site Y has had only two full generations since introduction, and although still small, has increased at approximately the estimated rate for its habitat each year. The populations at C and S have had one generation to date, with a similar increase to the predicted rate at the former site and a greater one at the latter (Table 13.2).

Taken together, these early results suggest that the scientific analysis on which the new conservation programme was based was broadly correct, although two potential long term problems of climate change and achieving consistent practical management of sites remain.

13.6 CONCLUSIONS

It is illogical, from a continental viewpoint, that such large resources have been spent on conserving *M. arion* in Britain, for this involves trying to maintain the hardest type of *Maculinea* in the most difficult part of its range. The same is true of the re-establishment of *M. teleius* and, to a lesser extent, of *M. nausithous*, in The Netherlands. Unfortunately, the will and means to conserve *Maculinea* (and other butterflies) tend to be generated in countries only after the trauma of approaching extinction, and far from competing for resources, the Dutch and British programmes have generated additional funds for butterfly projects elsewhere. They have also alerted other nations to problems they may face, while giving much pleasure at home. A scientific justification is that one learns more quickly about the minimum requirements of a species by trying to manage its populations in the harshest possible environment.

However, it is clear that future resources should mainly be spent on conserving existing populations of all species of *Maculinea* nearer the centres of their ranges. These areas also support their fascinating parasitoids, which are in even greater need of conservation than the butterflies. There is now an urgent need for many more reserves or national parks containing *Maculinea* to be established, particularly for the three wetland species in eastern Europe, Spain, southern France and northern Italy, and for the land then to be managed with the deliberate aim of maintaining populations. The information to achieve this is probably now adequate for all European *Maculinea* species, although there is much still to learn about the fine-tuning of management, about the minimum requirements of each species, and about the needs of the parasitoids.

More general lessons have also been learnt from attempts to conserve *M. arion*. One is the need to base conservation measures on ecological studies. For many years there were ample resources to conserve several British populations of *M. arion*, but no one knew quite what to do. In addition, research is more expensive, and success more elusive, if started when a species is already close to extinction. Restoring a habitat is also a difficult and expensive task compared to maintaining an existing one, and re-establishments are a costly and unsatisfactory substitute.

Fortunately, these lessons were heeded among British conservationists, and the extinction of *M. arion* was responsible for immediate funding of research into an endangered fritillary, *Mellicta athalia* (heath fritillary) and other declining butter-flies. As a result of Warren's (1991) timely research, *M. athalia* has been saved from almost certain extinction in Britain. It will be further consolation if continental conservationists successfully apply these lessons to their local *Maculinea* populations.

13.7 ACKNOWLEDGEMENTS

Research on *Maculinea* has been carried out in collaboration with too many colleagues to name them all, but I would especially like to thank G.W. Elmes, J.C. Wardlaw, D.J. Simcox, R. Prowse, R.T. Clarke, M.E. Hochberg, R. Mattoni, M. Munguira and J. van der Made. The programme to re-establish *M. arion* in Britain has variously been funded by English Nature, the Institute of Terrestriasl Ecology, Butterfly Conservation, the National Trust, the World-Wide Fund for Nature, Sir Terence Conran, the Somerset Trust for Nature Conservation and Bicton Agricultural College. I am grateful for their generous support, and to many individuals who have shared the practical, scientific and administrative burdens.

CHAPTER 14

Managing local microclimates for the high brown fritillary, *Argynnis adippe*

M.S. Warren

14.1 INTRODUCTION: THE DECLINE OF *ARGYNNIS ADIPPE* IN BRITAIN

Of all the butterflies that occur in Britain, perhaps none has declined as dramatically in recent years as *Argynnis adippe* (high brown fritillary). The species was once widespread throughout England and Wales, and many early books referred to it simply as being present in most large woods (e.g. Frohawk, 1934). As late as the 1950s it was reported to be abundant in many of its localities but it then underwent a rapid and severe decline (Heath, Pollard and Thomas, 1984).

Recent national and regional surveys organized by Matthew Oates and myself on behalf of Butterfly Conservation (BC), English Nature (EN) and the Joint Committee for the Conservation of British Insects (JCCBI) have shown that the number of 10 km grid squares with records has fallen from over 500 since records began to just 32 (with definite colonies) in 1992. This represents a decline in range of over 94%, mostly in the space of 30 years. Moreover, many current colonies are small, some with only sporadic sightings, and may not persist for long. Since 1970, *A. adippe* has become extinct in at least 13 counties and is now probably Britain's most endangered resident butterfly (Warren, 1992c). It has recently been listed (in 1992) as a protected species under the 1981 Wildlife and Countryside Act. Its present strongholds are in south-west England and the south Lake District of north-west England, with a thin scatter of colonies in other regions such as the West Midlands and south Wales (Warren, 1992c).

The rapid decline of *A. adippe* poses several serious questions for conservationists, particularly as its larval food-plants (*Viola* spp.) are still widespread in Britain and remain common on most former localities. Another puzzle is why it has declined so much more severely than the other four *Viola*-feeding fritillaries found

Ecology and Conservation of Butterflies Edited by Andrew S. Pullin.
Published in 1995 by Chapman & Hall. ISBN 0 412 56970 1

in Britain, some of which are still abundant. To answer these and other questions, EN commissioned a research project, starting in 1990, to examine the requirements of *A. adippe* and to identify the measures needed to conserve its remaining breeding habitats. The following account describes some of the preliminary results of the study, which is planned to run for at least another 2 years.

14.2   HABITAT SELECTION

Argynnis adippe currently breeds in two main habitats in Britain: *Pteridium aquilinum* (bracken)-dominated grassland (throughout its range) and limestone rock outcrops, usually where scrub or woodland has recently been cleared (north-west England only). The former habitats are usually at low altitudes (up to 200–300 m) on open but sheltered slopes, often facing south. Breeding areas tend to be dominated by *P. aquilinum*, although this is often interspersed with grassy areas. The rock outcrops used in Britain are all on the carboniferous limestone of the south lakes area in north-west England. Here, breeding areas occur on freely drained limestone soils, typically in sheltered clearings where woodland or scrub has recently been cleared, or where there are early successional stages in the colonization of bare rock. The larval food-plant in both habitats is normally *Viola riviniana* (common dog-violet), although *Viola hirta* (hairy violet) is also used regularly in limestone areas.

Information on egg positions is hard to obtain in the wild as females are highly mobile and difficult to observe when egg-laying. In *P. aquilinum* habitats they fly low over the tops of the fronds and suddenly drop down to search the ground. Once on the ground, females crawl around for several minutes repeatedly probing the vegetation with their abdomens. Egg-laying usually occurs after they have crawled over a leaf of the food-plant, but not always. Sometimes eggs are laid within seconds of the female landing, before touching any *Viola* plants.

During the extensive bouts of abdomen probing, eggs often appear to have been laid when subsequent searches prove otherwise. This problem arises because the tip of the ovipositor is the same pale brown colour as the egg and the two are difficult to separate unless observed within a few centimetres, which risks disturbance. From lengthy observations on egg-laying it seems clear that many egg sites are rejected and that 'false' egg-laying behaviour is common, making it even more difficult to obtain egg data. Information on its habitat selection has consequently been obtained partly from a small number of direct observations on egg-laying and partly by searching for larvae during the spring when they can be found basking on dead vegetation (section 14.3).

Data on the eggs located in 1990 and 1991 are shown in Table 14.1. Eggs were laid near the food-plant (within 50 cm) on firm substrates which were unlikely to rot down as the eggs overwintered. In woodland clearings, they were laid in very short vegetation (mostly 1–5 cm), often next to small outcrops of limestone rock, where there was a good cover of mosses (usually overlying rock) and where there was only a sparse cover of grass. In *P. aquilinum* habitats, they were laid in small gaps in otherwise dense *P. aquilinum* stands, where the ground

Table 14.1 *Argynnis adippe* egg site data 1990–91

Habitat	Height egg above ground (cm)	Turf/litter height (cm)		No. Viola plants	Viola sp.	% Cover						P. aquilinum cover		
		above egg	around egg			Leaf litter	P. aquil- inum	Moss	Grass	Bare earth	Bare rock	No. fronds	Average height (cm)	% Canopy
P. aquilinum habitats														
W. Midlands (n = 15)	4.6 (2–8)	5.3 (4–8)	6.2 (3–13)	10.8 (2–23)	6.8 (1–25)	–	69 (40–100)	0	23 (5–30)	0	0	18.8 (13–32)	82.1 (40–110)	50 (30–70)
NW England (n = 6)	5.2 (3–14)	5.8 (2–8)	6.5 (2–16)	24.3 (4–41)	8.2 (2–14)	–	68 (55–90)	3 (0–8)	32 (12–40)	0	0	9.5 (7–13)	59.5 (51–73)	51 (30–70)
Woodland clearings (on thin limestone soils in NW England)														
6-year-old coppice (n = 22)	1.2 (0.5–2)	–	–	10.5 (0–35)	7.6 (0–30)	19 (2–60)	0	57 (10–100)	6	0.5 (0–35)	12 (0–5)	0 (0–50)	–	–
1-year-old (n = 22)	1.9 (0.5–3)	2.2 (1–3)	2.7 (1–6)	6.7 (0–20)	6.7 (0–14)	48 (25–90)	0	36 (6–60)	4 (0–17)	8 (0–25)	2 (0–9)	0	–	–

Figures show averages (+ observed range) of measurements made within 40 cm² quadrats centred on the egg, or within 1 m² quadrats for *P. aquilinum* canopies. Turf height measurements were taken with a 10 cm diameter wooden disc dropped from a height of 1 m, and *P. aquilinum* height with a 30 cm diameter disc dropped from a height of 2 m (both four measurements per quadrat).

Violets in *P. aquilinum* habitats were all *Viola riviniana* and either *V. riviniana* or *V. hirta* in woodland clearings.

was largely covered by dead litter and where there was again little grass. Substrates chosen were typically dead *P. aquilinum* litter and moss, small twigs or dead leaves in woodland. Vegetation with fragmented, loose plant debris was usually rejected for egg-laying as females seemed to be unable to find a firm substrate which would be suitable for the eggs to overwinter.

Random quadrats sampled during the spring showed that habitat selection by *A. adippe* larvae was broadly similar to that suggested from egg positions. In both habitats, larvae were found in areas where the main food-plant *V. riviniana* was particularly abundant and where grass cover was low (Table 14.2).

In the woodland clearings studied in north-west England they selected areas with bare rock and short vegetation. In the *P. aquilinum* habitats of the West Midlands, dead *P. aquilinum* was abundant throughout the area sampled and the larvae showed only a slight selection for areas with higher than average cover.

14.3 LARVAL BASKING BEHAVIOUR

Argynnis adippe larvae hatch in March and become fully grown by the end of May or beginning of June, when they pupate amongst the leaf litter. The larvae are diurnal and, whenever the weather is cool, they bask on the upper surface of leaves or bare ground, apparently warming themselves in any weak sunshine. Similar behaviour has been observed in larvae of other *Viola*-feeding fritillaries including *Boloria euphrosyne* (pearl-bordered fritillary) and to a lesser extent in *Boloria selene* (small pearl-bordered fritillary) and *Argynnis aglaja* (dark-green fritillary) (Warren, 1992d).

The basking behaviour of the gregarious black larvae of *Eurodryas aurinia* (marsh fritillary) has been shown by Porter (1982) to allow them to raise their body temperature greatly above ambient temperature, thereby allowing rapid development during cool spring weather. Basking behaviour is less well developed in the solitary, brown-coloured larvae of *A. adippe* but presumably serves the same function. Also, the main period of larval development is later in the year than *E. aurinia* (which pupates at the end of April) when average temperatures are higher. During hot weather, the larvae of *A. adippe* do not bask but hide under dense vegetation where they remain still for long periods between frantic bouts of feeding. The ease with which they can be found in the wild therefore depends on the precise weather conditions and time of day.

The microclimate selected for egg-laying is known to be crucial for other *Viola*-feeding fritillaries, with each species having its own preference (Thomas, 1991; Thomas, Snazell and Moy, 1994). It is therefore highly likely that microclimate plays a crucial role in habitat selection by *A. adippe*, particularly in the light of its larval basking behaviour. Trial measurements of the temperature within the habitats show large differences between different vegetation types (Table 14.3).

In *P. aquilinum* habitats, the dead litter where larvae bask and where eggs were laid was 2–5 °C warmer than short, grassy vegetation (5 cm tall) early on a sunny day, and 7–10 °C warmer than the ambient temperature (Table 14.3a). Differences were rather less, about 1 °C warmer, on a cloudy day.

Table 14.2 Habitat selection by *Argynnis adippe* larvae in *P. aquilinum* habitats and woodland clearings

	% Viola	% Leaf litter	% Dead P. aquilinum	% Bare earth/rock	% Grass	% Moss	Depth P. aquilinum litter (cm)	Turf height (cm)
P. aquilinum habitats (West Midlands)								
Larvae (n = 9)	12.9	–	56.4	0.5	25.4	–	5.9	–
Random (n = 40)	6.8	–	47.4	2.4	33.8	–	5.8	–
t-values (47d.f.)	2.19*	–	0.76	1.15	0.66	–	0.10	–
Coppiced woodland (on thin limestone soils in NW England)								
Larvae (n = 13)	13.1	11.9	–	21.3	13.0	24.6	–	3.9
Random (n = 50)	2.8	9.4	–	8.6	41.3	25.2	–	11.4
t-values (61d.f.)	5.37***	0.93	–	2.21	-2.98**	-0.08	–	-4.84***

Figures show the means of various vegetation parameters measured in 40 cm^2 quadrats placed around the larvae found on independent searches, and in quadrats placed randomly within the habitat. Turf height measurements taken using a 10 cm disc dropped from 1 m (four per quadrat).

The significance of the differences between the means has been assessed by the Student's t-test: * $P < 0.05$; ** $P < 0.01$; *** $P < 0.001$.

Table 14.3 Micro-climate measurements in *Argynnis adippe* habitats, 1991, in (a) *P. aquilinum* habitats and (b) woodland clearings

(a) *P. aquilinum* habitats (West Midlands)

Date		21/5/91	22/5/91
Time		0900–0920	0940–1000
Weather conditions		100% sun, wind 3	0% sun, wind 2
Microhabitat			
Shade of nearby wood (ambient)		15.8	15.1
Grassy vegetation:	c. 1 cm tall	24.9 (top)	–
	c. 5 cm tall	20.6 (top)	16.8 (base)
P. aquilinum litter:	c. 1 cm deep	25.1 (top)	17.7 (top)
	c. 5 cm deep	22.8 (top)	17.8 (base)
	10–15 cm deep	21.4 (top)	17.0 (top)

(b) Coppiced woodland (6–7 years old on thin limestone soils in NW England)

Date		1/6/91	4/6/91
Time		1100–1130	0930–1000
Weather conditions		100% sun, wind 4 (vegetation dry)	100% sun, wind 1 (vegetation wet)
Microhabitat			
Shade of nearby wood (ambient)		15.0	8.8
Bare limestone rock		31.6	25.4
Moss over rock		41.8	26.4
3 cm deep moss/leaf litter		39.6 (top)	30.7 (top)
Grassy vegetation:	5 cm tall	35.9 (top)	–
		35.7 (base)	20.3 (base)
	10 cm tall	35.3 (base)	13.1 (base)

Figures show mean temperatures (°C) of five readings taken with Hygrotest 6200 probe. Measurements were taken in quick succession, alternating between microhabitats to even out variation in time. Windspeeds were measured on the Beaufort scale (0, still–6, very strong).

In woodland clearings, the 3 cm deep moss and leaf litter where larvae often bask was 10–17 °C warmer than surrounding grassy vegetation early on a sunny day, and about 4 °C warmer by mid-morning when the latter was dry (Table 14.3b). The moss over rock, where most *A. adippe* eggs were laid, was the warmest place measured under dry, sunny conditions, being 27 °C warmer than ambient, 2 °C warmer than the leaf litter, and 6 °C warmer than fairly short grassy vegetation. Thus *A. adippe* seems to breed in the warmest microhabitats in some of the warmest and most sheltered habitats where *Viola* spp. grow in Britain.

14.4 REASONS FOR DECLINE

The evidence gathered so far suggests that *A. adippe* is the most highly selective of the five *Viola*-feeding fritillaries in Britain (cf. Thomas, Snazell and Moy, 1994). This may explain why it has had a more restricted historical distribution than the others, reaching its northern limit in north-west England while *A. aglaja*, *B. euphrosyne* and *B. selene* are all quite widespread in Scotland. *Argynnis paphia* (silver-washed fritillary) has a similar British range to *A. adippe*, but extends just into Scotland and is widespread in Ireland (Heath, Pollard and Thomas, 1984). The information obtained on habitat selection by *A. adippe* also suggests some possible reasons for its decline and why this has been the most severe amongst this group of butterflies.

In the past, *A. adippe* was generally regarded as being a woodland species in Britain (e.g. Ford, 1945). Through most of England and large parts of Wales, woods were traditionally managed by coppicing, whereby portions of the wood were cut on rotation and allowed to regrow (Peterken, 1981). This regular cutting provided a sequence of sunny clearings where *Viola* and other ground flora were abundant in early successional woodland habitats (Warren and Key, 1991). New coppice clearings could therefore have provided highly suitable habitats in southern Britain, as they do for other fritillaries (Warren and Thomas, 1992). Although *A. adippe* seems to breed only in the very warmest vegetation around rock outcrops in woodland habitats in north-west England, it may have been slightly less restricted (possibly without the need for rock outcrops?) in the milder climates further south, as has been shown for other butterflies (J. Thomas, 1993). Alternatively, *A. adippe* may have bred predominantly in *P. aquilinum* stands that had developed, or had been opened up, following woodland clearance.

Like many other woodland butterflies (Warren and Key, 1991), the decline of *A. adippe* was probably caused primarily by the drastic decline of coppicing during the twentieth century combined with the switch to high forest or plantation management. Since the first accurate census of 1904, coppicing has declined by 94%, with the result that most modern woodland habitats are far more shady than they have been for hundreds, if not thousands, of years (Warren and Key, 1991). Even as late as the 1950s about 20% of English woods were still being coppiced, but the figure today has fallen to just 2% (Forestry Commission, 1984). Most of this is *Castanea sativa* (sweet chestnut) coppice, where conditions are often too acidic for the larval food-plant *Viola*.

The sudden decline of *A. adippe* since the 1950s may also have been related to other factors, notably the replanting of many ancient deciduous woodlands with conifers during the 1950s and 1960s (Warren and Key, 1991). This replanting would have provided similar open conditions to coppice in the early stages and may have caused a temporary surge in numbers over this period. However, plantations soon become too shady to support early successional woodland butterflies, unless numerous open rides are present. Moreover, plantations are cut on long rotations of 70–100 years and provide fewer canopy gaps, at less regular intervals, than traditional coppice that was cut on cycles of 5–20 years (Warren and Thomas, 1992).

At the same time that *A. adippe* was declining in woodlands, many colonies were disappearing from other habitats. The latter are less well documented but the butterfly was undoubtedly once present on many heathy commons as well as in woodland (Frohawk, 1934). Given our present knowledge of its requirements, it seems likely that such colonies bred in *P. aquilinum* stands and not in woodland.

These habitats have also suffered from changing land use and the abandonment of traditional practices in ways that are harmful to *A. adippe*. For example, most common land was formerly grazed by livestock and *P. aquilinum* litter was gathered in large quantities for animal bedding (Smith and Taylor, 1986; Tubbs, 1986). The latter practice has now almost entirely ceased throughout Britain and, due to economic pressures, many commons in lowland areas have not been grazed since the 1940s.

Although *A. adippe* seems to require fairly dense *P. aquilinum* for breeding, unmanaged stands can quickly become unsuitable due to the build-up of dense litter which eventually suppresses all other plant growth. Most existing colonies occur in areas still being grazed by cattle and/or ponies, such as Dartmoor in south-west Britain. These animals rarely eat *P. aquilinum* because it is toxic and unpalatable, but their trampling prevents total domination because it damages new fronds and helps break up the litter (Smith and Taylor, 1986). There is strong evidence from Herefordshire, where management has ceased more recently, that the local extinction of *A. adippe* on commons has been a direct consequence of abandonment.

Thus the decline of *A. adippe* in Britain appears to be linked to the changing management of both woodland and common land, which became particularly severe in the post-war period. An important feature of these changes with respect to woodland butterflies has been the loss of habitats with very warm microclimates. In addition to *A. adippe*, butterflies like *B. euphrosyne* and *Mellicta athalia* (heath fritillary) select early successional woodland habitats and have experienced major declines in recent years (Warren, Thomas and Thomas, 1984; Warren, 1987c; Thomas, Snazell and Moy, 1994). Moreover, J. Thomas (1993) suggests that such species may only have been able to survive in Britain because of the existence of warm man-made habitats such as those created by coppicing. These may have provided refugia in which the species could survive when the climate cooled 5000 years ago.

Evidence is now growing that several other butterflies commonly breed in the warm habitats provided by *P. aquilinum* and, with the decline in traditional woodland management, may be becoming increasingly reliant on these communities. They include *B. euphrosyne*, which is abundant on all *A. adippe* sites, and to a lesser extent *B. selene*. Moreover, several recent studies have shown that *P. aquilinum* habitats may also be important to other forms of wildlife (Pakeman and Marrs, 1992; Warren, 1993c).

14.5 EXPERIMENTAL HABITAT MANAGEMENT

14.5.1 Woodland management on limestone

Having identified some of the butterfly's main requirements, it has been possible to begin habitat management aimed at maintaining, and preferably enhancing, remaining populations. Such management has largely been experimental, partly because there are large gaps in our knowledge of the effects of management on

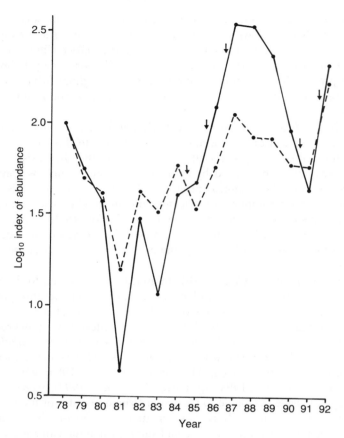

Figure 14.1   Annual changes in a population of *Argynnis adippe* in response to coppicing at Gait Barrows National Nature Reserve, NW England (solid line), 1978–92. Adults were recorded on a butterfly monitoring transect and are compared to the collated index on other monitored sites (dotted line) using the method of Pollard, Hall and Bibby (1986). The collated index was based on three sites in 1978 rising to 16 sites in 1992 (note log scale). Arrows indicate dates of coppicing. Note that most of the woodland on the site was cut during the 1960s and then left to regrow until conservation management started in 1984–85.

both flora and fauna, and because the response can be highly variable. Close monitoring has therefore been an integral part of the project.

In limestone habitats in north-west England, the positive response of *A. adippe* populations to coppicing has been known for several years (Pollard, Hall and Bibby, 1986). At Gait Barrows National Nature Reserve (NNR), there have been two periods of coppicing since the site was acquired as a NNR in 1977, and both have led to rapid increases in *A. adippe* compared to the overall trend on monitored sites (Figure 14.1).

The woodland had nearly all been cut down during the 1960s and most clearings were becoming very overgrown by the time transect recording began in

1978. Numbers declined until 1983 but increased rapidly after strips of woodland (amounting to *c.* 3 ha) were coppiced alongside the transect route from 1984–86. These had become dominated by tall grassland and tree regrowth by the time the present study started in 1990, and numbers had begun to decline relative to other sites (Figure 14.1). Although the coppice was still quite open and sunny even 6 years after coppicing (due to the slow tree regrowth on the thin soils), breeding was found to be restricted to a few rock outcrops where there was still sparse vegetation. Two new panels (*c.* 0.35 ha each) were then cut in 1991 and 1992, and numbers began to increase again (Figure 14.1).

A crucial result of egg-laying studies in the newly cut coppice was that eggs were only laid around a small but prominent rock outcrop, while the majority of *Viola* plants growing in patches of bare earth were ignored. The latter were used by *B. euphrosyne*, pointing to an important difference in the requirements of the two species which otherwise seem to be quite similar (cf. Thomas, Snazell and Moy, 1994). Fortunately, the woodland on this reserve overlies very thin, rocky soils and it is relatively easy to produce good breeding conditions. Elsewhere coppicing or scrub cutting will have to be specifically targeted on rocky areas to maximize the benefits to *A. adippe*.

14.5.2   The role of *P. aquilinum* management

Far less is known about possible ways of managing *P. aquilinum* habitats for *A. adippe* and the approach has been even more experimental. Although there is a vast amount of information on the ecology and control of *P. aquilinum* (e.g. Smith and Taylor, 1986), most studies have been concerned with heathland or upland communities where it is usually regarded as an invasive weed (Marrs, 1987; Lowday and Marrs, 1992). Few studies have examined habitats that contain a rich vernal flora such as those used by *A. adippe* and other fritillaries (Pakeman and Marrs, 1992; Warren, 1992d, 1993c).

The experiments conducted so far have fallen into three main categories:

1. basic research on the dynamics of *P. aquilinum* communities supporting a vernal flora;
2. enhancement of existing breeding areas (where *Viola* spp. are usually abundant under a variable but dense *P. aquilinum* cover; and
3. restoration of neglected *P. aquilinum* stands (i.e. with deep litter and few or no *Viola*).

As most potential forms of management are likely to have a slow and progressive effect on the vegetation, and the response of *A. adippe* populations may be delayed even further, these experiments are long term in nature and only a brief description is given here.

Experimental plots were established in 1990 in major breeding areas of *A. adippe* in the West Midlands to look at yearly fluctuations in vegetation structure and composition; the effects of summer cutting (end of July); and the raking of *P. aquilinum* litter (in autumn, simulating traditional cutting for animal bedding). The preliminary results show that there is considerable yearly variation in

P. aquilinum height, presumably in response to weather factors, although frond density has remained more or less constant. As expected from other studies (e.g. Lowday and Marrs, 1992) frond height has been reduced considerably in the cut plots and slightly reduced in the raked plots, although frond density showed little change after 3 years. *Viola* spp. density declined in both the control and raked plots in 1992, perhaps in response to increased *P. aquilinum* height that year, but increased markedly in the cut plots, presumably as the *P. aquilinum* has been weakened. The cover of dead litter has dropped greatly in both the cut and raked plots, so breeding conditions for *A. adippe* may not necessarily be improved while these treatments continue.

In a separate experiment, also in the West Midlands, studies are being conducted on the effects of autumn flailing, using a tractor-drawn swipe which macerates the litter and hastens its decomposition. This treatment has been applied to a large area of *P. aquilinum* to reduce the fire risk on a hillside where grazing levels had dropped over recent decades. *Argynnis adippe* was formerly abundant here but has now become rare, possibly because eggs laid on *P. aquilinum* are damaged by flailing (or survive badly afterwards) or because of changes in the microclimate. Flailed areas have become dominated by grasses, with low cover of *P. aquilinum* litter during the spring and summer. Consequently, they may provide few suitable warm basking areas for the larvae despite *Viola* spp. being abundant. The experiment has involved establishing permanent plots in flailed areas and comparing these to plots in a small area where flailing was stopped temporarily in 1991. If the results show substantial benefits for *A. adippe*, then flailing may be put on to a 4 or 5 year cycle to produce some suitable breeding patches while preventing the build-up of deep *P. aquilinum* litter.

Due to the current lack of accurate information, it has been thought unwise to change existing management where this has been relatively stable over recent years and where no obvious problems are developing. On such sites, the general approach has been to establish baseline data on adult population levels and to identify trends in numbers. New butterfly monitoring transects (using the method of Pollard, 1977) have been established on several *A. adippe* sites by BC volunteers, and timed counts of adults are conducted each year on many others (using method 2 of Warren, Thomas and Thomas, 1984).

Experimental management has been restricted to cutting narrow paths (*c.* 1 m wide) during late June in dense *P. aquilinum* stands where *Viola* spp. are abundant, thereby simulating animal tracks in areas where grazing levels were generally low. The aim was to break up the dense stands and provide numerous canopy gaps where *A. adippe* females could lay eggs. Observations on two sites showed that females used these paths frequently and several eggs were laid within the *P. aquilinum* on either side. However, it is difficult to equate this with breeding success as the butterfly normally breeds at low densities and survival rates are very difficult to assess. Egg or larval survival appeared to be poor where eggs were laid in paths through very tall, dense *P. aquilinum* as there were few signs of characteristic larval feeding damage the following spring.

Another pressing problem in the conservation of *A. adippe* is whether stands that have developed a very deep litter can be restored. Studies in heathland habitats have shown that *P. aquilinum* control often leads to colonization by 'weedy' species, such as *Urtica dioica* (stinging nettle) and *Holcus mollis* (creeping soft grass), and not those of the former vegetation which are preferred for nature conservation (Pakeman and Marrs, 1992). The underlying problem is that *P. aquilinum* enriches the soil and allows colonization by competitive, nutrient-demanding species. In the case of heathlands, which occur on very poor acidic soils, regular cutting and removal of *P. aquilinum* or soil stripping is necessary to aid revegetation by characteristic plants (Marrs *et al.*, 1992).

The full extent of this problem on *A. adippe* sites is not clear as the soils that support a vernal woodland flora are less acidic and sandy compared to heathland. However, there have been several instances where *P. aquilinum* adjacent to *A. adippe* breeding habitats has been controlled for general conservation purposes with very disappointing results. In nearly every case the flora is now dominated by *H. mollis* and/or *Hyacinthoides non-scriptus* (bluebell) with few or no *Viola*.

Current attempts to restore neglected *P. aquilinum* stands for *A. adippe* include the re-introduction of cattle or pony grazing (e.g. on two sites in the West Midlands, one funded by EN, the other by BC). The intention of these operations is to reduce or break-up the cover of *P. aquilinum* and reduce the depth of litter, while maintaining its overall dominance. The length of the rotation is based on guesswork at this stage, and will initially be about once every 4–5 years.

14.6 SITE PROTECTION AND FUTURE PROSPECTS

Of the 50 *A. adippe* breeding areas known in Britain in 1992, a large proportion are protected in some way. About 40% are notified as Sites of Special Scientific Interest under the 1981 Wildlife and Countryside Act, and many others are nature reserves. The latter include three NNRs (one with a very large colony) and eight County Trust reserves. In addition, many important localities are owned by the National Trust and several are being managed under agreements with the Dartmoor National Park and BC.

The future of *A. adippe* in Britain depends not so much on conservation designations, but on how many sites can be brought into favourable management. This will be affected by two main factors: the identification of appropriate management techniques for each habitat; and the implementation of recommended management. As many organizations have shown a strong interest in conserving the species, much will depend on the former. In the limestone habitats of north-west England, the butterfly appears to be relatively secure because many populations are responding well to rotational cutting of woodland or scrub, and successional changes are very slow owing to the thin, rocky soils. However, the majority of colonies (perhaps 80%) breed in *P. aquilinum* habitats where comparatively little is known about their conservation. Ecological research is beginning to improve our understanding of *P. aquilinum* habitats but there is still a long way to go before practical measures can be implemented with any certainty.

Looking further ahead, it may eventually be possible to reinstate suitable habitat management on selected former sites. If these exist reasonably close to existing colonies, they may well be colonized naturally. Recent mark–recapture studies have shown that *A. adippe* is fairly mobile, with individuals regularly moving 2 km between colonies and occasional sightings 5 km from known breeding areas. However, because its decline has been so drastic, any recovery can only be anticipated within its present core areas (south-west England and the south Lake District). Throughout most of this former range, it is unlikely to take advantage of any improvements in habitat management, including those that may ensue from general conservation initiatives. Consequently, any widescale recovery in Britain will only be possible by resorting to man-made releases. These have never been attempted seriously with *A. adippe* and are only likely to be successful if preceded by careful planning.

14.7 ACKNOWLEDGEMENTS

The ecological study of *A. adippe* was commissioned by EN as part of its programme of research into nature conservation. The surveys were organized or funded by BC, EN and the JCCBI. I would like to thank all the surveyors and recorders who have contributed to our knowledge of this species, in particular Matthew Oates. I am grateful to the Institute for Terrestrial Ecology for supplying data from the Butterfly Monitoring Scheme and to David Sheppard for his comments on the text.

PART FOUR

European and Global Perspectives

Implications of biogeographical structures for the conservation of European butterflies

R.L.H. Dennis and W.R. Williams

15.1 INTRODUCTION

The biogeographical structure of organisms has at least two points of relevance for their conservation. First, species differ in geographical status, influencing regional faunal composition and species-dynamics. Secondly, human pressures of population growth, of industrial and urban development, and from agriculture on the one hand, and of conservation resources, public awareness and sympathy, and legislation on the other, have a geographical bias, much of it accounted for by state boundaries. These geographical patterns determine the nature of problems and solutions for a regional fauna.

Conservation policy and programmes have arisen where there is a threat to wildlife. It is entirely reasonable that this should be most advanced where the greatest losses have been observed, both of populations and species. Such action is usually initiated within political units (countries, etc.), if only because state boundaries have facilitated the enumeration of species and their mapping, and form the vehicle for policy making. It is not surprising that butterfly conservation developed first in the industrialized countries of north-west Europe. In these regions, wildscape has been most threatened by modern agricultural techniques; the size of the fauna is relatively small and easily documented, as therefore are losses, and the political status and relative wealth of these areas has provided the leisure time and freedom for the study of wildlife. As awareness of the need for conservation and the facilities for carrying it out depend largely on wealth distribution, the concentration of butterfly conservation in north-west Europe may be an inadequate measure of the need for it elsewhere over Europe. In this chapter, a

Ecology and Conservation of Butterflies Edited by Andrew S. Pullin.
Published in 1995 by Chapman & Hall. ISBN 0 412 56970 1

series of biogeographical issues are addressed on the European fauna under three broad questions. First, if data on population losses, extinctions and past distributions of species over many states are poorly known, are there other aspects of the geography of European butterflies that can be used to guide conservation policy? In fact, might there not be better indicators for conservation policy? Secondly, what serious environmental changes threaten European species over what space and time scales? Thirdly, can these issues be contained within state boundaries or do they have a wider significance that requires cross-state co-operation?

15.2 BIOGEOGRAPHICAL MEASURES OF THE NEED FOR CONSERVATION

15.2.1 A European data bank; an essential tool for conservation

It is not possible to conserve what is not adequately mapped. Nor is it possible to analyse biogeographical structures of a continental fauna without such a data-base. Although distribution maps at high resolution ($10 \, km^2$, $2 \, km^2$) are increasingly emerging for European countries (e.g. Luxembourg: Meyer and Pelles, 1981; Switzerland: Gonseth, 1987; Yugoslavia: Jaksic, 1988; Britain: Emmet and Heath, 1989; The Netherlands: Tax, 1989; Austria: Reichl, 1992), for many states the fauna have not been mapped to Cartesian co-ordinates and accurate mapping of Lepidoptera over Europe has just begun (Faunistica Lepidopterorum European Project, Svendsen and Fibiger, 1992). In the absence of detailed accurate data, analyses carried out for this paper have been restricted to a set of 85 'natural' primary areas overlaid on the base maps in Higgins and Riley (1983) and Higgins and Hargreaves (1983). These data have some disadvantages. The number of species may be underscored (cf. Higgins, 1975; Kudrna, 1986). The 'natural' areas are large and vary in dimensions. The distributions of species are almost certainly inaccurate. Natural primary areas based on relief were selected for two reasons. Use of Cartesian co-ordinates (i.e. $100 \, km$ UTM grid) would give a spurious impression of accuracy. Larger Cartesian units (e.g. $250 \, km$ UTM grid) would result in the overlap of distinctive physical units without improving accuracy. As it is, text detail on distributions in the sources refer to physical regions (i.e. mountain and inter-montane areas) usually without specifying localities. The data comprise dichotomous scores (0, absence; 1, presence) on 393 species for 85 regions.

For analyses, the primary area ($Q = 85$) × species ($R = 393$) matrix has been used directly and converted into square matrices defining associations between the primary areas (Q matrix) and between the species (R matrix). Simple similarity measures have been applied (i.e. PHI correlation coefficient; Jaccard similarity coefficient S_J; Siegel, 1956; Sneath and Sokal, 1973). Their comparative use and the reasons for using them are given elsewhere (Dennis, Williams and Shreeve, 1991). A variety of multivariate techniques have been used to process the matrices, including:

1. non-hierarchical ordination by principal component analysis, factor analysis (VARIMAX, orthogonal rotation), principal co-ordinate analysis and non-metric scaling (ALSCAL);
2. hierarchical (SAHN) clustering by complete linkage and single linkage; and
3. classification by a polythetic divisive algorithm (TWINSPAN) (Gower, 1967; Sneath and Sokal, 1973; Hill, Bunce and Shaw, 1975; Hill, 1979).

The use of a variety of techniques is important (Hengeveld, 1990). Ordination techniques are effective in representing distances between major clusters but are less good than SAHN clustering in reproducing affinities between close neighbours. Further details of the techniques are given in Dennis, Williams and Shreeve (1991). Analyses were conducted via the AMDAHL 58/60 computer at Durham University, using routines in SPSS and TWINSPAN (Nie, Hull and Jenkins, 1975; Hill, 1979). Data from 254 stations in the west Palaearctic have been used for comparison of faunal units on climatic parameters.

15.2.2 Species richness gradients: 'hot-spots' for European butterflies

If species are regarded as equivalent in status, then it may be considered that areas deserving most attention for conservation would be those having maximum species richness. Compared to a policy whereby conservation priority is given to areas of greatest proportional loss, this is almost an inverse bias, since areas of greatest proportional loss tend to be those with limited species diversity. What is the pattern of species richness over Europe and where are the areas of maximum species richness? Only a generalized picture of species abundance is available (Figure 15.1a).

Highest species diversity (z-score > 1) straddles montane southern central Europe from north-central Spain, through the Pyrenees, Massif Central of France, the Alps, the Dinaric alps and the Carpathians and throughout mainland Greece. Peak values (z-score > 2) occur in the Alps proper. Diversity declines systematically northwards across northern Europe to arctic Scandinavia (z-score < -1.5) and southwards, though less dramatically throughout peninsular Italy and Iberia, but only attaining negative standard scores in southern Portugal. Particularly low diversities are expected for, and occur on, islands in the Mediterranean and Britain, the values reflecting island area and isolation as well as geographical position.

15.2.3 Endemicity: measures of scarce European butterflies

If the status attributed to species is inverse to their geographical range (weight inverse to abundance), then rare species, and areas where they aggregate, may be regarded as most deserving of conservation efforts. Certainly species which are both uncommon numerically and spatially are regarded as most susceptible to extinction (Dennis, 1993). But what is an endemic species and what is the pattern of European butterfly endemics? Endemicity has relevance only to a specified geographical area. Thus, at one extreme, some 140 of the 393 butterfly species

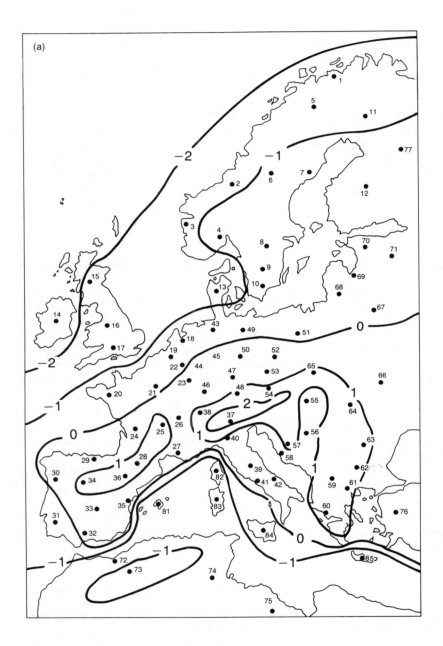

Figure 15.1   Species richness and endemicity in European and north African butterflies. (a) Standard scores of species richness based on 85 primary 'natural' areas standardized by regression for 250 km² units; (b) number of endemic species for 10 contiguous and non-contiguous areas (solid line) and zone of primary area endemics (dashed line). (After Dennis, 1993.)

Figure 15.1 Continued.

occurring in the mapped zone (Higgins and Riley, 1983) are endemic to it. At the other, a number of species are also endemic to the smallest mapping units referred to in this assessment ($n = 41$). However, there is no reason why endemicity should not be accounted for in increments of base units; such calculations may be considered for contiguous and non-contiguous areas separately or together. When this is done, it is found that the number of new primary areas with 'endemics' stabilizes for seven contiguous primary areas and for 10 non-contiguous primary areas. Thus, the pattern of endemicity is illustrated for 10 primary areas.

Defined in this way, endemicity is a southern European and north African phenomenon (Figure 15.1b). The virtual absence of endemics north of the southern European mountain belt is striking. Only one endemic species is found in northern Europe, having both a north Scandinavian and alpine distribution (i.e. *Pyrgus andromedae* (alpine grizzled skipper)), and this may eventually be found to have a wider distribution in Asia (Kudrna, personal communication). Peak endemicity occurs in the Alps proper but high values are found in other mountain areas; the Atlas chain in north Africa is especially rich in endemics. The region of single primary area endemics is slightly different (Figure 15.1b). It envelopes areas of lower as well as of higher cumulative regional endemicity, such as the Aegean islands and Mediterranean islands such as Sardinia. Naturally, the map of single spatial unit endemics will change most readily as new finds are made (e.g. *Maniola chia* on Chios; Thomson, 1987).

15.2.4   Is there a regional structure to European butterflies?

Neither species abundance patterns nor endemicity patterns may account for spatial bias among butterfly taxa. Conservation policies based on species richness or endemicity may omit specific taxa or faunal elements, particularly if geographical bias for these various measures does not coincide. To the extent that components of a butterfly fauna are unique to different geographical parts of Europe, conservation policy needs to make some regional appraisal of faunal elements and faunal units. What faunal patterns occur over Europe? Does taxonomic bias exist and can distinctive faunal elements and faunal units be discerned? Do the different faunal units convey an equivalent amount of significance for conservation?

Two matrices (Q and R matrix) have been explored for spatial structure in European butterflies. Q mode analysis determines more generalized structures (faunal regions, geographical areas differing significantly from others in taxa) from links among more cosmopolitan species. R mode analysis draws out smaller spatial structures by way of a two-stage process; first of identifying species which share similar distributions, then of deriving spatial units defined by them on their conjoint occurrence (faunal elements, species sharing a distinctive geographical distribution). Both stages of analysis revealed evocative geographical structures to European butterflies based on the taxonomic species concept (see Dennis, Williams and Shreeve, 1991).

Regionalization of the European fauna is best summarized by the factor analysis ordination (Figure 15.2a). This corresponds closely with the product of polythetic division (TWINSPAN) and homostats from complete linkage clustering. Europe north of the Alps is dominated by three zonal regions and the British Isles; south of the mountain chain, the Mediterranean is divided into western and eastern sectors, and separated from north Africa and the Mediterranean islands. The Italian peninsula has equivocal allegiance but factor loadings place it nearer to the western Mediterranean region. Some primary areas with low communalities (cumulative loadings on extracted factors) have substantive unique variance, indicative of endemism or of unusual combinations of species.

There is also close agreement between the ordination and clustering products for the *R* matrix. The faunal elements are displayed from basal stems in complete linkage (Figure 15.2b); an arbitrary cutoff of a 70% species inclusion level is applied to highlight the core of elements. As expected, many more faunal elements are derived than faunal regions. A preponderance of them emerge in southern Europe, in the mountain chain, the Mediterranean zone and north Africa. Only four occur for Europe north of the alpine chain, two of which envelop large portions of continental Europe.

Detailed results of these analyses are given elsewhere (Dennis, Williams and Shreeve, 1991). In summary, the results disclose:

1. There are clear geographical patterns of faunal regions and faunal elements.
2. Faunal regions are generally matched by faunal elements, although a multiplicity of faunal elements often exists for particular faunal regions, as in the case of the west and east Mediterranean.
3. Some faunal regions do not have comparable faunal elements (e.g. British Isles) and very distinctive faunal elements exist for geographical zones which did not clearly emerge as faunal regions (e.g. the southern mountain chain). The difference in the number of spatial structures produced from the *R* and *Q* matrices results primarily from the excess in the numbers of species over regions, therefore from the additional information in the *R* matrix compared to the *Q* matrix. The determination of faunal regions from the *Q* matrix is also constrained by the numerical dominance of species with cosmopolitan distributions over those with distributions restricted to smaller parts of Europe. Analysis of the *R* matrix effectively filters out relationships at different spatial scales. The absence of a British faunal element to match a faunal region is explained by the fact that all but one of the British species (*Erebia epiphron* (mountain ringlet)) are root species of the prime European extent element. The British faunal region is characterized by being a zone of faunal impoverishment; it lacks other species which collectively describe the European extent region (Figure 15.2a, region I).
4. Regional structures in northern and southern Europe have a latitudinal zonation and southern Europe divides into west–east components.
5. Regional structures in northern Europe are fewer, spatially larger and open-ended compared to those in southern Europe. The majority of species belonging to northern structures extend across the Palaearctic. Regional structures south of the northern edge of the Alps are mainly restricted to the mapped area and comprise high frequencies of endemic species.

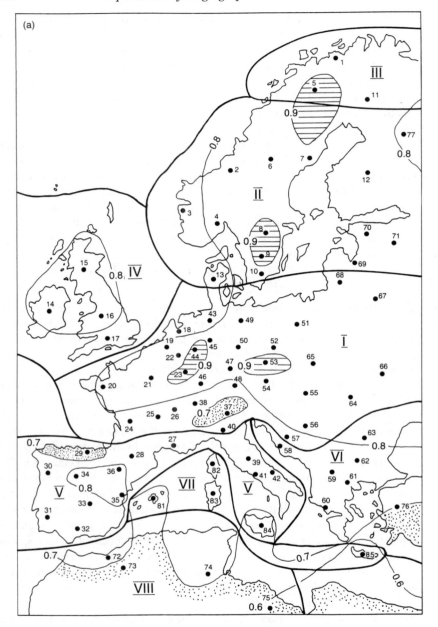

Figure 15.2 Faunal regions and faunal elements in European and north African butter-flies. (a) Regionalization from factor analysis of the Q matrix for 85 regions (nine factors); regions are numbered I–VIII; isolines give communalities; areas scoring low (<0.7) are stippled and high (>0.9) are shaded. (b) Faunal elements based on complete linkage analysis of the R matrix for 393 species (70% inclusion level); thick continuous line, extent elements; thin continuous line, southern elements. (After Dennis, Williams and Shreeve, 1991.)

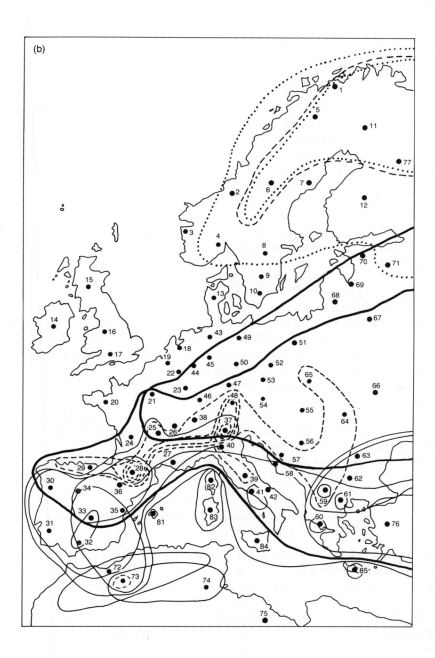

Figure 15.2 Continued.

It is important to understand that despite the disclosure of these spatial units, there are substantial overlaps of species ranges and that there are few absolute boundaries separating these units. For example, some 10% of species belonging to the extent group cover much of the mapped area. All faunal regions and the geographical areas depicted by faunal elements are characterized by some, occasionally a high degree of, internal diversity. But several features point to the faunal spatial structures as being much more than a random collection of species and the outcome of purely stochastic events: consistencies in pattern between analyses, latitudinal banding of zones, multiplicity of east–west units in the Mediterranean, geographical bias of endemics to southern Europe and what must amount to the highly significant discrimination in faunal content among European regions. Moreover, subdivisions within taxa associated with the application of the biological species concept (e.g. *Pontia daplidice* and *P. edusa* (bath whites); references in Dennis, Williams and Shreeve, 1991) parallel the geographical zonation of faunal regions and elements among taxa.

The contribution of taxa to regional faunas also varies geographically (Dennis, unpublished). The relative frequencies of Papilionidae and Pieridae decline northwards throughout Europe (Papilionidae from >6% to <2%; Pieridae from >20% to <10%) whereas that of Nymphalidae (excluding Satyrinae) increases with latitude (from 15% to >25%). Among Satyrinae, Hesperiidae and Lycaenidae more complex patterns emerge. Satyrinae have peak proportional abundance in montane areas (>25%), Hesperiidae in southern Iberia and Turkey (>15%) and Lycaenidae in Iberia and the Balkans but also, more surprisingly, in Denmark (>35%). The Satyrinae are also highly unusual inasmuch as it is the only large taxonomic group for which the 'global' frequency of species (30.8% of European taxa) exceeds the regional mean frequency of species (mean 19.82%, standard deviation 3.4, $n = 83$) indicative of the multiplicity of species with restricted distributions. In the residual Nymphalidae the bias is reversed ('global' frequency 17.6%; regional mean 24.9%; standard deviation 4.9; $n = 83$); these species have wide distributions. This observation has important implications for the spatial dynamics and the adaptability of taxa in the face of environmental changes, and suggests that a closer inspection of spatial and ecological attributes is warranted.

15.2.5   Conservation of European species: can appropriate measures be found?

A direct approach to determine priorities for conservation of all European species was initiated by Kudrna (1986) using two measures, a chorological index and a vulnerability index. The former is a measure of a species geographical disposition based on ranked scores for its geographical range and distribution (density of sites). The vulnerability index is similarly constructed but on the basis of range contraction, habitat loss and collecting. It is less successful than the chorological index, since for many European species insufficient data exist. However, the chorological index allows the direct evaluation of European species. Moreover, mapped units or sites can be directly compared from the sum of index values for species or from average scores. Subsequently, more detailed vulnerability measures, which can be applied for conservation policy, have been constructed by Dennis and Shreeve (1991) and Dennis (1993), but only for a limited regional fauna for which adequate data exist. Even then, there are important biological attributes for which data are lacking. Kudrna (1986) emphasized the importance of a comprehensive conservation programme, aimed at the conservation of whole communities rather than the protection of single species; whole habitats not just the insects. He realized that the essential criteria for site evaluation are species richness and the number of scarce species. Clear effective quantitative measures are described and documented for part of central Spain by Baz (1991). They are:

$$I_i = \sum_{j=1}^{m} S_{ij}Q_j / \sum_{j=1}^{m} Q_j$$

$$Q_j = \sum_{j=1}^{m} S_{ij}Q_j / \sum_{j=1}^{m} S_{ij}$$

where the conservation priority of each species is

$$Q_j = 1 - P_j$$

and the proportional occurrence of species$_j$ is

$$P_j = \sum_{i=1}^{n} S_{ij} / n$$

when n is the number of sites, m is the number of species and S_{ij} is the presence (1) or absence (0) of species$_j$ at site$_i$. The behaviour of such measures needs to be scrutinized, as does the effect of irregular spatial units and arbitrary regional bounds, but it holds promise for ranking species and sites for conservation.

Applied to the data for the 85 'European' primary areas, clear patterns are disclosed. I_i, the importance index (Figure 15.3a), has its highest values in the southern montane belt, the Alps and their eastern extension into the Balkans, the Pyrenees and northern Spain.

Values for I decline northwards to Scandinavia, to a minimum in Britain, and into the Mediterranean islands. This is not entirely surprising, as the denominator,

$$\sum_{j=1}^{m} Q_j,$$

will favour areas with high species diversity. Q_j, the average priority of areal faunas (Figure 15.3b), has a very different geographical pattern. High values are found not only in the southern montane belt but also in the arctic and north Africa. The lowest values are again in Britain; the Mediterranean islands have relatively higher values of Q_j than of I_i. High values of Q_j for the arctic do not correspond with any endemicity measures, but then Q_j is only assessed for areas within the sampling frame and does take account of wider distributions. Since individual states and larger political aggregations can cater only for the conservation of their own fauna, this may be a more realistic evaluation of the importance of arctic fauna. Baz (1991) suggests summing the two indices; on this basis, conservation priority is indicated for southern and eastern Europe relatively to the north and west.

More sophisticated indices based on the information content of cladistic classifications, and which take account of taxic diversity (Vane-Wright, Humphries and

Figure 15.3 Conservation indices for European butterflies; (a) importance index (I_i); (b) average priority index (Q_i). See text for definitions.

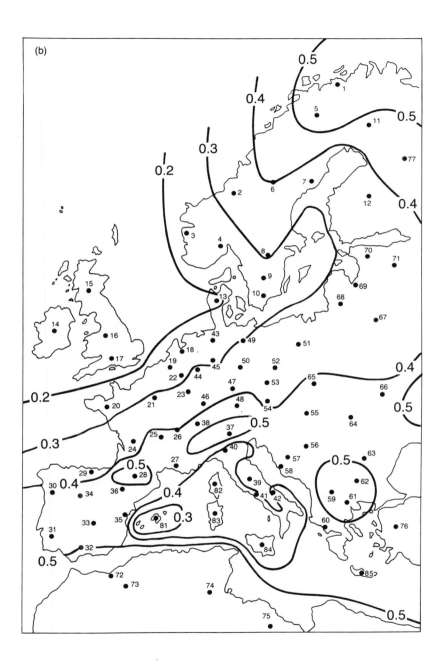

Figure 15.3 Continued.

Williams, 1991), cannot be applied to the European fauna as the essential taxonomic work has not been done.

15.3   ENVIRONMENTAL THREATS TO EUROPEAN BUTTERFLIES

Threats to the survival of European butterflies may be described as natural and human in origin. Natural factors would include such influences as climate, floods, fires and vegetation succession; human factors are perhaps more numerous but are broadly categorized into land-use changes (e.g. agriculture, forestry, industry, recreation and urban growth), pollution, collecting and trade. There is no strict division as human agencies can be responsible for changes that are also attributed a natural basis (e.g. greenhouse climate changes; flooding land for reservoirs, accidental or purposeful generation of fires, etc.). Natural environmental changes are generally scale dependent and are well described by events in log denary axes of space and time. For example, changes in whole vegetation and climate belts (10^6 km^2) occur at the scale of 1000–10 000 years, whereas those within the regeneration complex (10^{-2}–10^{-3} km^2) occur at the scale of 10–100 years. Modern human influences have overturned this relationship. Since AD 1800, human population growth and industrialized agriculture, both developing at an exponential rate, have threatened changes on spatial scales of 10^6 km^2 in time scales of 100–100 years. There is insufficient space to review these changes in any detail, but a brief discussion of natural long-term climatic cycles as well as short-term anthropogenically induced changes in climate and land use have direct relevance for conservation policy.

15.3.1   Milankovitch cycles, northern European *tabula rasa* and southern European refuges

Stratigraphic analysis of ice cores and deep-sea cores (Bartlein and Prentice, 1989; Cronin and Schneider, 1990) indicates that there have been some 46 glacial–interglacial cycles during the past 2.5 million years. These have been modelled by Milankovitch cycles (temporal cycles in the Earth's radiation receipt) in the Earth's orbit and axial inclination, and thus global insolation receipt (Berger and Loutre, 1991). Because of the markedly different climatic zones occupied by different faunal elements in Europe, these climatic changes must have had a profoundly different impact on butterflies of different regions. For distinctly northern elements (viz. arctic and boreal habitats) limiting conditions and refuges occur during shorter interglacials; for more southern elements, including the species-rich extent group, refuges are occupied during longer glacial phases. Apart from the far northern elements, most, if not all, populations of species north of the alpine mountain chain are erased in a glacial phase (Dennis, Williams and Shreeve, 1991). Most species currently north of the alps depend on the maintenance of glacial refuges in southern Europe. It has also been argued that

Mediterranean Europe and the southern mountain chain provide suitable conditions for ongoing evolution (i.e. stability, allopatry, diverse environments) during the Pleistocene, for example for *Erebia* species in the mountains and Lycaenidae in Iberia and the Balkans (Dennis, Williams and Shreeve, 1991; Dennis, 1993; see also Bennett, Tzedakis and Willis, 1991). It follows that the survival of species in southern Europe is fundamental to the long-term maintenance (10^4–10^7 years) of species in northern Europe.

15.3.2 Implications of increased output of greenhouse gases for European butterflies

The implications of the increased output of greenhouse gases for European butterflies are potentially serious (Dennis and Shreeve, 1991; Dennis, 1993). Even the most restrictive controls, in which contributory gas emissions are reduced to 50% of the 1985 levels by AD 2050, predict a doubling of radiative forcing by AD 2100, but by a factor of five for a 'business-as-usual' scenario (Houghton, Jenkins and Ephraums, 1990). The consensus from general circulation models (GCMs) of global mean temperature rise with doubling in CO_2 atmospheric concentration (ΔT_{2x}) is 1.5–4.5 °C, with a best estimate of 2.5 °C and *c.* 1.8 °C by AD 2030. All models demonstrate much greater rises in temperature for high latitudes; $\geqslant 8$ °C for arctic habitats (Woodwell, 1989; Mitchell *et al.*, 1990; Mitchell and Zeng, 1991). Accompanying these changes in temperature will be changes in precipitation and water balance, although regional predictions are less easily made. Increases in precipitation are more probable in northern Europe and decreases in southern Europe (Jones, 1990; Warrick and Barrow, 1991).

The fate of faunal elements is closely tied up with biomes to which they closely correspond. Particular concern has been expressed about the tundra and alpine communities which are predicted to contract at the expense of boreal taiga and montane coniferous forests; coniferous habitats lack butterfly associations. Topographic and edaphic variability (i.e. slopes with varied aspects, shading and drainage) will form an important component in the survival of arctic and montane butterflies. The effect of climate change on faunal elements depends largely on the vulnerability of component species and the speed of projected changes, which could rival the maximum rates of climatic change during the last glacial hemicycle (Dennis and Shreeve, 1991; Dennis, 1993). The faunal unit most likely to benefit from climatic change is the extent group which includes *c.* 40% of the entire European fauna. Many species belonging to this unit have the facility to spread north and west towards Scandinavia and Britain. Faunal units least likely to benefit are subalpine and Mediterranean zone endemics, for which temperature changes may cause further elimination and isolation of habitats, in the former case due to elevational shifts in vegetation zones and in the latter case due to drought. The UK Meteorological Office GCM projection of climate under ΔT_{2x} scenario for the Mediterranean region, reveals a large reduction in the growing season (i.e. months with mean monthly temperature >5 °C and rainfall $>$ potential evapotranspiration) (Brouwer, 1989) and advancing aridity. Although arctic and northern elements have an apparently wide resource base, they could easily contract, particularly as these areas become more suitable for cultivation and afforestation with conifers.

15.3.3   The erosion of butterfly habitats over Europe

Current trends would lead to a firm projection of continued habitat loss over Europe, for butterflies and for other organisms. As human activity varies spatially, in magnitude and in kind, it should be possible to harness knowledge from the human sciences, particularly geography, to predict where changes are most likely to occur and thus to establish the degree of threat to butterfly populations. Changes to the land, use or abuse, are largely the product of the potential for change. This will depend on indigenous human demographic parameters and stimuli in the form of wealth acquisition and creation. Certain land-use changes follow highly predictable patterns (e.g. urban sprawl around cities and along coastlines for tourism; new road networks circumventing towns and linking major settlements; changes in agricultural practices towards increased technology and intensity, larger units and fields); moreover much change is intended policy and planned, as in the European Community (EC) directives for developing countries in southern Europe (COM 1984, 1985). There is insufficient space to discuss this issue at length, but three points make this topic of paramount importance:

1. significant 'land-use' changes have already occurred in zones of high species richness and endemicity in southern Europe (Kudrna, 1986);
2. steep terrain is not an obstacle to some land uses and can positively encourage it (e.g. tourism, recreation); and
3. relatively low gross national product (GNP) does not bar growth and change, much EC funding and foreign investment has been placed in Mediterranean Europe (Wallis, 1986).

Thus butterfly populations in southern Europe are not secure by being placed in a relatively underdeveloped zone. The reverse, in fact, may be the case as it may be argued that development has a higher priority in this region. Higher priority for development usually equates with lower priority for wildlife, and over much of southern Europe conservation has a lower profile in the public and political domain than it has in the north-west.

15.4   THE NEED FOR A CO-ORDINATED POLICY TOWARDS CONSERVING EUROPEAN BUTTERFLIES

A number of compelling reasons exist for a more co-ordinated wildlife conservation policy and programme for Europe beyond initiatives within state boundaries. It should facilitate the standardization of taxonomy for organisms; it should reduce ambiguity as to the status of species, biogeographical and conservational; it should rationalize the legal requirements as regards land-use changes, collecting and trade. It could, through a pooling of experience and expertise, and the application of site selection techniques over wider regions (see Margules, 1989;

Margules and Stein, 1989; Pressey and Nicholls, 1989a,b), determine more effective 'rescue' programmes for threatened habitats and taxa. Clearly, conservation needs to develop on the same plane as economic development in the European Community, so that potential for change can be challenged and matched by appropriate conservation measures.

There are at least two aspects of European butterfly biogeography that make a co-ordinated policy towards conservation highly desirable. First, is the obvious point concerning the mismatch of species' ranges and state boundaries; piecemeal surveys, policies and programmes within state boundaries are likely to be less successful, at least less efficient, for the effective conservation of taxa and habitats than those that account fully for their wider variability and geographical status. The second point is less obvious but is fundamental to the long-term survival of the European fauna. The bulk of populations and genetic variability belonging to north European taxa, that is of the numerically large extent element, have a life span restricted to each interglacial phase (20 000–40 000 years); the replenishment of populations and variability depends on the long-term maintenance of taxa in southern European refuges. These have persisted throughout the Quaternary (2.5 million years) (Dennis, Williams and Shreeve, 1991). Irrespective of this issue, the southern part of Europe also has the greatest species diversity, the majority of endemics, the most distinct faunal elements and scores highest on conservation indices. Yet, most applied research for the conservation of butterflies is undertaken in northern Europe not in the south. It seems entirely reasonable that the allocation of resources for conservation of fauna over Europe should correspond with measures of taxonomic priority and that the interchange of assistance should equate with support that is given to economic development.

15.5   ACKNOWLEDGEMENTS

We extend our grateful thanks to Otakar Kudrna and Andrew Pullin for reviewing this paper and for their helpful comments.

CHAPTER 16

Measuring changes in butterfly abundance in The Netherlands

C.A.M. van Swaay

16.1 INTRODUCTION

The Dutch Butterfly Foundation (DBF, De Vlinderstichting) was founded in 1983. Its main aim is to maintain and restore the Dutch butterfly fauna. Starting with a small group of enthusiastic volunteers, it soon grew to become the focus for butterfly research in The Netherlands (together with the Agricultural University of Wageningen). At present the foundation employs 12 staff. The activities can be divided in two main categories (Veling and van Swaay, 1992):

16.1.1 Research and advice

At this moment three general research projects are being carried out by the Dutch Butterfly Foundation:

1. Collection of detailed distribution data of all Dutch butterflies and day-active moths. Ecological data are also gathered, particularly for rare and characteristic species.
2. Dutch Butterfly Monitoring Scheme (DBMS): monitoring the changes in the abundance of Dutch butterflies. This project can be compared with the British Butterfly Monitoring Scheme (BMS).
3. First sightings: by collecting the first three observations of common butterflies a lot of people are stimulated to look for butterflies. The data can be used to study the effect of weather on the start of the flight period of butterflies.

The data collected in these research projects are immediately used to give advice in nature management and urban and rural planning to achieve maintenance and restoration of butterfly fauna. Advice is given in the following forms:

Ecology and Conservation of Butterflies Edited by Andrew S. Pullin.
Published in 1995 by Chapman & Hall. ISBN 0 412 56970 1

1. Courses in 'butterfly-friendly management' are given to planners and managers of nature reserves, public spaces in towns and road verges in the countryside.
2. With the help of the detailed information on distribution and ecology of butterflies, the DBF can make precise management plans for nature reserves. These plans use the butterfly fauna as example species, to achieve a better management for all flora and fauna.
3. The butterfly data are used in 'environmental impact studies', in which the effect of new developments on the environment is studied. If necessary, fieldwork is done to test the data.

16.1.2 Information and education

To achieve a change in planning and management, information must be made available to politicians, planners, landowners and the general public. The DBF produces guides, leaflets, posters, travelling exhibitions, slide-series and brochures on butterflies in different Dutch landscapes. Lectures are given for many different groups, the general public is informed on how they can create their own butterfly gardens and the DBF has frequent contacts with the press.

By interesting schoolchildren in butterflies, many people, including parents, can be reached. Lessons are produced on the ecology and biology of butterflies, with emphasis on their behaviour and how we can take concrete action for their conservation. Live material (eggs, larvae, pupae) is sent to schools to be bred in class situations. Common and widespread species such as *Aglais urticae* (small tortoiseshell), *Inachis io* (peacock) and *Pieris brassicae* (large white) are used, and can be released by the children after successful breeding.

16.2 RESEARCH METHODS

16.2.1 Calculating butterfly abundance

In 1981 the Dutch Butterfly Mapping Scheme began at the Nature Conservation Department of the Agricultural University of Wageningen. The whole country was investigated for the occurrence of butterflies, resulting in the Dutch butterfly atlas (Tax, 1989). Not only new observations were collected. Almost all museum and private butterfly collections were visited. Also, all the literature on Dutch butterflies was searched for observations (van Swaay, 1991). This resulted in a database with 106 000 'old' records (until 1980) and 238 000 'new' records. In Figure 16.1 the number of records for each 5 year period is given.

This material offered a tool for investigating the historical changes in the Dutch butterfly fauna (van Swaay, 1990).

A new method was developed by van Swaay (1990) to study changes in the abundance of butterflies, which enabled analysis of the period to 1980. Until that date there were only minor and slow changes in the way butterfly abundance was recorded and the method could easily correct these changes. One of the greatest

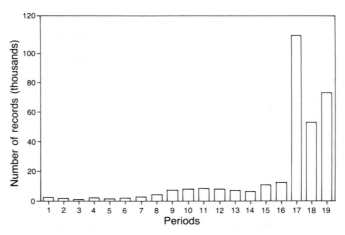

Figure 16.1 Number of records for each 5 year period studied (1 = 1901–05; 2 = 1906–10 . . . 18 = 1986–90; 19 = 1991–92).

disadvantages of this method was that it could not be used for the periods after 1980, when the DBMS began (see Figure 16.1). This meant that a good comparison of the changes after 1980 with the changes before 1980 was very difficult.

Latour and van Swaay (1992) produced a new and relatively simple method to calculate and compare changes in abundance before and after 1980 in a part of The Netherlands. This method is now used for the whole country. Common species with distributions that were more or less unchanged between 1901 and 1990 were selected. The average abundance per 5 year period was then determined for each by calculating the average number of squares where these species were seen. This is a measure of the investigation intensity. The number of squares per 5 year period were then calculated for all other species and results were corrected for the investigation intensity, giving the relative abundance.

The criteria for the common species which were selected to determine the investigation intensity were that they had to be recorded in more than 50% of the squares equally distributed over the whole country. From these species the results of van Swaay (1990) were used to calculate the regression line over the period 1 (1901–05) to 16 (1976–80). The results are given in Table 16.1.

The species also had to be common at the beginning of the century: the intercept should be greater than 10 (which means that the species was seen in more than 10% of the investigated squares). Thirdly, they should not have fluctuated too much: $R^2 > 0.10$; and, lastly, they shouldn't have increased or decreased notably – the direction coefficient should be between -1 and $+1$.

Only three species fulfilled these criteria: *Coenonympha pamphilus* (small heath), *Lycaena phlaeas* (small copper) and *Maniola jurtina* (meadow brown). Table 16.2 gives the number of squares where these species have been recorded per period.

The average is used as a correction factor for the other species. Table 16.3 gives the results.

Table 16.1 Regression line of the results of van Swaay (1990) in the periods 1–16; only resident species that occurred in The Netherlands in more than 50% of the squares equally distributed over the country are treated

Species	R^2	Intercept on y-axis	Direction
Thymelicus lineola	0.43	6.0	0.35
Pieris brassicae	0.93	−2.9	1.27
Pieris rapae	0.89	−3.7	1.51
Pieris napi	0.90	−1.9	1.29
Gonepteryx rhamni	0.75	4.3	1.03
Lycaena phlaeas	0.29	14.9	0.39
Polyommatus icarus	0.05	17.7	0.15
Celastrina argiolus	0.09	11.8	0.18
Inachis io	0.71	−3.5	1.42
Aglais urticae	0.78	−5.3	1.67
Araschnia levana	0.51	−5.1	1.55
Coenonympha pamphilus	0.57	11.9	0.71
Lasiommata megera	0.57	6.29	0.78
Maniola jurtina	0.28	13.7	0.30

Table 16.2 Number of squares where *Coenonympha pamphilus*, *Lycaena phlaeas* and *Maniola jurtina* have been observed per 5 year period; the calculated average is used to correct for the investigation intensity

Period	C. pamphilus	L. phlaeas	M. jurtina	Average
1901–05	31	37	30	33
1906–10	18	21	19	19
1911–15	18	21	17	19
1916–20	31	27	25	28
1921–25	21	21	22	21
1926–30	36	40	35	37
1931–35	28	41	39	36
1936–40	65	56	60	60
1941–45	90	85	77	84
1946–50	98	90	80	89
1951–55	74	102	82	86
1956–60	104	112	97	104
1961–65	87	81	84	84
1966–70	109	93	74	92
1971–75	174	145	120	146
1976–80	145	133	110	129
1981–85	896	762	692	783
1986–90	437	383	374	398
1991–92	486	460	475	474

Table 16.3 Relative abundance of butterflies in The Netherlands over the 5-year periods 1901–1905 (1) to 1986–1990 (18) plus 1991–1992 (19)

Species	1	2	3	4	5	6	7	8	9	10	11	12	13	14	15	16	17	18	19
Heteropterus morpheus	18	47	43	4	23	14	14	7	10	7	7	4	2	2	4	5	3	2	2
Carterocephalus palaemon	67	93	54	33	66	43	39	48	46	11	15	17	19	23	12	5	5	11	7
Thymelicus sylvestris	49	67	27	58	33	38	61	51	45	39	51	35	55	32	50	46	31	27	38
Thymelicus lineola	6	5	5	58	9	11	8	7	5	57	57	54	60	48	55	56	62	68	74
Thymelicus acteon		5	5	7	47	22	39	20	27	9	5	2	4	2	1	0			
Hesperia comma	52	52	54	54	94	62	75	73	63	40	38	21	37	24	14	15	9	15	11
Ochlodes venata	77	83	86	87	42	14	14	15	12	65	57	66	101	78	58	60	59	63	59
Erynnis tages	58	52	32	47	84	3	6	5	2	21	20	11	5	4	2	1	0	0	0
Spialia sertorius	3	72	11	4	42	43	36	48	52	2	2	3	7	4	2				
Pyrgus malvae	64	67	70	94	70	35	111	93	86	35	43	31	30	21	11	10	5	6	3
Papilio machaon	40	41	96	94	52	73	81	65	48	96	56	26	26	41	42	32	9	14	34
Gonepteryx rhamni	40	98	64	61	28	22	28	41	37	90	62	66	90	93	96	127	100	117	137
Aporia crataegi	34	10	38	65	38	19	33	20	37	38	31	19	12	4	7	1	2	2	1
Pieris brassicae	3	5	11	22	28	22	25	30	33	49	56	58	70	88	79	97	105	115	114
Pieris rapae	6	36	11	29	70	32	33	36	36	47	76	67	86	90	105	107	125	141	140
Pieris napi	6	129	11	25	42	57	81	68	63	46	60	68	86	92	78	106	108	127	134
Anthocharis cardamines	67	72	59	87	33	38	50	48	37	53	59	47	51	96	76	91	50	67	68
Callophrys rubi	43	57	54	61	66	14	22	25	11	41	50	42	65	64	31	41	25	27	29
Thecla betulae	24	83	16	29	75	49	50	46	20	13	24	8	10	9	10	5	3	3	4
Quercusia quercus	64	72	38	72	5	32	53	58	39	30	47	31	52	48	31	49	27	32	35
Nordmannia ilicis	83	10	54	69		11	3	3	1	31	50	28	46	36	20	30	9	11	12
Satyrium w-album	3	10	4	4				3		1	1	2	2			5	0		
Satyrium pruni										2					1				
Lycaena phlaeas	113	109	112	98	98	108	114	93	101	101	119	107	96	101	99	103	97	96	103
Lycaena dispar			11	14	14	11	17	8	5	9	13	12	12	12	8	6	2	1	1
Heodes tityrus	86	119	64	108	38	43	61	63	63	77	84	59	60	35	28	30	22	23	13
Palaeochrysophanus hippothoe	21	16				3	11		5	2			1				0	0	
Cupido minimus	3	5	5	25	28	3	6	2	4	2	1		6	2		2	1		
Plebejus argus	46	41	64	36	14	43	44	43	31	41	2	37	63	55	39	30	19	27	20
Lycaeides idas	9	5	5	4	47	8	3	3	2		55	4		2	1	4	0		
Vacciniina optilete	3	10								2	3	1	1	3	2		1	1	0
Aricia agestis	52	57	48	47	28	24	25	35	43	53	51	31	20	23	18	19	18	20	16
Polyommatus icarus	107	93	102	152	131	124	122	136	100	114	122	97	89	96	98	78	86	83	77
Cyaniris semiargus	31	52	32	61	23	16	33	18	10	21	12	1	2	4	5		0	1	
Celastrina argiolus	67	93	86	108	84	70	89	80	52	62	80	93	61	108	44	73	65	85	95

Table 16.3 Continued

Species	1	2	3	4	5	6	7	8	9	10	11	12	13	14	15	16	17	18	19
Maculinea alcon ericae	31	10	5	18	19	3	14	17	14	28	24	20	33	25	27	19	11	6	10
Maculinea alcon arenaria				11	5		6	2	2	1	2	3	1	1	1	1			
Maculinea arion	9	5	5	18	5	3	3	2	1	2	7	10	4	3		21			
Maculinea teleius	21	26	16	18	9	5	8	5	6	11	13	6	4	3	1	1	3	1	
Maculinea nausithous	12	21	11	18	14	5	8	3	5	10	43	22	21	17	14	12	11	12	
Ladoga camilla	43	47	21	14	33	41	67	51	43	27		1		1	1	1	3	1	0
Limenitis populi		5	16	11	5	8	11	3	2	1		8	5	1	3	1		1	0
Apatura iris	24		11	29	5	16	17	12	11	10	8	8	5	9	3	12	3	5	3
Aglais urticae	49	114	27	61	14	19	28	38	46	41	47	63	64	88	124	124	123	133	155
Nymphalis polychloros	40	47	64	43	84	51	36	80	55	40	41	14	19	5	9	17	2	6	1
Nymphalis antiopa	9	21	23	33	66	32	31	36	31	32	31	27	7	5	4	9	3	6	2
Inachis io	49	16	32	14	5	11	28	36	26	44	55	47	55	85	100	139	117	130	147
Polygonia c-album	6		11	4	23	59	67	63	58	92	78	32	58	65	44	37	33	54	87
Araschnia levana	86	10	5	7	5	5	8	12	24	71	199	58	48	80	57	118	75	90	96
Boloria aquilonaris	24	124	91	90	84	3	6	60	5	2	2	2	2	8	3	5		1	1
Boloria selene	55	36	38	29	47	38	69	2	60	66	60	56	25	37	19	10	6	9	6
Boloria euphrosyne	31	88	129	76	108	3	6	65	98	132	3	2	73	1					
Issoria lathonia	52	36	11	25	33	35	61	22	27	31	80	79	13	62	33	19	11	12	14
Fabriciana niobe	24	47	27	18	30	30	33	22	27	27	28	29	11	18	10	10	5	8	5
Argynnis aglaja	43	21	16	14	61	38	28	22	31	27	28	19	18	12	9	3	3	5	3
Argynnis paphia	43	62	27	47	42	22	25	17	20	22	14	10	5	14	5	9	3	2	3
Eurodryas aurinia	21	109	16	14	38	24	11	20	17	15	7	7	5	3	1	1	0		
Melitaea cinxia	46	10	48	11	14	24	6	5	18	15	21	4	1	5	5	2	0		
Melitaea diamina	24	83	5	11	9	5	3	2	1	3	2	2				5		2	3
Mellicta athalia	64	36	32	36	61	27	50	18	21	24	29	16	21	9	8	5	2	2	3
Pararge aegeria	61	52	59	54	33	19	50	36	46	69	80	55	50	53	51	62	51	65	69
Lasiommata megera	12	83	48	101	52	35	83	85	69	71	72	61	73	99	100	96	106	103	116
Aphantopus hyperantus	15	27	27	58	33	51	42	73	49	51	51	47	79	53	53	39	50	51	57
Coenonympha hero	12	5	5	7	14	3	3	8	1	1	2	1		3	3	2		2	
Coenonympha arcania	15	21	16	112	98	97	78	108	107	110	86	100	104	118	119	112	114	110	97
Coenonympha pamphilus	95	93	96	33	38	24	22	23	19	31	16	11	30	20	10	12	5	3	1
Coenonympha tullia	28	62	38	65	70	32	39	43	35	60	59	48	56	46	43	53	50	54	57
Pyronia tithonus	61	72	75	90	103	95	108	99	92	90	95	93	100	80	82	85	88	94	100
Maniola jurtina	92	98	91	76	89	70	75	80	73	74	63	60	87	66	44	58	33	40	40
Hipparchia semele	67	83	48		5			3	6	2	7	3	4	4	3	2	1	2	3
Hipparchia statilinus	6	5																	

16.2.2   Effects of climate change

The relative abundance of the 5 year periods between 1901 and 1980 is also used to study the effects of the climate on butterflies this century. The weather over the 5 year periods was calculated from data from KNMI (1985) for each season (spring: March–May; summer: June–August; autumn: September–November; winter: December–February). For spring and summer the average maximum temperature, the number of days with a maximum temperature higher than 20 °C and the precipitation were used; and for autumn and winter, the average maximum temperature and the precipitation.

For each period of 5 years a Pearson correlation was calculated to compare these parameters to the abundance of the butterflies.

16.2.3   Monitoring scheme

In 1989 the DBF began a butterfly monitoring scheme using recording methods essentially the same as those used in Britain. Only part of the 1989 season was recorded, but from 1990 onwards the complete recording seasons (April–September) were covered: in 1990 there were 93 sites, in 1991 162 sites and in 1992 224 sites. Sixty-six sites were monitored in all years.

16.3   RESULTS

16.3.1   Butterfly abundance

Table 16.3 clearly shows that the decrease of many species has not stopped. A few examples can be used to illustrate this.

(a)   Heodes tityrus *(sooty copper)*

This butterfly was fairly common in The Netherlands in the first part of the twentieth century (Figure 16.2). The relative abundance was always between 43 (1926–30) and 119 (1906–10). After 1950, the species disappeared in the south and west, with the exception of a few dune populations. After 1986, *H. tityrus* also became rare in the north and east. The relative abundance has now sunk to 13. Not only is the number of sites is decreasing: on the monitoring sites the number of individuals has also decreased every year since 1990 (Table 16.4).

(b)   Boloria selene *(small pearl-bordered fritillary)*

This used to be a typical species of wet, nutrient-poor grasslands and could be found all over the country (Figure 16.3). With the modernization of agriculture many of these grasslands have been 'improved', making them unfit for butter-flies. Nutrient-poor grasslands only remain in nature reserves. This slow process resulted in a steady decline of this species. Now only a few good

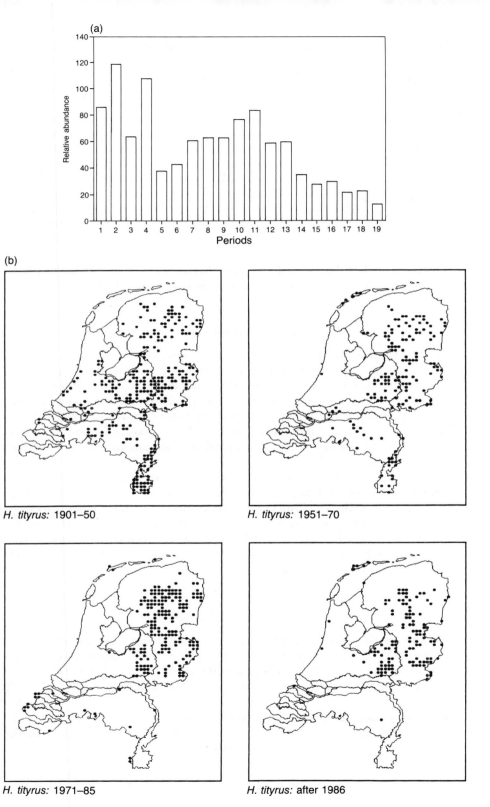

Figure 16.2 (a) Relative abundance and (b) distribution of *Heodes tityrus* in The Netherlands (periods as in Figure 16.1).

Table 16.4 Number of butterflies seen on the 66 sites that were monitored in 1990, 1991 and 1992

	1990		1991		1992	
Species	*sites*	*individuals*	*sites*	*individuals*	*sites*	*individuals*
Heodes tityrus	14	230	9	34	8	17
Boloria selene	3	359	3	283	3	414
Argynnis aglaja	3	8	3	61	5	197
Coenonympha tullia	2	26	1	3	1	1
Araschnia levana	48	1102	37	622	43	1145

populations are left. The relative abundance has decreased from 124 (1906–10) to six after 1990. The numbers on the monitoring sites are still high.

(c) Argynnis aglaja *(dark-green fritillary)*

This was never as common as *H. tityrus* or *B. selene*, but could still be found in many locations at the beginning of the twentieth century (Figure 16.4). The species occurred in two types of habitat: dry, sandy grasslands and heathlands, and wet, nutrient-poor grasslands (together with *B. selene*). By the middle of the twentieth century it had disappeared from the latter habitat and could only be found in the dunes and the Veluwe (middle of the country). Since then it has disappeared from the dunes on the mainland and can now only be found on the islands in the Wadden Sea.

Interestingly, the numbers on the monitoring sites have risen considerably. This is because it can occur in large numbers on thistle-fields on heathlands. In some years there are hardly any thistles on the monitoring sites, and in other years they are very abundant. This automatically means that *A. aglaja* is very numerous in some years and can be rare (on a site) in other years, but just a few metres away from the site it can be very common. Therefore the DBMS is not a good monitoring tool for this species.

(d) Coenonympha tullia *(large heath)*

The number of locations where this species can be found is declining year by year (Figure 16.5). Until recently, peat bogs, the habitat of this butterfly, were being converted to agricultural land. In the past 10 years this species has suffered from the drying out of the remaining peat bogs, which are nature reserves.

Nevertheless not all species have declined. *Araschnia levana* (European map butterfly) colonized The Netherlands in less than 10 years (Figure 16.6). Nowadays it can be found all over the country.

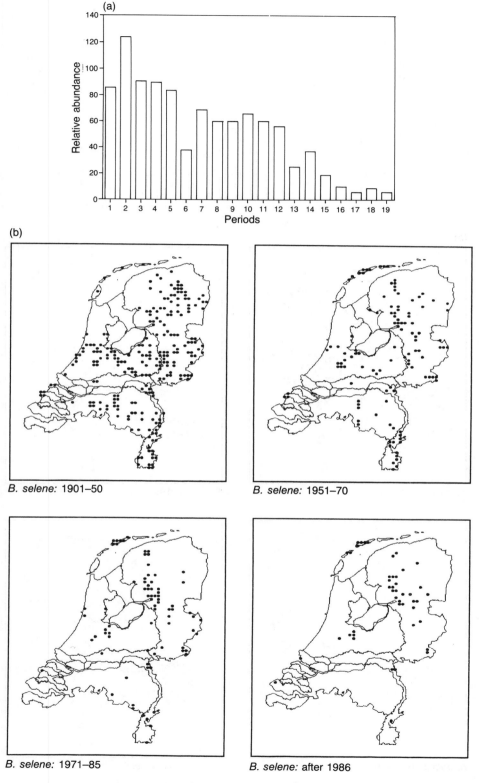

Figure 16.3 (a) Relative abundance and (b) distribution of *Boloria selene* in The Netherlands (periods as in Figure 16.1).

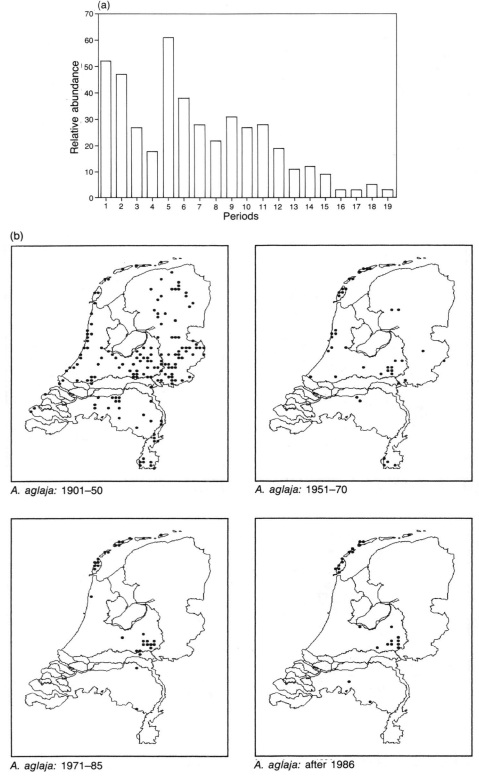

Figure 16.4 (a) Relative abundance and (b) distribution of *Argynnis aglaja* in the Netherlands (periods as in Figure 16.1).

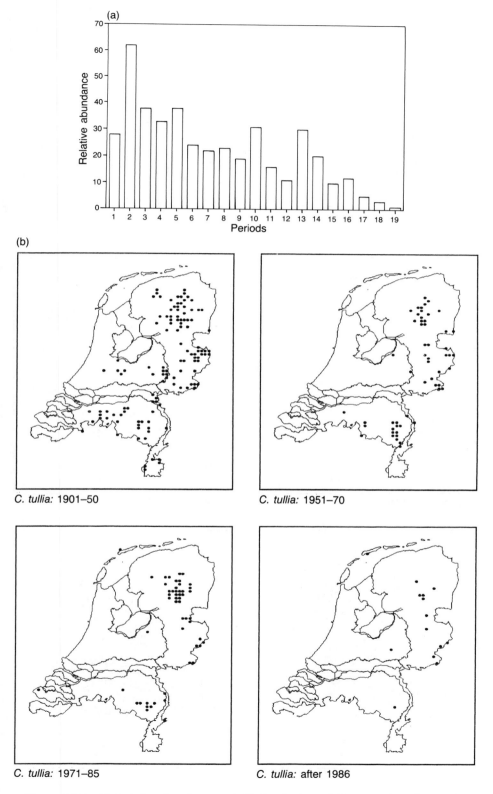

Figure 16.5 (a) Relative abundance and (b) distribution of *Coenonympha tullia* in The Netherlands (periods as in Figure 16.1).

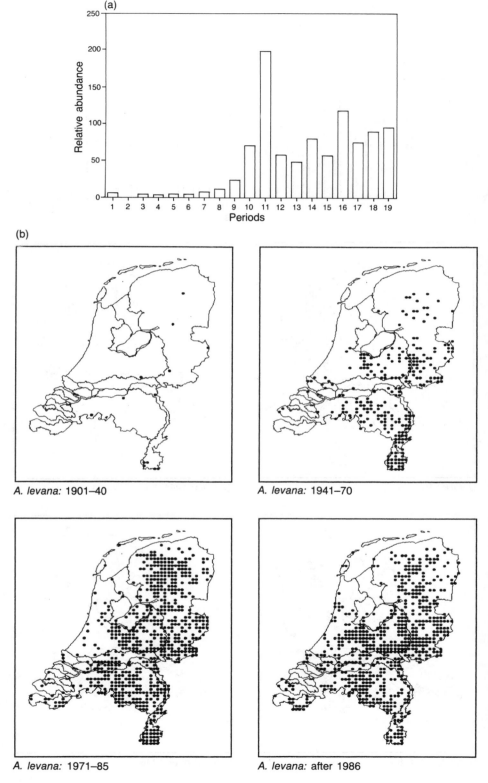

Figure 16.6 (a) Relative abundance and (b) distribution of *Araschnia levana* in The Netherlands (periods as in Figure 16.1).

16.3.2 Climatic change

The results of the correlation between butterfly abundance and climatic parameters are given in Table 16.5. Six groups of species can be distinguished:

(a) *Species correlated with a warm or dry spring (group I)*

Thymelicus acteon (Lulworth skipper) is also correlated with the number of days in summer with a temperature higher than 20 °C. Only *Fabriciana niobe* (niobe fritillary), *A. aglaja* and *Nymphalis antiopa* (Camberwell beauty) are correlated with dry spring conditions, the latter also with warm spring weather.

(b) *Species correlated with a warm summer (group II)*

Thymelicus acteon, *Pontia daplidice* (bath white) and *Papilio machaon* (swallowtail) (all also correlated with a warm spring). *Colias croceus* (clouded yellow), a migrant species in The Netherlands, can be very abundant when a warm autumn follows a warm summer. No species are correlated with summer precipitation.

(c) *Species correlated with the autumn precipitation (group III)*

Melitaea cinxia (Glanville fritillary) favours a dry, and *Polyommatus icarus* (common blue) a wet, autumn.

(d) *Species correlated with a cold or dry winter (group IV)*

Hipparchia semele (grayling) is the only species correlated with a cold winter.

(e) *Species correlated with a cold or wet spring (group V)*

Except for *Callophrys rubi* (green hairstreak) and *Ochlodes venata* (large skipper) spring precipitation is not very important. All of these species are abundant in The Netherlands and the rest of western Europe, or have even expanded their range.

(f) *Species not correlated with any of the climatic parameters (group VI)*

The reasons for their changes in abundance are obviously not the (small) climatic changes of this century.

16.4 DISCUSSION

16.4.1 Butterfly abundance

Compared to the situation as described by van Swaay (1990) the deterioration of the Dutch butterfly fauna still goes on. From the 71 native species 51 (72%) have declined, of which 15 species are probably extinct. Two other species, *M. cinxia* and *Coenonympha arcania* (pearly heath), which have not been recorded since 1986 and 1987 respectively, will soon have to be considered extinct. In 1993 a very small population of *Cyaniris semiargus* (mazarine blue) was found, but it is too early to say if this population will be able to maintain itself.

Table 16.5 Correlations between the abundance of butterflies in The Netherlands from 1901–1980 and three climatic parameters

Species	Spring T	Spring D	Spring P	Summer T	Summer D	Summer P	Autumn T	Autumn P	Winter T	Winter P
Group I										
Nymphalis antiopa	+	+	+							
Thymelicus acteon	+				+					
Issoria lathonia	+									
Aporia crataegi		+								
Pyrgus malvae		+								
Cyaniris semiargus		+								
Argynnis aglaja			+							
Fabriciana niobe			+							
Group II										
Pontia daplidice	+				+	+				
Papilio machaon	+				+	+				
Colias croceus				+				+		
Ladoga camilla					+					
Polyommatus icarus					+				−	
Group III										
Melitaea cinxia								+		
Group IV										
Hipparchia semele									−	
Mellicta athalia										+
Melanargia galathea										+
Thymelicus sylvestris										+
Group V										
Callophrys rubi	−	−							−	
Anthocharis cardamines	−									
Quercusia quercus	−									
Inachis io	−									
Thymelicus lineola	−	−								
Pieris brassicae	−	−								
Pieris rapae	−	−								
Pieris napi	−	−								
Aphantopus hyperantus	−	−								
Gonepteryx rhamni	−									
Aglais urticae	−									
Cynthia cardui	−									
Vanessa atalanta	−									

Table 16.5 Continued

Species	Spring			Summer			Autumn		Winter	
	T	D	P	T	D	P	T	P	T	P
Group V – continued										
Lycaena phlaeas		−								
Plebejus argus		−								
Maniola jurtina		−								
Ochlodes venata		−	−							
Group VI										
Hesperia comma										
Erynnis tages										
Carterocephalus palaemon										
Colias hyale										
Thecla betulae										
Nordmannia ilicis										
Lycaena dispar										
Heodes tityrus										
Aricia agestis										
Cupido minimus										
Celastrina argiolus										
Maculinea arion										
Maculinea nausithous										
Maculinea teleius										
Apatura iris										
Limenitis populi										
Araschnia levana										
Polygonia c-album										
Nymphalis polychloros										
Argynnis paphia										
Boloria selene										
Melitaea diamina										
Eurodryas aurinia										
Pararge aegeria										
Lasiommata megera										
Pyronia tithonus										
Coenonympha tullia										
Coenonympha arcania										
Coenonympha pamphilus										

T, temperature.
D, number of days with a temperature higher than 20 °C.
P, precipitation. In this case + means a positive correlation with a low precipitation (dry season).

This makes the butterflies one of the most threatened group of animals in The Netherlands.

16.4.2 Climate change

In the period 1901–80 the weather varied considerably (even when combined to periods of 5 years). A correlation of a species in Table 16.5 with one of the climatic parameters does not necessarily mean that the climate was the direct cause of the changes. It may well be possible that the climate influenced the habitat of the butterfly, which in turn caused the changes in the abundance of the butterfly. Another important factor to keep in mind is the fact that a change in climate may coincide with a change in land use. In The Netherlands many butterfly habitats have declined or even disappeared since the spread of modern intensive agriculture in the 1950s and 1960s (van Swaay, 1990). In this period some climatic parameters changed considerably: the temperature both in spring and summer was considerably lower than in the previous period and the precipitation much higher. Therefore we cannot be sure if the change in abundance was induced by the changes in climate or by the changes in land use.

On the other hand, there could be species that are very sensitive to climatic changes in The Netherlands, but which are not correlated to one of the investigated parameters, perhaps because another factor, such as the changed land use, was overwhelming.

The results of this investigation are only applicable in The Netherlands. Of the 65 investigated species, 30 (46%) are correlated with at least one of the climatic parameters investigated. Many species reach the edge of their range here, and this means that these species will react very quickly to changes in climate. Towards the centre of their range this response may be greatly reduced. In other areas of Europe different species may be sensitive to climate changes.

The weather in spring seems to be especially important for many species. The species of group I are all species with a more continental distribution, except for *A. aglaja* which is found over almost the whole of Europe. Notably *N. antiopa* needs a warm and dry spring. This butterfly, which hibernates as an adult, reproduces during this period. Nevertheless, other species hibernating as adults, such as *I. io*, *A. urticae* or *Gonepteryx rhamni* (brimstone), are not so sensitive. *Argynnis aglaja* and *F. niobe* can be found in approximately the same habitat in The Netherlands, especially the coastal dunes and inland heathlands. These species, which hibernate as small larvae, are sensitive to wet spring conditions.

The species in group V are negatively correlated with warm spring conditions. All these species are still fairly common in The Netherlands, and almost all of them also in the British Isles. This means that they are quite resistant to the cool springs of the atlantic climate. *Vanessa atalanta* (red admiral) and *Cynthia cardui* (painted lady) are immigrants from the Mediterranean or even Africa. Maybe cool spring conditions in western Europe coincide with optimal weather for the development of these species in the south of Europe. Another reason for the negative correlation of all the species of group V with warm spring conditions is that in a cool spring the larval food-plants are in a better condition for the larval development. This caused the decline of *Aphantopus hyperantus* (ringlet) after the warm, dry year 1976. In subsequent years, which were cooler and wetter, the population recovered quickly (Pollard, Hall and Bibby, 1986).

No species is positively correlated with a cool summer, but some butterflies show a very strong correlation with warm summer conditions, especially *P. daplidice* and *P. machaon.* These two species also need warm spring conditions. This means that they were relatively abundant in the warm 1940s, but almost disappeared in the cool 1960s. The correlation of *Ladoga camilla* (white admiral) with good summer weather was also established by Pollard (1979) in England, who found a good correlation between the early summer weather and the timing of the spread of *L. camilla* this century. The June temperature could be used as an indirect measure of the speed of development. This relationship was believed to be between the duration of the late larval and pupal stages and the mortality of these stages. Successive years of low mortality would result in a rapid increase in population (Pollard, 1979).

Only three species show a correlation with autumn weather. In periods with many warm summers *C. croceus* can be abundant, and when September and October are also warm the population can get very large, before it crashes after the first cold nights. Why *P. icarus* is correlated with a wet autumn is unclear. In autumn *M. cinxia* forms larval nests on *Plantago lanceolata* (ribwort plantain) near to the ground. This species is apparently favoured by dry autumn weather.

Three species show a correlation with dry winters. It is surprising that no species is negatively correlated with wet winters, since in some entomological papers it is stated that disease of the larval stages is favoured by moisture. *Hipparchia semele* and *C. rubi* are correlated with cold winter temperatures.

Group VI contains those species for which no correlation could be found. Of course this doesn't mean that they are not influenced by a changing climate. More likely other factors will also have played a major role in their change of abundance. Most of these have already been discussed by van Swaay (1990).

16.5 CONCLUSIONS

Biotope destruction has been the most important reason for the strong decrease in the Dutch butterfly fauna (van Swaay, 1990). Species that were typical of nutrient-poor grasslands have declined most conspicuously. In the beginning of the twentieth century this type of biotope was still commonly used as agricultural grassland. Nowadays these can only be described as 'green deserts', where no butterflies can survive. Therefore most rarer species are now restricted to nature reserves. Here they are threatened by acid rain and the sinking of the ground water level.

Almost half of the investigated species are correlated with a climatic parameter. More towards the centre of Europe this percentage will be lower, since many species do not live so close to the edge of their distribution as in The Netherlands. But it is clear that climate can play an important role in the distribution of butterflies. These data can be used to predict the effects of the expected future climatic changes.

CHAPTER 17

Conservation of butterflies in central Europe

O. Kudrna

17.1 INTRODUCTION

For the purpose of this chapter, the present central Europe consists of the following countries: the former Federal Republic of Germany, Federal Republic of Austria, Czech Republic, Republic of Hungary, Slovak Republic and the Republic of Poland (Figure 17.1).

It is an arbitrary designation and one could argue about the exclusion of north-western Germany or about the inclusion of the Republic of Slovenia. Historically and politically the latter belongs to central Europe, but biogeographically it belongs to southern Europe or to the group of alpine countries. Austria, being an alpine country, will be referred to only briefly here.

The geographical term 'central Europe' has changed its meaning many times throughout history and, in addition to this, has its political, cultural and perhaps even religious aspects. The Iron Curtain of the cold war period cut across the former central Europe and only the Federal Republic of Germany and Austria became politically and economically integrated parts of the Western world. The remaining countries were then usually referred to as the eastern or communist block. The disintegration of the communist block in 1989 put an end to the bipolar division of Europe and the term central Europe regained full meaning.

Aware of the extremely serious state of environmental pollution in the central European countries of the former Soviet block, the Gesellschaft für Schmetterlingsschutz organized a small conference 'Schutz der Tagfalter im Osten Mitteleuropas: Böhmen, Mähren, Slowakei, Ungarn' for invited experts in Oberelsbach (Germany: NW Bavaria) on 1 October 1990 (Kudrna, 1991), within the 3. Rhöner Symposium für Schmetterlingsschutz, with the aim of discussing the situation and of trying to find solutions to outstanding issues in the form of a 'first-aid' programme.

Ecology and Conservation of Butterflies Edited by Andrew S. Pullin.
Published in 1995 by Chapman & Hall. ISBN 0 412 56970 1

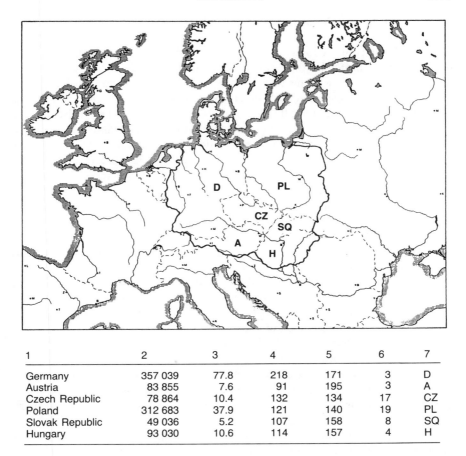

1	2	3	4	5	6	7
Germany	357 039	77.8	218	171	3	D
Austria	83 855	7.6	91	195	3	A
Czech Republic	78 864	10.4	132	134	17	CZ
Poland	312 683	37.9	121	140	19	PL
Slovak Republic	49 036	5.2	107	158	8	SQ
Hungary	93 030	10.6	114	157	4	H

Figure 17.1 Central European countries. In the table below, values are given for: (2) area in km^2; (3) population in millions; (4) population density per km^2; (5) number of butterfly species present; (6) number of extinctions. (7) Abbreviations used on the map.

It is a long-established fact that butterflies have been seriously declining in diversity of species and in the number of their populations in most of Europe. The reasons for their decline in central Europe are well known (Blab and Kudrna, 1982; Kudrna, 1986) and need not be discussed here. Nor is the aim of this chapter to provide general information on the organization of and legislation concerning the conservation of nature. The chief aim is to show what is being done in central Europe for the conservation of butterflies.

Some readers might find this review biased; it is not exhaustive and concentrates on aspects deemed more important by the author. I do not believe that glorification of moderate successes to please some readers can positively contribute towards the advance of butterfly conservation in central Europe, or anywhere. It is more constructive to point out significant deficiencies.

17.2　FORMER FEDERAL REPUBLIC OF GERMANY

Germany is the largest country of central Europe by a huge margin. Conservation of nature is, in principle, the responsibility of the individual federal states, the federal administration being responsible chiefly for framing legislation, co-ordination and co-operation with other countries. This means that the federal states are free to decide about their own state-specific conservation legislation and priorities to a degree, as long as they follow certain legislative guidelines set up by the federal government. Germany is traditionally a federal state: the modern 'Deutsches Reich' was set up in the second half of the nineteenth century. Federalism is generally more developed and valued in the more conservative states, like the Freistaat Bayern, the former Kingdom of Bavaria, than in the states created after the Second World War, partly following the shape of the occupation zones. For the conservation of butterflies this brings advantages and disadvantages alike. There are no common federal minimum requirements, which means that conservation attempts can be deplorably low in some states, and that there is plenty of room for initiative and enterprise. Unfortunately, initiative and enterprise are not exactly features characteristic of the civil service in any country, and the conservation of nature is dominated by the state administration everywhere in Germany. It must not be overlooked that the ministries and state agencies responsible for the conservation of nature form only one component of a complex mechanism called state administration, and that their activities are limited by the amount of money allocated by the government. Last, but not least, the choice of priorities and activities is still being influenced negatively by some very old-fashioned 'bio-bureaucrats' of the 'generalist' or 'ornithologist' variety, fossilized in many institutions at all levels of administration and often passing their experiences on to younger colleagues.

As far back as 1936 four butterfly species were formally given legal protection in Germany: *Iphiclides podalirius* (scarce swallowtail), *Parnassius apollo* (apollo), *P. phoebus* (small apollo) and *P. mnemosyne* (clouded apollo); their habitats were not given adequate protection. In the early eighties a federal bill on the conservation of nature was passed by the government. In this bill all European butterfly species except the genus *Pieris*, and later a few migratory species such as *Cynthia cardui* (painted lady) *Vanessa atalanta* (red admiral), *Aglais urticae* (small tortoiseshell) and *Inachis io* (peacock) were listed as protected, and the catching and breeding of these species for any purpose, including research, became prohibited except if specifically allowed by the nature conservancy authority concerned. For instance this law forbids the possession in Germany of specimens of nearly all European butterfly species collected in other countries after 1980, even if they are common and not protected or threatened in the country of origin, research purposes not excepted. To obtain a collecting permit, a written application stating the reasons must be made. The collecting permit, if granted by the authority, may often incorporate ridiculous conditions. The legislation antagonized butterfly collectors instead of educating and integrating them; butterfly collection can become a valuable 'database'. A few years ago, Bavaria took the rare initiative of changing this law in such a way that it can now be considered abolished for all genuine research purposes. The damage this law has caused, especially among the amateur lepidopterists, is immense and will take years to overcome. The situation in the former German Democratic Republic was, in this respect, much easier.

The main reasons behind the above law were probably capricious authorities exercising their power over a small group of scientists unable to protect themselves against the pressure, partly because of the lack of a really independent advisory scientific body (entomology in Germany has always been organized regionally, with no strong all-German entomological society; the division of Germany after the Second World War increased the regionalization). It was much easier for any authority to blame entomologists than to prevent the destruction of biotopes. In some ways, this law was very convenient for certain large nationwide nature conservation organizations, like, for instance, the well-known Bund für Umwelt und Naturschultz Deutschland (BUND). In general such organizations combine simple enthusiasm of their membership with considerable political influence of their leaders and a hopeless lack of specialists and expertise, a very dangerous mixture altogether. A layman can easily observe an entomologist catching a butterfly, the only one he happened to see, not being trained to find butterflies; the entomologist is thus for the layman the cause for the decline of butterflies. The layman is not often aware that the real cause of the decline of butterflies is the destruction of their habitats. These were abundantly known to the decision-makers at the time of drafting the legislation, but conveniently 'overlooked': it is much easier to control entomologists than farmers and developers.

The 'Federal Committee for Entomology', composed of seven members, was set up in 1992 by the Naturschutzbund Deutschland (NABU, over 140 000 members); the NABU originated by joining together two societies for the conservation of birds and nature in the former West and East Germany. It is the first step in the right direction.

The German approach to the conservation of nature differs from the English; with the exception of mammals, birds, reptiles and amphibians, conservation efforts in Germany are very rarely species or species-group specific.

The first German treatment of butterfly conservation was presented by Blab and Kudrna (1982); the book received much attention but most recommendations have never been fully implemented. It is hardly surprising that the recent 'Aktion Schmetterling' organized chiefly by BUND has brought no positive results worthy of reference. The book published specifically for the occasion (Blab *et al.*, 1987) soon became popular, but it gives the readership an impression that everything is known, light and easy, and can thus even damage prospects of obtaining research funding to solve outstanding problems.

After the publication of the first federal *Rote Liste*, red lists of threatened species of butterflies and moths were published for most federal states. They are not exactly comparable to the red data books as they usually contain lists of 'naked names' of species believed to be threatened in the area, without species-specific comments. The listing of a species does not mean that species-specific steps for its conservation will be taken; the red lists are essentially political documents. The categories used do not correspond with the International Union for the Conservation of Nature (IUCN) guidelines, and differ from state to state.

In general, butterflies are rarest in north-west Germany (which also has most local extinctions), and most common in the south-east. Apart from this, the south is generally richer in butterflies than the north, i.e. Bavaria has approximately 160 species and Nordrhein-Westfalen about 95 species. Only three 'northern' butterfly species do not live in the south: *Argyronome laodice* (Pallas's fritillary), *Heteropterus morpheus* (large chequered skipper) and *Carterocephalus silvicolus* (northern chequered skipper). The lack of butterflies in the north-west correlates with the atlantic climate, the high concentration of human population and the high concentration of heavy industry, as well as industrialized agriculture and the ecologically uniform countryside. The low number of nationwide extinctions of resident species recorded to date (i.e. *P. phoebus*, *Brenthis daphne* (marbled fritillary), *Coenonympha oedippus* (false ringlet)) is misleading; Germany is a large country and many regional extinctions have been recorded. Several species are now confined to one or a few restricted localities with a fair chance of extinction within the next 5 or 10 years if no species-specific conservation measures are taken. The exact status of a few of them (extant or extinct in Germany) is unknown.

The most conspicuous German butterfly project is the publication of a two-volume, lavishly illustrated monograph of the butterflies of Baden-Württemberg (Ebert and Rennwald, 1991); since 1977 the project has received long-term generous financial support by the state.

The Gesellschaft für Schmetterlingsschutz e.V. was founded in 1988. It organizes annually the 'Rhöner Symposium für Schmetterlingsschutz' and publishes a serial *Oedippus* devoted to butterfly conservation; six issues have been published to date, including Kudrna and Mayer (1990), Kudrna and Seufert (1991), Balint and Fiedler (1992), Kudrna (1992, 1993).

Specific plans for the conservation of selected threatened and rare species are most urgently needed; only about four papers, dealing with *I. podalirius* (Kinkler, 1991), *P. apollo* (Richarz, 1989), *P. mnemosyne* (Kudrna and Seufert, 1991) and *Colias myrmidone* (Danube clouded yellow) (Kudrna and Mayer, 1990) have been published so far. It is generally very difficult to persuade the authorities that species-specific conservation is necessary, and even more so to put species-specific measures into practice. This is in sharp contrast with certain general measures purporting to serve all species of a given habitat, as long as these are paid for in a form of indirect subsidies to farmers.

Last, but not least, the successful re-establishment of *Colias palaeno* (moorland clouded yellow) in the Rhön (Kudrna, 1992) must be mentioned; it is probably the first successful re-establishment of this species and the first project of this kind in Germany.

A distribution atlas is badly missed; the one compiled nearly 20 years ago (Schreiber, 1976) is incomplete, unreliable and totally out of date. Unfortunately it is very unlikely that an atlas could be completed within the next decade or two: there is no interest in official places in Bonn to start such a project. As Bavaria is by far the richest German state from the lepidopterological point of view, a Bavarian butterfly conservation project would not only secure the survival of most species but could also set new standards, at least for Germany. Long overdue and badly needed, too, is monitoring of selected butterfly species.

17.3 FEDERAL REPUBLIC OF AUSTRIA

Although Austria belongs geographically in central Europe, it is, in the first place, an alpine country from the point of view of butterfly conservation. Suffice to say that with nearly 200 butterfly species it is by far the richest central European country. With the setting up of a zoological databank 'Zoodat' in Linz, the chances for the publication of a distribution atlas of the butterflies were excellent. Unfortunately, the atlas published by Reichl in 1992 is unlikely to completely satisfy the needs of nature conservation, since all records made after 1959 are treated as recent although many populations and localities probably no longer exist. The obsolete nomenclature makes the use of the work very difficult to anyone who does not possess the outdated and long-superseded first edition of Forster and Wohlfahrt (1955).

With the exception of a successful re-establishment of *Zerynthia polyxena* (southern festoon) and the management of its habitats in parts of Styria (Gepp, 1981), to my knowledge little has been done in way of species-specific butterfly conservation projects. Red lists of threatened species have been compiled for various regions and a few species are protected by law.

17.4 CZECH REPUBLIC

The Czech Republic is the western, larger part of the former Czechoslovakia, and before that for some thousand years, the Kingdom of Bohemia, later a part of the Austro-Hungarian Empire. From the historical, cultural and economical points of view it is quite different from Slovakia, and except for the communist period typically a Western country, with rich cultural traditions going back to the Middle Ages, when in 1348 Prague became the city with the first university in central Europe. Whereas Slovakia was an underdeveloped agricultural country in 1918 (when Czechoslovakia was created as an independent state), Bohemia and Moravia were among the rich, advanced and industrialized European countries.

Lepidopterology, and indeed entomology in general, has a long tradition in the Czech Republic. Many earlier papers were conveniently published in German, the leading scientific language, well understood throughout the central Europe of the period. The most important critical accounts of the distribution of the Lepidoptera in Bohemia (Sterneck, 1929) and Moravia (Skala, 1912, 1932) were published in German. After the Second World War Schwarz (1948–49) published an excellent treatment, for its time, of butterflies of Czechoslovakia, followed by a zoogeographical study by Moucha (1959).

The first consistent attempt to inform the general public about insect conservation issues, including butterflies (Novak and Spitzer, 1982), was followed by the publication of a Czechoslovak red data book, the third volume of which is devoted to Invertebrata (Skapec, 1992), listing about 10 threatened butterfly species living in the Czech Republic. This was preceded by a draft of a chapter on Lepidoptera for a planned Czech red data book (Soldat, 1987). Important information on the changes in distribution of butterflies in Moravia was published by Kralicek and Povolny (1980).

In their report on the conservation of butterflies in the Czech Republic, resulting from the conference 'Schutz der Tagfalterfauna im Osten Mitteleuropas', Kudrna and Kralicek (1991) concluded that 17 butterfly species have probably become extinct in the country during the twentieth century: *P. apollo, Leptidea morsei* (Fenton's wood white), *Pieris mannii* (southern small white), *Lycaena helle* (violet copper), *L. thersamon* (lesser fiery copper), *Cupido alcetas* (Provençal short-tailed blue), *Polyommatus eroides* (false eros blue), *Hypodryas maturna* (scarce fritillary), *Limenitis reducta* (southern white admiral), *Neptis sappho*, (common glider), *Polygonia l-album*, *Nymphalis xanthomelas* (yellow-legged tortoiseshell), *Coenonympha hero* (scarce heath), *Carterocephalus flocciferus* (tufted marbled skipper), *C. lavatherae* (marbled skipper), *Pyrgus alveus* (large grizzled skipper) and *P. armoricanus* (Oberthür's grizzled skipper).

Most of the extinct species can be placed in two categories: isolated populations of rare species or populations on the edge of their range. These facts have to be taken into account when planning a future conservation strategy. The fact that the regional extinction of some species of the second group can also be interpreted as natural losses or as unsuccessful attempts at range expansion or even range dynamics (e.g. *P. l-album*, *N. xanthomelas*) does not make much difference.

Much damage to butterfly fauna has been done by state-financed land-improvement schemes under the communist regime. The coming privatization and restructuring of agriculture and withdrawal of state subsidies can positively contribute towards the saving of the remaining unimproved meadows. Provisions concerning their maintenance as habitat conservation are being considered by nature conservation authorities.

An important success is the re-establishment of *P. apollo* near the town of Stramberk in north-eastern Moravia after complex habitat restoration (J. Lukasek, personal communication).

Kudrna and Kralicek (1991) made the following priority recommendations for butterfly conservation in the Czech Republic:

1. recording and mapping of butterfly distribution;
2. study of threatened species (ecology and population biology);
3. monitoring of threatened species.

The Gesellschaft für Schmetterlingsschutz, in co-operation with the members of the Lepidoptera Section of the Czech Entomological Society, started a butterfly mapping project in 1992. The response was excellent and at the time of writing this chapter there are already over 140 lepidopterists engaged in voluntary recording. A preliminary distribution atlas of butterflies of the Czech Republic is scheduled for publication in September 1994.

A butterfly conservation society was set up in October 1993, and currently a butterfly conservation foundation for the Czech Republic is in preparation, as a registered charity supported by the Ministry of Environment. There is a fair chance that ecological studies of the first threatened species can start in 1994–95, to be followed by a monitoring of selected species on a national scale in the not too distant future.

17.5 REPUBLIC OF POLAND

Poland is the second largest central European country, with large, undisturbed areas situated chiefly in the north-east of the country; land use still includes much traditional agriculture. Indigenous butterflies have always enjoyed much attention from Polish lepidopterists. All Rhopalocera families have been dealt with in the acknowledged series *Klucze do Oznaczania Owadow Polski* (Krzywicki, 1959–70). In spite of this, considerable losses in butterfly populations have been recorded during the past three or four decades. This led to the publication of a comprehensive report by Dabrowski and Krzywicki (1982), summarized later in German by Krzywicki (1982). The authors analysed the reasons for decline of the rare and threatened species and made recommendations for the conservation of their habitats. As a follow-up, recording of butterflies in Poland was started and a distribution atlas is scheduled for publication in 1996 or 1997 (Buszko, 1992). In addition to the above, some earlier experiments on the re-establishment of *P. apollo* have taken place (Palik, 1980).

Although many important papers on the distribution of butterflies in Poland have been published since the Second World War, large areas of the country remain very little known, such as the almost untouched country along the Biebrza River, probably rich in butterflies.

It seems that Polish lepidopterists, who were very much left alone and allowed to travel relatively freely abroad by the former communist regime, are losing touch with the West at the time of the freedom of their country.

17.6 SLOVAK REPUBLIC

The presence of high mountains and post-glacial immigration from south-eastern Europe (e.g. Pannonia) are chiefly responsible for the high species diversity in Slovakia. This was surely aided by the late industrialization and intensification of agriculture, compared with any other central European country, except perhaps Poland.

Slovakia gained independence as from 1 January 1993. For over 70 years it formed a part of federal Czechoslovakia, and prior to 1918 it was governed as a part of Hungary (the former 'Upper Hungary') for almost 1000 years, since the early Middle Ages. As the conservation of nature was largely the responsibility of each of the two federal republics and earlier federal states forming the former Czechoslovakia, Slovakia was represented at the butterfly conservation conference in Oberelsbach (Kulfan and Kulfan, 1991).

According to Kulfan and Kulfan (1991) the following eight species have become extinct in Slovakia: *C. lavatherae, Iolana iolas* (iolas blue), *L. helle, Argynnis pandora* (cardinal), *Boloria eunomia* (bog fritillary), *P. l-album, C. hero* and *Pyronia tithonus* (gatekeeper). None the less, I doubt very much that all the above species were ever really established in Slovakia.

The red data book for the whole former Czechoslovakia (Skapec, 1992) lists 12 threatened butterfly species from Slovakia; the more detailed list of threatened species compiled by Kulfan and Kulfan (1991) is longer and possibly more reliable.

Since the publication of Hruby's (1964) critical account of the distribution (with lists of localities) of the order Lepidoptera in Slovakia, recently updated (Reiprich and Okali, 1989), the distribution of butterflies is probably better documented, although not necessarily better known, than that of any other central European country. Unfortunately neither of the two publications directly fulfils the needs of butterfly conservation.

No specific major projects for the conservation of butterflies have taken place to date, despite recommendations (Kulfan and Kulfan, 1991). Even the attempts to map the distribution of at least the butterflies (if not all Lepidoptera), using existing data of Hruby (1964), Reiprich and Okali (1989) and others, are not very likely to succeed in the near future.

17.7   REPUBLIC OF HUNGARY

Balint (1991) compiled a comprehensive report on butterfly conservation in Hungary. According to his report, 35 species have been formally protected by law since 1982: it is forbidden to collect them. Among these are some common species (e.g. *I. io* and *Papilio machaon* (swallowtail)) which need no protection, as well as species not present in Hungary (e.g. *L. helle*).

A red data book (Rakonczay, 1989) was recently published and is aimed more at the general public than at professional biologists.

Although some of the nature reserves and other protected areas are rich in butterflies, their active conservation, especially concerning acutely threatened species, leaves much to be desired.

Four species have become extinct or 'disappeared' (Balint, 1991) in the twentieth century, three of them during the past decades: *L. morsei, Melanargia russiae* (Esper's marbled white), *N. xanthomelas* and *P. l-album*. A plan for the conservation of *Plebejus pylaon sephirus* (zephyr blue) was recently published (Balint and Fiedler, 1992) but apparently not put into operation as yet. Several threatened species are in need of urgent conservation measures; these have not been worked out as yet.

Balint's (1991) recommendations include the publication of a modern book, in Hungarian language, on Hungarian butterflies, with special reference to their ecology and conservation; setting-up a scheme for the mapping of all species and, if possible, monitoring of some threatened species and intensification of applied ecological research; and, last but not least, to improve public awareness.

17.8 CONCLUSIONS AND OUTLOOK

A massive decline of butterflies has taken place in all countries in central Europe since the Second World War. Only species-specific conservation measures can put an end to the decline of ecological specialists confined in certain restricted areas and/or to specific rare habitats. Before such measures can be taken, the applied ecology and population biology of the species must be studied. This needs time, generally at least 2 or 3 years for every species. The present status of many species deemed threatened is very inadequately known; possibly some of them are already extinct. It is therefore imperative to set up simple monitoring of at least the most acutely endangered species. The only reliable method of assessing the butterfly fauna of a region is recording, resulting in the publication of a distribution atlas. With the support of keen lepidopterists, a preliminary atlas for a region or a small country can be completed within 2 or 3 years; it is unlikely to answer detailed questions but can serve as a base for further study and for the first conservation measure alike. The ideal answer would be to set up regional centres for the conservation of butterflies.

We know from experience that successful projects aimed at butterfly conservation have mostly been carried out outside of nature conservation authorities, but usually supported by them. As nature conservation authorities rarely employ specialists, they should support, financially and otherwise, the work of lepidopterists working on conservation-relevant topics, and give them a free hand to identify their priority tasks. Nature conservation authorities are more useful as agents negotiating with land owners, farmers, etc., and in implementing practical conservation measures. Nature-conservation-oriented foundations could help greatly by supporting worthwhile scientifically based projects, carried out or supervised by acknowledged specialists and experts. Without species-specific measures, the lists of threatened and extinct species will be getting longer from year to year.

At present, the most urgent primary tasks for the conservation of butterflies in central Europe are the compilation of distribution atlases (at least for important regions), working out plans for the management of (at least) critically endangered species (based upon field studies of their ecology and population biology), and monitoring of selected species and communities. Results of scientific studies must be implemented without undue delay by the authorities concerned; it is their responsibility.

In view of the imminent changes of climate, experiments with the establishment of threatened butterfly species in suitable habitats need more support. Applied rhopalocerological conservation-orientated research deserves the full support of conservation authorities and private foundations. The success of our present work can significantly influence the long-term presence of many butterflies in central Europe.

CHAPTER 18

Ecology and conservation of alpine Lepidoptera

A. Erhardt

18.1 INTRODUCTION

Although there are indications that Lepidoptera had started to decrease early in the twentieth century (Entomologischer Verein Alpstein, 1989), they have declined throughout Europe most markedly during the past few decades. This decrease is so strong that it has been noticed not only by lepidopterists but even by lay-persons. According to red data lists, about 40–50% of all Lepidoptera species occurring in Germany and Austria are endangered, and 2–5% of the species occurring in these countries have already become extinct (Ebert and Falkner, 1978; Wagener *et al.*, 1979; Pretscher, 1984; Gepp, 1981; Huemer, 1994; Huemer, Reichl and Wieser, 1994). In Switzerland, 39% of all butterfly species are endangered (Gonseth, 1987), although no recent extinctions have been reported so far. However, in some areas of Switzerland the percentage of endangered butterfly species is distinctly higher. According to Bryner (1987), 53% of the butterfly species originally occurring in the Seeland (area between the lakes of Biel and Murten) have already become extinct, 27% are endangered, leaving only 20% of mainly trivial species unthreatened. The figure for moths, although much less well known, is probably similar to that of butterflies.

In comparison to the deplorable situation in the central European lowlands, Lepidoptera occurring in the Alps appear to be less threatened (Blab and Kudrna, 1982; Kappeler, 1987). This is due to a number of factors: the Alps include inaccessible areas which remain undisturbed by humans; the higher altitudes prevent the most intense forms of agriculture; landscape is less altered and destroyed by human activity, and consequently primordial habitats such as steep gorges, floodplains or moorlands, sites cleared by avalanches and alpine areas above the tree line still remain in fairly natural condition.

Table 18.1 shows that with increasing elevation both the absolute number and

Ecology and Conservation of Butterflies Edited by Andrew S. Pullin.
Published in 1995 by Chapman & Hall. ISBN 0 412 56970 1

Table 18.1 Elevational range of all Swiss butterfly species and of most endangered Swiss butterfly species (after Schweizerischer Bund für Naturschutz, 1987)

Elevational range	Number and % of all species		Number and % of most endangered species	
Lowland	57	(32.9%)	20	(69%)
Lowland to sub-alpine	80	(46.2%)	8	(27.6%)
Sub-alpine to alpine above timber line	36	(20.8%)	1	(3.4%)
Total number of species	173		29	

the percentage of highly threatened Swiss butterfly species decreases (Schweizerischer Bund für Naturschutz, 1987).

The only threatened species in subalpine to alpine areas is *Erebia christi* (Ratzer's ringlet), which is geographically restricted to an extremely small area in the southern Swiss Alps (Laggin Valley). This small range is most likely a relic from periods previous to the last glaciation(s) when *E. christi* was probably more widespread, and coincides with a southern nunatak, i.e. a mountain which, during the glaciation periods, remained unglaciated within the surrounding ice cover and thus served as a refuge for animals and plants (Merxmüller, 1952).

The Alps still provide refugial habitats for *c.* 80 lowland species whose elevational range extends also into subalpine areas. However, the less serious situation regarding alpine Lepidoptera has resulted in only limited efforts being made to protect them. In fact, besides the few long-established nature reserves (Duffey, 1982), no conservation efforts for alpine Lepidoptera have been undertaken so far. In addition, the reserves are mostly unmanaged except for keeping Cervidae (deer) populations at tolerable levels, and are not aimed specifically to protect Lepidoptera. They are also far from including all of the highly diverse alpine habitats and can thus only harbour a limited, although at times high, number of species (Pictet, 1942). Alpine Lepidoptera, especially species that live in subalpine, traditionally cultivated grasslands, are therefore under increasing threat, and we are in danger of loosing some of our most diverse Lepidoptera communities.

Recent ecological work on alpine Lepidoptera has been limited, possibly because they are less threatened than lowland species which have consequently drawn most of the attention. However, lowland species have hardly been studied in Switzerland either, obviously due to lack of money and interest from the government and institutions (Geiger and Gonseth, 1992).

This chapter reviews recent work on the ecology and conservation of alpine Lepidoptera. The impact of post-war changes in grassland use and the role of adult resources are emphasized.

18.2 ECOLOGICAL WORK ON EUROPEAN ALPINE LEPIDOPTERA

18.2.1 General

Recent investigations on the ecology of alpine Lepidoptera are limited. Geiger and Scholl (1981) have studied the dispersal behaviour of *Colias phicomone* (mountain clouded yellow) in an alpine meadow near Grindelwald in the Swiss Alps by mark–release–recapture techniques. The investigated meadow extended over a length of 2.5 km and a width of 600 m. Over 80% of the male and female butterflies were recaptured within less than 50 m of the last point of capture. This indicates that in spite of their fast and powerful flight, individual butterflies do not disperse over the whole meadow but reside at small subsites, and may even form subpopulations within a large continuous suitable habitat.

Loertscher (1991) investigated the population dynamics of *Erebia meolans* (Piedmont ringlet) and *E. aethiops* (Scotch argus), again in the area of Grindelwald, also using capture–recapture techniques. He found that both species flew over a period of 3 weeks, that males emerged before females and that *E. aethiops* males showed a distinct habitat preference for a steep meadow with tall grass bordered by woods and interspersed with bushes and small trees.

Balletto and co-workers (Balletto, Barberis and Toso, 1982; Balletto *et al.*, 1988) investigated butterfly communities in grasslands of the Italian Alps and the Apennines and found that discrete butterfly communities inhabit different vegetational units.

I have investigated the effects of post-war changes in grassland use on diurnal Lepidoptera in the Swiss Central Alps (Tavetsch Valley) (Erhardt, 1985a,b,c 1992b; Erhardt and Thomas, 1991). Since this study is still the only major one that has been conducted in the Alps in recent years and since it proved to be relevant for both the ecology and conservation of Lepidoptera, its results are summarized here in somewhat greater detail (section 18.2.4).

18.2.2 New species and rediscoveries

The Lepidoptera fauna of the Alps is still rather poorly known. This applies equally to their distribution, biology, ecology and taxonomy. However, due to the work of a group of lepidopterists during the 1980s in Switzerland, our knowledge of alpine butterflies has greatly improved (Schweizerischer Bund für Naturschutz, 1987; Gonseth, 1987).

Nevertheless, much has still to be discovered, and a new species of blue butterfly, *Agrodiaetus huemedusae* (Piedmont anomalous blue), was described as recently as 1976 (Toso and Balletto, 1976). As in most other regions of Europe, previously unrecorded species of Heterocera are being found in the Alps each year, some of which are even new to science (e.g. Tarmann, 1984; Huemer, 1986; Huemer and Tarmann, 1989, 1992, 1994a,b; Huemer *et al.*, 1992). An additional problem in the high Alps is the inaccessibility and the unpredictable weather. For instance, in the very touristic region around Zermatt, two high alpine species, the geometrid *Psodos wehrlii* and the psychid *Dahlica wehrlii* have been rediscovered after gaps of 50 and 75 years, respectively.

18.2.3 Adult nectar resources

The role that ecological conditions play in the development of the early stages of Lepidoptera is not doubted and has been shown by many striking investigations (e.g. Thomas, Chapter 13 this volume). Adult resources of Lepidoptera have received much less attention, although it has been shown that they have profound effects on the population dynamics of Lepidoptera (Boggs, 1987). The most striking example of these effects is that of the tropical pollen-feeding *Heliconius* butterflies (Gilbert, 1972). For temperate butterflies, the role of adult resources appears to be more controversial. While often called opportunistic (Courtney, 1986; Wynhoff, 1992), it has also been shown that temperate butterflies have distinct flower preferences (Erhardt and Thomas, 1991; Porter, Steel and Thomas, 1992) and that the distribution of their nectar plants can even affect oviposition sites (Murphy, Menninger and Ehrlich, 1984). This suggests that adult resources could be more important for the population dynamics of temperate butterflies than so far assumed. A further indication of this comes from the findings of Loertscher, Erhardt and Zettel (1994), that the microdistribution of several butterfly species was determined by the distribution of their preferred nectar plants in an abandoned meadow in the southern Alps of Switzerland (Figure 18.1).

During my research in subalpine meadows (Erhardt, 1985a,b,c) I recorded over 15 000 flower visits of diurnal Lepidoptera and subsequently analysed the nectar of the most frequently visited flowers. The detailed findings of this study would exceed the scope of this chapter, but the general results and trends can be summarized as follows:

1. Out of *c.* 170 potential nectar plants in the study area, only *c.* 20 were visited regularly by the observed Lepidoptera.
2. Most preferred were purple and/or yellow capitula of Compositae and Dipsacaceae, but white capitula were also preferentially visited by a few Lepidoptera (e.g. *Erebia melampus* (lesser mountain ringlet) and *Heodes virgaureae* (scarce copper)).
3. Over 50% of the observed visits were to the flowers of only three plant species (*Knautia arvensis* (field scabious), *Centaurea scabiosa* (greater knapweed) and *Arnica montana* (arnica)). These nectar plants were especially preferred by many butterfly species, in spite of greater abundances of other potential nectar plants.
4. When faced with the same choice of nectar plants, different Lepidoptera species differed in their flower preferences.
5. Flower preferences of single species may vary according to the flowers present in a particular habitat.
6. Nectar of the preferred flowers was characterized by either:
    (a) high quantities of nectar per flower or floret in an inflorescence,

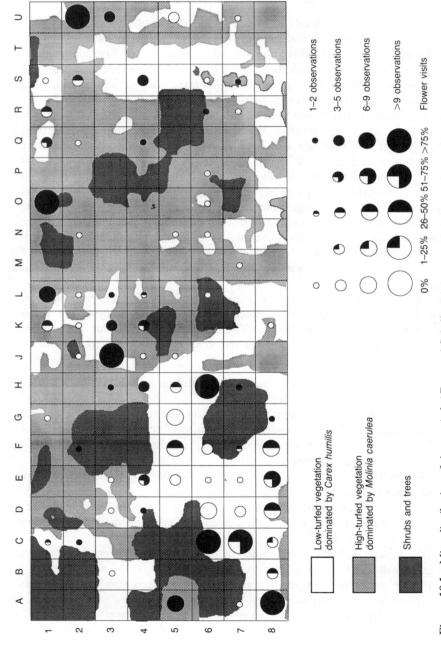

Figure 18.1 Microdistribution of *Agrumenia loti* Den. and Schiff., with percentage of flower visits in an abandoned meadow at Monte San Giorgio (elevation 1097 m) in the southern Swiss Alps (Loertscher, Erhardt and Zettel, 1994).

moderate to high proportions of sucrose as compared with fructose and glucose (sucrose, fructose and glucose are the main sugars found in floral nectar; Baker and Baker, 1983), and by relatively low concentrations of amino acids; or

    (b) low quantities of nectar per flower or floret, low levels of sucrose but high levels of fructose and high concentrations of amino acids.

7. There was a weak, but statistically significant, correlation between proboscis lengths of Lepidoptera and corolla tube lengths of the visited flowers.

8. Some species, e.g. the Lycaenids *Eumedonia eumedon* (geranium argus), *Cyaniris semiargus* (mazarine blue) and *Polyommatus icarus* (common blue), not only visited the flowers of their larval host, but even seemed to prefer them. A precondition for visitation of the flowers of the larval host by adult Lepidoptera is, of course, that the host-plant is in bloom and that the flowers produce nectar during the flight period of the adults. However, once this condition is fulfilled, there are strong advantages for Lepidoptera to use the flowers of their larval host as nectar resources:

    (a) If a Lepidoptera species also pollinates the flowers of its larval host, it enhances the reproductive success of its host-plant and thus also increases its own potential for reproductive success.

    (b) If the host plant flowers profusely enough, it also covers the needs of the adults. Consequently, a species may depend only on its larval host-plant and not on additional plants as adult resources.

    (c) Only one search image for one plant species for oviposition and feeding has to be formed by females.

    (d) Secondary compounds that larvae acquire from the host-plant might also be accessible in the nectar for the adult insects.

9. Factors responsible for the observed flower preferences include the proboscis length and energy requirements of Lepidoptera, and corolla tube length, colour, shape, odour, nectar quantity and quality of different flowers. The ability of Lepidoptera to produce saliva in order to dilute nectar, which can be highly viscous or even crystalline, as in certain Umbellifers, may also be an important factor.

10. The clear preference for a small, restricted number of nectar plants suggests that these nectar plants could play a particularly important role for the population dynamics of nectar-feeding Lepidoptera, and may thus function as 'key-stone nectar plants' (Mills, Soulé and Doak, 1993).

18.2.4 Changes in grassland management: a major threat

Methods in grassland management have changed profoundly during the past decades. This change began after the Second World War, when the pressure for higher yields and the rising costs of labour started to cause an increasing industrialization in agriculture. The impact has been greatest in lowlands but has also affected grassland management in the Alps. Before this, grasslands had been managed in a similar manner for hundreds of years. They were man-made, mainly unfertilized, mown and lightly grazed meadows, and had become important

Figure 18.2 North-facing slope in the Tavetsch Valley near the village of Selva. In the valley bottom, hay-meadows are fertilized. The steeper parts of the slope have been abandoned and then invaded by dwarf shrubs (*Vaccinium myrtillus*, *Vaccinium uliginosum*, *Empetrum nigrum*), shrubs (*Rhododendron ferrugineum*) and in most places by green alder (*Alnus viridis*). Lightly grazed meadows can only rarely be found on the flatter parts of the slope.

secondary habitats for butterflies. This development peaked at the turn of the century. However, these traditionally cultivated grasslands have decreased enormously since the Second World War due to two major changes in cultivation. Easily accessible meadows in the valley bottom were heavily fertilized and most areas, especially steep slopes which brought only small returns, were abandoned. Today, abandoned grasslands are very common in the Alps (Figure 18.2).

They play not only an important role for forestry but also greatly influence landscape and are a real problem (Surber, Amiet and Kobert, 1973). Between 1954 and 1972 *c.* 200 pastures in the Bavarian Alps were abandoned, leaving an area of 8000 ha unmanaged (Plachter, 1991). In the Swiss Alps, *c.* 80 000 ha of grassland were already abandoned by the beginning of the 1980s. By the end of the twentieth century, 29% of the grasslands in the Central Alps and 41% in the Southern Alps of Switzerland are expected to be abandoned (Surber, Amiet and Kobert, 1973). In the Austrian Alps, abandoned grassland amounts to *c.* 270 000 ha (Plachter, 1991). In contrast, only about 5% of all cultivated grassland is still managed in traditional ways in the Alps of central Switzerland (Zoller and Bischof, 1980). This figure is probably similar, if not even more extreme, in alpine areas of the neighbouring countries (Balletto, 1992; Descimon, personal communication). In the Jura mountains, the figure is even more dramatic: less than 1% of traditionally cultivated grassland remains (Zoller and Bischof, 1980). I have investigated the impact of these changes in grassland management on diurnal Lepidoptera (i.e. all species of Rhopalocera, Hesperiidae and Zygaenidae, and day-active species of Bombycidae, Sphingidae, Noctuidae and Geometridae). The detailed methodology and results of this investigation are given by Erhardt (1985b,c) and are summarized briefly here. Not unexpectedly, these changes in grassland management had a strong impact on Lepidoptera.

(a) Cultivated grassland

The different investigated types of cultivated grasslands included lightly grazed meadows, unfertilized mown meadows and three different types of fertilized meadows with increasing intensity of fertilization. Extensively grazed meadows were inhabited by most Lepidoptera species (mean = 41 species, 39 autochthonous, i.e. completing their life cycle in these meadows, south-facing slope, $n = 2$). Unfertilized, mown meadows were also species rich (mean = 34.5, 31.5 autochthonous, south-facing slope, $n = 4$), but increasing fertilization caused a severe decline in the species richness of Lepidoptera. Thus the most heavily fertilized meadows in the valley bottom were inhabited by only eight species, of which but five were autochthonous and only one, *Palaeochrysophanus hippothoe* (purple-edged copper), was not a trivial species (see below).

In general, the diversity of Lepidoptera was strongly correlated with plant diversity in all investigated vegetation types ($P < 0.05$, Spearman's rank correlation). This correlation is clearly due to the dependence of the phytophagous larvae on vascular plants. However, there was a discrepancy in the case of unfertilized, mown meadows. This is because the act of mowing produces well-balanced conditions for competition between growing herbs, and results in a sward that is especially rich in herb species (Figure 18.3).

On the other hand, the sudden uniform breakdown of the structure of the vegetation by mowing reduces Lepidoptera populations by destroying the flowers used by the adults and, more harmfully, by killing many of the immature stages. This disturbance is more harmful to Lepidoptera in subalpine meadows than is light grazing and trampling by cattle, because cattle graze selectively and leave patches of sward undisturbed. As a consequence, butterfly diversity is lower in unfertilized, mown meadows, although plant diversity is higher there than in lightly grazed meadows.

(b) Abandoned grassland

When cultivated grassland is abandoned, a succession of different vegetation types occurs, each with its own fauna. The plant succession is complex and not fully explained in many parts of Europe (Bischof, 1980). However, in the subalpine region of the Swiss Central Alps, the principal trends can be summarized as follows (Bischof, 1980, 1984): on unfertilized mown meadows early successional stages are dominated by grasses and herbs. These are then invaded either by dwarf

Figure 18.3 Detail of unfertilized, mown meadow. Note the marvellous aspect of *Paradisea liliastrum*, and the high diversity of other herbs. In the foreground *Pulsatilla apiifolia*.

shrubs (*Vaccinium myrtillus* (bilberry), *V. uliginosum* (northern bilberry), *Calluna vulgaris* (heather)) or by young trees (*Betula pendula* (silver birch), *Alnus viridis* (green alder)). Later stages may eventually revert to forest climax vegetation dominated by *Picea abies* (Norway spruce).

There was again a good correlation between the number of species of plants and Lepidoptera over the range of successional stages sampled ($P<0.05$, Spearman's rank correlation), although the deviation in Lepidoptera in unfertilized, mown meadows was again evident. Thus although the diversity of plant species was lower in abandoned areas that were still dominated by grass and herbs, diversity of Lepidoptera was higher than in unfertilized, mown meadows. In fact, early successional stages dominated by grasses and herbs were the habitats richest in Lepidoptera of all investigated vegetation types. Several circumstances favour Lepidoptera in these habitats:

1. the vegetation is still rich in plant species;
2. no disturbance by cultivation measures occurs, which allows larvae to complete their life cycles and adults to feed on flowers during the whole growing season;
3. on south-facing slopes, a shallow humus layer can prevent rapid succession and create rather stable grassland habitats with microclimates which are especially warm and favourable to Lepidoptera.

Such sites can therefore support an especially rich Lepidoptera fauna (mean = 48.5 species, 46.5 autochthonous, *n* = 2), including some rare species such as the diurnal noctuid moth, *Chloridea ononis*. This species was resident in the study area, in contrast to former statements that it does not breed north of the southern border of the Alps, and that specimens found further north are always of migratory origin (Erhardt, 1990).

In later successional stages dominated by dwarf shrubs, reduced plant diversity leads to a decrease in Lepidoptera diversity. Even so, slightly more species occur in these stages than in the unfertilized, mown meadows (mean = 37.7, 35.7 autochthonous, south-facing slope, *n* = 3).

Species richness of Lepidoptera falls rapidly with the arrival of shrubs and trees. Indeed, successional stages dominated by *A. viridis* are practically devoid of diurnal Lepidoptera. The climax vegetation of woodland dominated by *P. abies* is inhabited by few, characteristic species of Lepidoptera (mean = 7, all autochthonous, south-facing slope, *n* = 2).

Similar trends were recorded on north- and south-facing subalpine slopes, although species richness was generally lower on the former. The species composition also differed on the two aspects. Many species were confined to south-facing slopes but a few were restricted to northerly aspects. The latter include *Erebia pharte* (blind ringlet), *E. eriphyle* (eriphyle ringlet) and a rare endemic of the European Alps, *Eurodryas intermedia wolfensbergeri* (asian fritillary) (see below).

Factors that determine the species richness of plants and Lepidoptera in the investigated cultivated and abandoned grasslands are summarized in Figure 18.4.

(c) Changes in species and families of Lepidoptera

The trends in total species richness described in the previous two sections mask a more rapid turnover in the presence of individual species and families of Lepidoptera in fertilized or abandoned grassland (Figure 18.5).

At a family or generic level, Papilionids (*Papilio machaon* (swallowtail), *Parnassius apollo* (apollo)) were only recorded regularly in early successional stages. *Erebia* species were mainly eurytopic (i.e. autochthonous but not restricted to specialized habitats), as expected of a genus near its centre of distribution (Kühnelt, 1943), and showed the greatest species diversity in unfertilized, mown meadows and recently abandoned grassland. Lycaenidae, on the other hand, include a high percentage of stenotopic species (i.e. autochthonous species restricted to a few specialized habitats) in the Alps. They occurred mainly in lightly cultivated meadows and decreased steeply with increasing fertilization, disappearing in the later stages of abandonment. In contrast to other Lepidoptera families, the species number of Lycaenidae closely paralleled that of vascular plants in the different categories of vegetation.

In the following section, five selected species are presented as examples and are briefly commented upon.

Palaeochrysophanus hippothoe deserves some attention because it was the only stenotopic species found in fertilized mown meadows. Furthermore, its montane subspecies *P. h. eurydame* is endangered in central Europe (Pretscher, 1984;

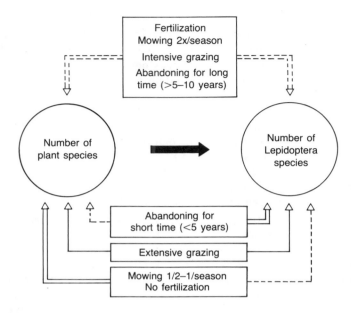

Figure 18.4 Factors affecting plant and Lepidoptera diversity in cultivated and abandoned grassland. Nature of effect: --- , negative effect; ===, strongly negative effect; ——, positive effect; ===, strongly positive effect.

Schweizerischer Bund für Naturschutz, 1987; Ebert and Rennwald, 1991). It generally occurs in moist grasslands and moorlands (Bergmann, 1952). Its occurrence in fertilized meadows, especially in the moist meadows in the valley bottom, may actually be a relic from times when these meadows were mown but not yet fertilized. However, it may also have profited from fertilization, since fertilization favours its larval host, *Polygonum bistorta* (bistort).

Palaeochrysophanus hippothoe also shows how a butterfly species may survive the impact of mowing. At the time when meadows are mown, many of the adults have already emerged and are flying in the fertilized meadows, taking nectar from some of the meadow flowers (most notably *P. bistorta*, the larval host). When meadows are mown, the adults leave the meadow and can be found visiting flowers in the close surroundings, in unfertilized, mown meadows or early successional stages (in both, the larval host *P. bistorta* is absent). However, I have observed females returning to the mown meadows when hay-making was finished to lay their eggs on the remaining leaflets or stalks of the larval food-plant. This shows that the surroundings of mown meadows have an important function to support the adults during the time of hay-making.

Lysandra dorylas (turquoise blue) is not only endangered in montane areas (Riess, 1978; Schweizerischer Bund für Naturschutz, 1987; Ebert and Rennwald, 1991), it must also be considered as endangered in the study area. It was restricted to unfertilized, mown meadows with a shallow humus layer on south-facing slopes and to corresponding early successional stages. Clearly, this species has suffered from both fertilization and abandonment. This is even more threatening than for other species, since in the study area it has probably never been abundant.

The complex life cycle and the associated complex ecology of *Maculinea arion* (large blue) have been revealed by the admirable investigations of Thomas and co-workers (see Thomas, Chapter 13 this volume). In the study area, the (sub)alpine subspecies *M. a. obscura* was stenotopic in extensively grazed meadows. This observation agrees very well with the habitat requirements of the species in lower elevations (Thomas, 1980, 1984a).

Colias palaeno (moorland clouded yellow) is much threatened in lowlands. However, the subalpine race (f. *europomene*) has profited from abandonment, since it could colonize successional stages dominated by dwarf shrubs with *V. uliginosum*, its larval host. These successional stages have become an important secondary habitat for *C. palaeno* in subalpine areas. However, neither cultivated meadows nor other successional stages are suitable habitats for this species.

Eurodryas i. wolfensbergeri is endemic to subalpine areas of the European Alps, where it is quite rare, but may be locally abundant (Schweizerischer Bund für Naturschutz, 1987). Again, the sibling species *Eurodryas maturna* (scarce fritillary) is severely threatened in the lowlands of central Europe. In the study area, *E. intermedia* was confined to north-facing slopes, where it was restricted to lightly grazed meadows and to early successional stages. Although it probably also suffers from abandonment, it was a pleasant surprise to discover this species in the study area, since it was previously not known to exist in this part of the central Alps.

(d) Primordial habitats

An aspect more and more relevant for conservation is the investigation and the status of the natural primordial habitats of Lepidoptera. Cultivated meadows and most stages of abandoned grassland are secondary habitats caused by human activity in historical times. The members of these communities must therefore have had natural primordial habitats. Prior to human intervention, the landscape was dominated by woods (climax vegetation) after climatic conditions had become similar to those of today following the last glaciation. Because Lepidoptera and plant species found in today's secondary habitats do not belong to the flora and fauna of forests, they must have been much rarer and were probably restricted to sites without trees, or may even have been absent. Potential primordial habitats are the floodplains of rivers, moorlands, steep gorges and rocks, and, in the subalpine region, sites cleared by avalanches or alpine areas above the tree line. An investigation of possible primordial habitats of the Lepidoptera species found in the investigated cultivated grasslands and successional stages in the study area showed that a large proportion of these species also occurred in a gorge of the upper reaches of the River Rhine and/or in alpine meadows above the tree line (Table 18.2).

A smaller but still significant fraction has obviously invaded the subalpine man-made vegetation types from lowland areas, whereas only one migratory species, *Autographa gamma* (silver-Y moth), could be recorded. This general situation

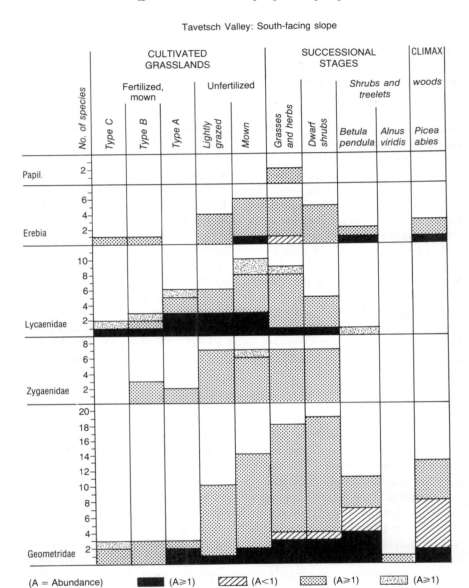

Figure 18.5 Species number and ecological status of different Lepidoptera families in cultivated meadows and in different stages of abandoned grassland. Abundance $\geq 1 : \geq 5$ observed individuals per flight period and area of $2500\,m^2$. Abundance $<1 : <5$ observed individuals per flight period and area of $2500\,m^2$. Stenotopic: autochthonous species (i.e. species completes its entire life cycle in the same vegetation type) restricted to a few specialized habitats; eurytopic: autochthonous species present in several similar vegetation types; xenotopic: species present in a vegetation type without completing its life cycle there (larval food-plant lacking), visitors, neighbours. Fertilized meadows: type A, south-facing,

Table 18.2 Relative occurrence in primordial habitats of Lepidoptera species found in cultivated and abandoned grasslands (Tavetsch Valley)

Primordial habitat	South-facing slope	North-facing slope
Gorge of Rhine (A)	43.2%	38.6%
Alpine areas above timber line (B)	19.8%	22.7%
A and B	16.0%	22.7%
Lowlands (from literature)	19.8%	13.6%
Migratory species	1.2%	2.2%
Total number of species	81	44

raises a point of special concern: today, primordial habitats have been vastly destroyed (e.g. floodplains, moorlands and others), especially in the lowlands. As a consequence, many Lepidoptera species are by now completely dependant on secondary, man-made habitats over large geographical areas. These secondary habitats are extremely fragile since their existence depends completely on the maintenance of traditional cultivation practices, which notably included abandonment for short periods of time. Today, these sites have not only become very rare, but are also extremely prone to changes. If these few remaining sites cannot be preserved, species might be lost from large geographical areas or even become extinct. The extinction of *M. arion* in Britain is a warning example.

18.3   THREATS AND CONSERVATION MEASURES

Although Lepidoptera are less threatened in the Alps than in the lowlands, they are far from being safe in the future. As already shown in the previous section, a number of increasing threats will affect them (Erhardt, 1985a,b,c; Schweizerischer Bund für Naturschutz, 1987; Plachter, 1991; Balletto, 1992).

18.3.1   Increasing intensification in agriculture

Traditionally lightly cultivated grasslands which are inhabited by a rich and diverse Lepidoptera fauna (see above) have strongly decreased and are heavily fertilized, especially in easily accessible areas.

---

mown 1×/season ($n=1$); type B, south-facing, mown 2×/season ($n=1$); type C, valley bottom, mown 2×/season ($n=1$); lightly grazed ($n=2$); unfertilized mown, mown 1×/season or 1× every second or third season ($n=4$); grasses and herbs ($n=4$); dwarf shrubs ($n=3$); shrubs and treelets of *Betula pendula* ($n=1$) and of *Alnus viridis* ($n=1$); woods ($n=2$).

18.3.2 Abandonment

As mentioned in the previous section, on less easily accessible areas such as steep slopes, which were hard to cultivate and brought only small returns, traditionally cultivated grasslands were, and still are, being abandoned. The succession occurring on abandoned grassland favours Lepidoptera for short periods of time (less than 5–10 years), but in the long run strongly reduces their diversity to rather trivial species which inhabit wooded vegetation.

18.3.3 Afforestation

Abandoned grasslands are often afforested with *Picea* spp. because such grasslands provide gliding surfaces for snow and increase the danger of avalanches. However, these afforestations, besides being expensive, prevent natural succession of the associated plant and animal communities or possible recultivation. In addition, trees in afforestations are usually planted too close together to provide suitable habitats for species that occur regularly in *Picea* forests.

18.3.4 Other threats

1. Destruction of floodplains in valley bottoms has significantly increased.
2. Moorlands are under increasing pressure.
3. Constructions of dams for reservoir lakes have drowned whole valleys, especially in untouched remote areas, and associated diversions of alpine rivers cause desiccations of alpine river beds.
4. General effects of civilization, although less severe than in lowlands, have not left alpine areas untouched. Construction of buildings and housing areas and associated road construction have significantly reduced natural areas.
5. Tourism and all of its associated effects increasingly affect alpine areas. Development of recreation industry in the Alps is associated with high demands for infrastructure (restaurants, hotels, parking lots, etc.). The whole range of the Alps is inhabited by about 7 million residents. However, at peak vacation times there can be 10 million tourists in the Alps. During the whole year, *c.* 40 million tourists need a total of 250 million overnight accommodations. In addition, there are *c.* 60 million tourists who visit the Alps on day trips. Yearly numbers of overnight accommodations for tourists in the Austrian Alps increased from 18.1 million in 1950–51 to 93.5 million in 1974–75 (all information from Plachter, 1991).
6. Constructions of ski runs reduce large natural areas and destroy the humus layer, which is thin but often several thousand years old. This increases soil erosion and reduces the water-carrying capacity of the soil by a factor of 5–10. In contrast, restoration of alpine grassland above the tree line is extremely difficult. Already in 1983 there were 12 000 ski-lifts and cable cars and over 40 000 ski runs in the Alps. In Switzerland, the number of transportation systems for tourists increased from 140 in 1950 to 1700, with a length of *c.* 1800 km, in 1980. In Austria, *c.* 2000 ha of woods were cleared for ski runs from 1969 until 1978. The area used for ski runs in Austria equals roughly the

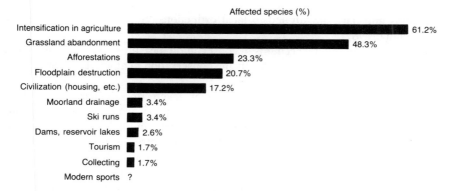

Figure 18.6 Threats affecting butterfly species in the Swiss Alps (total 116 species, Hesperiidae excluded; after Erhardt, 1985b, unpublished; Schweizerischer Bund für Naturschutz, 1987).

area used for traffic (Plachter, 1991). In addition, cable cars, chair- and ski-lifts bring people and the associated disturbances even into remote high alpine areas.

7. Last, but not least, modern sports such as mountain biking, moto cross, hang gliding, parachute flying and rollerskiing pose additional increasing threats to alpine habitats.

In Figure 18.6 the different threats are plotted in decreasing order according to the percentage of alpine butterfly species they affect.

Since there was no comparable information available from Italy, France or Austria, this figure is based on statements about threats to the different alpine butterfly species in Switzerland in Schweizerischer Bund für Naturschutz (1987), on data in Erhardt (1985b) and on field surveys (Erhardt, unpublished). However, the figure is most likely similar for the neighbouring countries. Although only nine butterfly species are considered as severely threatened in the Swiss Alps (Table 18.1), Figure 18.6 shows that intensification in agriculture and abandonment of grassland have detrimental effects on over half of all species. This illustrates again that, particularly in subalpine areas, we are facing severe declines in Lepidoptera. Afforestations, floodplain destruction and civilization affect *c.* 20% of all butterfly species, whereas the other mentioned threats seem at present to be less severe. However, moorland destruction is threatening three highly endangered species (*C. palaeno*, *Boloria aquilonaris* (cranberry fritillary) and *Lycaena helle* (violet copper)). In addition, these minor threats may well be underestimated, because their effect is difficult to assess and has not been investigated in detail so far. This is particularly true for the effects of modern sports. It seems safe, however, to state that collecting has so far had only a minor detrimental effect.

Except for collecting, all of these effects reduce and fragment suitable habitats for Lepidoptera in the Alps. Even though most of the affected species are not yet severely threatened today, they will be so, even in the not too distant future, if current developments in the Alps proceed unhalted. The above-mentioned factors are especially threatening isolated populations of alpine Lepidoptera in the Jura mountains and have already led to local extinctions (Schweizerischer Bund für Naturschutz, 1987).

There are many general accounts of conservation measures for Lepidoptera, but only measures that concern specifically the conservation of alpine Lepidoptera are addressed in the following section.

Until now, no real measures have been taken to safeguard particularly Lepidoptera in the Alps. The few established national parks can by no means guarantee protection for all alpine Lepidoptera species in the long run. Additional measures are therefore definitely required. Some countries, such as Germany and Austria, have introduced extensive bans on collecting, but such measures have, at best, little effect on the decline in populations, give conservationists a false sense of security and thus could even be counter-productive (see Kudrna, Chapter 17 this volume).

In general, measures should keep the mentioned threats within defined boundaries to guarantee a long-term survival of alpine Lepidoptera species. The optimal strategy would be to protect undisturbed areas large enough to allow natural vegetation dynamics to provide sufficient suitable primordial habitats for alpine Lepidoptera. However, this might not be possible due to the increasing pressure of civilization and tourism, and more specific measures are therefore needed.

Since traditional, lightly cultivated grasslands and early successional stages are especially rich in Lepidoptera, and since they also provide refugial habitats for a number of lowland species, these vegetation types need special protection. Because differently cultivated meadows and different stages of abandoned grassland are inhabited by different characteristic Lepidoptera faunas, the optimal strategy would be to preserve a balanced combination of all possible vegetation types. Consequently, the different successional stages need care, because otherwise their vegetation will change and develop towards woodland. For earlier successional stages, a rotational system of cultivation, either by mowing or light grazing, could be a useful method for their conservation (Morris and Thomas, 1991). However, unfertilized, mown and lightly grazed meadows are the most endangered vegetation types. They can be preserved only if traditional cultivation practices are maintained. A restoration of lightly cultivated grassland from fertilized meadows or from later successional stages would probably need considerable amounts of time.

Hope-Simpson (1965) estimated that the restoration of characteristic mature chalk grassland in Britain would require a full century. Similar, if not even greater amounts of, time are likely to be required for the restoration of subalpine, lightly cultivated grasslands. The regeneration of unfertilized, mown meadows by mowing and cutting of shrubs from successional stages abandoned for *c.* 20 years is currently being studied in the southern Swiss Alps in a project of the Swiss National Science Foundation. Controlled burning, sometimes also used in the past, could probably only locally be a useful measure, since it destroys all early stages of Lepidoptera and since the smoke would drive away the adults.

For species with restricted geographical distributions, reserves large enough to guarantee their persistence should be created.

Further measures include preventing further intensification in grassland use, protecting floodplains (today at particularly high risk), leaving sufficient amounts of water in rivers diverted for reservoir lakes to prevent desiccations of river beds, protecting alpine moorlands from trampling and fertilization by cattle, allowing for afforestations only where urgently needed (e.g. for protection from avalanches), controlling touristic development (road construction, housing, etc.) and prohibiting the destruction of alpine areas for the construction of ski runs (Schweizerischer Bund für Naturschutz, 1987).

Since Lepidoptera are particularly sensitive indicators of vegetation structure and vegetation changes (Erhardt, 1985b,c; Erhardt and Thomas, 1991), they are an important group of animals to consider when selecting and evaluating general nature reserves.

18.4   DESIDERATA AND CONCLUSIONS

Although less endangered than in lowlands, Lepidoptera in alpine areas are far from being unthreatened. Many factors reduce butterflies and moths in the Alps, among which changes in grassland use (fertilization and abandonment) appear to be particularly severe, besides a host of other man-made threats. Consequently, measures that limit these threats are definitely needed to conserve Lepidoptera in the Alps.

True ecological work on alpine Lepidoptera appears still to be in its infancy; new species, especially among Microlepidoptera, are still to be described, larval hosts are often only incompletely known, and food requirements of adult Lepidoptera are even less well studied. Long-term studies on the population dynamics of selected alpine Lepidoptera species would be particularly useful, and studies on the minimum viable populations of Lepidoptera are almost nonexistent, as is a monitoring scheme for endangered Lepidoptera species in the Swiss lowlands and in alpine areas. The minimum area required to support a population, and the effects of different types of management regime are further research topics which are urgently needed if conservationists are to stem current declines. Detailed ecological research should be focused on species that are already threatened today, or which are likely to be threatened by further developments in the Alps.

However, the Alps still provide opportunities to study Lepidoptera under natural, undisturbed conditions. General ecological studies of single species and communities of Lepidoptera under natural conditions should not be neglected as they carry a high potential to provide results which are also important for conservation.

18.5   ACKNOWLEDGEMENTS

I am grateful to Dr Jacqui Shykoff and Steven Whitebread for comments on and corrections of earlier drafts of the manuscript. I also thank Professor Henri

Descimon, Dr Johann Gepp, Dr Peter Huemer and Dr Diethart Matthies for valuable information and discussions. Research was supported by the Swiss National Science Foundation (project no. 3.643–0.75 and 'Nachwuchs stipendium' to Andreas Erhardt).

CHAPTER 19

Conservation of butterfly habitats and diversity in European Mediterranean countries

M.L. Munguira

19.1 INTRODUCTION

The biological history of the past few million years has made the Mediterranean area one of the richest areas of Europe in terms of species diversity. Three-quarters of the total European insect fauna are found in the Mediterranean (Balletto and Casale, 1991). The structure of the three peninsulas surrounded by this sea, with mountains in the north isolating them, has resulted in a high level of endemic species and very rich butterfly communities.

On the other hand, the area is the target of important tourist activity, being one of the principal income sources of countries such as Spain, Italy and Greece. Nevertheless, most of the land is unsuitable for agricultural or industrial activity and this has made the whole area a sanctuary for wildlife.

Awareness of butterfly conservation did not develop until quite recently, appearing in Spain in the 1970s after the publication of the book on Spanish butterflies (Gómez Bustillo and Fernández-Rubio, 1974) and the first Lepidoptera red data book (Viedma and Gómez Bustillo, 1976). In Italy the topic has had a similar development and became important after a paper on the conservation of Italian Lepidoptera with recommendations for future actions (Balletto and Kudrna, 1985) and the publication of the Italian red data book (Prola and Prola, 1990). Unfortunately there seems not to have been a similar development in Portugal, the former Yugoslavia and Greece, and this is particularly sad in Greece because of the very interesting endemic fauna of this country, discussed later.

The study of butterflies was at first particularly devoted to faunistics, then community ecology studies (e.g. Balletto *et al.*, 1977; Viejo, 1986) and at the

Ecology and Conservation of Butterflies Edited by Andrew S. Pullin.
Published in 1995 by Chapman & Hall. ISBN 0 412 56970 1

moment research is beginning to produce the first results as far as single-species ecology is concerned (e.g. Munguira, 1989; Jordano, Fernández and Rodríguez, 1990). Thus butterfly studies are beginning to make conservation ecology easier through the knowledge of the requirements for each species, although the same geographical dissimilarity mentioned above is evident.

19.2 BUTTERFLY DIVERSITY IN EUROPE

It is generally assumed that animal diversity increases in the northern hemisphere as we move to the south (see, for example, the thorough study on Palaearctic Papilionoidea: Kostrowicki, 1969; or García-Barros, 1988, for the genus *Hipparchia*). In some areas, such as the Iberian Peninsula, this tendency is reversed due to the peninsular effect (Martín and Gurrea, 1990), and in southern Europe species richness increases towards the Alps and decreases again in the Mediterranean (J. Martín, personal communication; see also Dennis and Williams, Chapter 15 this volume).

Using data from Kudrna (1986) the only four countries with more than 200 Papilionoidea species are Spain, France, Italy and the former Yugoslavia. In Bretherton (1966) the same four countries plus Switzerland have more than 200 'Rhopalocera'. Figure 19.1 shows that most of the countries with 100–200 species are central European, and most of the ones with less than 100 species are northern European.

France is the only central country with more than 200 species, because of its size and diversity of habitats created in part by its Mediterranean area and the mountain ranges found in its territory. These data show the butterfly richness in Mediterranean countries and therefore their importance for the conservation of overall butterfly fauna in Europe.

The differences between north and central countries are statistically significant, but those between the centre and the Mediterranean are not significant because of the increased richness in the Alps. This analysis uses political entities (i.e. countries) as a source. This is certainly not suitable from the biogeographical point of view, but as conservation is not only a biological but also a somewhat political issue, the approach is not totally wrong.

In an attempt to exclude the bias in the analysis caused by the differing areas of countries, the number of species in each country was plotted against the logarithm of the area of that country (ln area). Mediterranean countries also score high in this analysis (Figure 19.2), although very large countries (i.e. Spain) are clearly underrated, whereas small countries (e.g. Luxembourg) are overrated.

Differences are again significant between northern and central countries, but not when comparing central and southern countries.

19.3 ENDEMIC SPECIES

The number of species found in only one of the European countries is shown in Table 19.1.

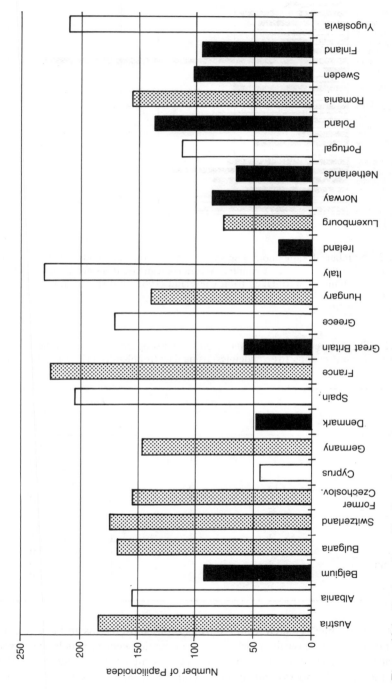

Figure 19.1 Total number of Papilionoidea recorded in European countries. Countries in northern, central and southern Europe are shown in black, grey and white, respectively. (Data taken from Kudrna, 1986.)

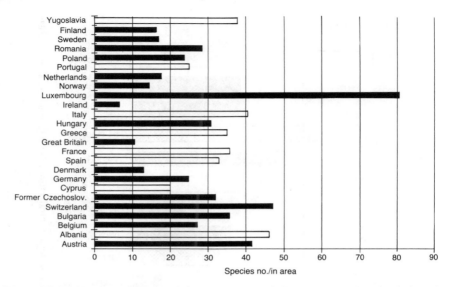

Figure 19.2 Number of Papilionoidea species in each European country, divided by the natural logarithm of the area of that country. Mediterranean countries are shown in white. The areas of the countries were taken from Cleveland (1987) and species numbers from Kudrna (1986).

Table 19.1 The number of species found in only one European country, and the number really restricted to one country (true endemics) using data from Kudrna (1986) and Higgins and Hargreaves (1983)

Country	Species in one country		True endemics	
	Kudrna	*Higgins*	*Kudrna*	*Higgins*
Russia	30	–	–	–
Spain	30	21	16	10
Greece	21	14	13	8
Italy	13	5	13	5
Portugal	6	1	0	0
Cyprus	5	–	4	–
Bulgaria	2	0	2	0
Austria	1	1	1	1
Yugoslavia	1	0	1	0
Great Britain	1	0	1	0
France	1	2	1	0
Switzerland	0	1	0	0

Data from this table are taken from Kudrna (1986) and Higgins and Hargreaves (1983), and do not necessarily show the true endemic species, which are mixed with species at the edge of their distribution range. True endemic species are also shown in the table. It must be taken into account that many of Kudrna's species are treated as just subspecies by Higgins (1975) and Higgins and Hargreaves (1983). By any standards, Mediterranean countries have the largest number of endemic species, accounting for 69% of the species present in only one country and 94% of the true endemic species (the last figure does not include the former Soviet Union). Only three species are endemic to countries outside the Mediterranean: *Aricia artaxerxes* (northern brown argus) from the UK, *Erebia orientalis* from Bulgaria and *Erebia claudina* (white speck ringlet) from Austria.

The importance of endemic species for an overall conservation strategy in Europe is evident. They represent 12% of the total number of species of Papilionoidea, and their disappearance from one country implies their total extinction. They are also products of a very peculiar biological history and are indicators of extremely rare animal and plant communities, whose conservation should be held as a priority. The main countries for endemic species in the Mediterranean are Spain, with 16 species, and Greece and Italy, with 13, confirming that the three peninsulas have had a very important role in the formation of endemic species in Europe and that some care should be taken to preserve this important heritage.

19.4 ENDANGERED HABITATS

This topic has been reviewed by Balletto and Casale (1991) when dealing with conservation of Mediterranean insects. Munguira, Martín and Balletto (1993) pointed out the relevant habitats for lycaenid conservation, and Balletto (1992) dealt with the Italian situation. Most of the ideas that follow are based on these previous reviews, although neither of them specifically deals with Mediterranean Papilionoidea.

Wet grassland is probably the most endangered habitat in Europe (Munguira, Martín and Balletto, 1993). The Mediterranean basin is not excluded from this reality, but the main values of the area as far as conservation is concerned are centred in mountain areas to which some priority should be given.

19.4.1 High-altitude grasslands and screes

Some of the higher Mediterranean mountains have several endemic butterflies and have very peculiar butterfly communities. Some of these high-altitude species are shared with central European or northern countries, and sometimes they tend to form distinct subspecies. Climax alpine or Mediterranean high-altitude grasslands are not particularly rich in butterfly species (Balletto, Barberis and Toso, 1982). Nevertheless some of the species living on these extreme habitats have a very restricted distribution range and are sometimes very rare.

These habitats are not severely threatened as a whole, but the communities living on them are so sensitive to slight changes that very strict measures should be taken to preserve them. Human influence on these areas is almost entirely related to tourism, mainly involving the building of ski courses in the places where butterflies live. The screes and orophilous grasslands of Sierra Nevada in southern

Spain are an example of these habitats. Here populations of two endemic species; *Lysandra golgus* (Nevada blue) and *Agriades zullichi*, and three endemic subspecies are threatened by redevelopment of a ski station (Munguira, 1989; Munguira and Martín, 1989). Other examples include Mont Pollino in southern Italy, the habitat of *Agrodiaetus galloi*, and the Chelmos or Taígetos Mountains in Greece, to which several endemics are restricted. *Polyommatus menelaos* (Grecian eros blue) lives in grassland habitats on Mt. Taígetos, and is another example of a butterfly which is vulnerable because of a very restricted range.

19.4.2   Mountain shrublands

Some climax shrublands in Mediterranean mountains are habitats for an extremely interesting butterfly fauna. The plant communities have a typical 'vegetable hedgehog' aspect (Polunin and Smithies, 1973) with spiny legumes (especially *Astragalus* species), crucifers and umbellifers, all with a cushion-like shape. These formations can be found in mountains of southern Europe and Africa, such as Sierra Nevada and Sierra de los Filabres in southern Spain or the Idhi and Dhíkti Mountains in Crete, the latter being the habitat of the narrowly endemic *Kretania psylorita* (Cretan argus).

Threats to these habitats in the Spanish sierras come from the plantation of *Pinus* spp., with subsequent vulnerability to fires, causing loss of the characteristic plant communities. Examples of habitat reductions after *Pinus* plantations on these areas can be found in Sierra de los Filabres, in the habitat of *Parnassius apollo* (apollo) and *Pseudochazara hippolyte* (Nevada grayling) (Kudrna, 1986), and Sierra Nevada, in places where the following species listed in the Spanish Lepidoptera red data book (Viedma and Gomez Bustillo, 1976) can be found: *Cupido lorquinii* (Lorquin's blue), *Aricia morronensis* (Spanish argus), *Lysandra nivescens* (mother of pearl blue), *P. apollo* and *P. hippolyte*. In Crete the type locality of *K. psylorita*, a mountain range where *Zerynthia cretica* (Cretan eastern festoon), *Coenonympha thyrsis* (Cretan small heath) and *Hipparchia cretica* (Cretan grayling) also fly, has been partially destroyed by a ski course and the road leading to it (Leigheb, Riboni and Cameron-Curry, 1990). Similar problems are shared by the heathlands found at the highest altitudes in Corsica, Sardinia and Elba which form the habitat of the restricted endemic, *Lycaeides corsica* (Balletto, Toso and Lattes, 1989).

19.4.3   Mountain grasslands

Mountain grasslands are the richest habitats for butterflies, both in Spain and in Italy. In our area these habitats are always the result of interaction between human activities and nature, the richest grasslands are those that result from transformation of former oak forests, such as *Quercus pubescens* in Italy (Balletto, Barberis and Toso, 1982) and *Q. faginea* and *Q. pyrenaica* in Spain (Viejo, Viedma and Martínez, 1989). In Italy overgrazing has been pointed out as a threat to these habitats, and the same can probably be said for some of the mountain grasslands in southern Spain and Greece, but the situation is probably going to change dramatically with EC agricultural policy. Although overgrazing may have been a real threat for some butterfly habitats, abandoning stock grazing is not going to favour their conservation. After abandoning pastures, natural succession will proceed and most of the grassland habitats will disappear, giving way to secondary forests.

Among the species affected are the *Maculinea* (large blues) which are near the limit of their distribution in the Mediterranean, but which have strong populations in most countries. The very local species of the genus *Agrodiaetus* (anomalous blues) typically live in this kind of grassland and some members of the group are certainly endangered: *A. humedasae* (Piedmont anomalous blue) in the Vall d'Aosta (Italy), *A. fulgens* (Catalonian anomalous blue) from just a few localities in Catalonia, and probably *A. nephohiptamenos* (Higgins's anomalous blue) from northern Greece. Many other species will be negatively affected by changes in mountain grassland management, the following are just some examples: *Papilio hospiton* (Corsican swallowtail), *Melanargia arge* (Italian marbled white), *M. pherusa* (Sicilian marbled white), *Fabriciana elisa* (Corsican fritillary) and *Pseudophilotes barbagiae* in Italy (Balletto, 1992); *C. lorquinii*, *Eumedonia eumedon* (geranium argus), *Erebia palarica* (Chapman's ringlet) and *E. zapateri* (Zapater's ringlet) in Spain (Munguira, Martín and Viejo, 1988; Munguira, 1989).

The use of pesticides and fertilizers in hay meadows, which has had a negative effect in most northern countries (Erhardt and Thomas, 1991), is probably irrelevant in Mediterranean countries, because grassland management has normally followed traditional land uses, and chemical treatments are not a common practice.

19.4.4 Dry grasslands

Dry grasslands are frequent at low or medium altitudes in Mediterranean countries. Although they can not be classified as steppes, the vegetation structure and some of the plant species are very similar to these formations peculiar in Europe to eastern countries. Poor soils, low rainfall and grazing by rabbits or domestic stock preserve the grassland plant communities. The presence of grasses and crucifers makes these habitats especially suitable for satyrines and pierids, some of which are not at all abundant in the Mediterranean.

Some of the Italian species listed as rare by Balletto (1992) occur in dry grasslands on the edge of their distribution range. For this reason their ecological requirements are different from those in their main strongholds (e.g. *Glaucopsyche melanops* (black-eyed blue), *Melanargia occitanica* (western marbled white) and *Hipparchia fidia* (striped grayling)). In Spain, dry grasslands are habitats for many species only found in the Iberian Peninsula and north Africa: *Pseudophilotes abencerragus* (false baton blue), *Melitaea aetherie* (aetherie fritillary), *Chazara prieuri* (southern hermit), *Euchloe belemia* (green-striped white) and *Zegris eupheme* (sooty orange tip, also present in Asia). The African migrant, *Colotis evagore* (desert orange tip), has established itself in the southern coast of the Iberian Peninsula, and lives in hot, dry habitats. It has, nevertheless, been unable to establish itself permanently on inland colonies due to host-plant availability and unfavourable climatic conditions (Jordano, Retamosa and Fernández Haeger, 1991).

Two other species, *Iolana iolas* (iolas blue) and *Plebejus pylaon* (zephyr blue), are endangered lycaenids typical in dry grasslands with some shrub cover, especially the first species. They are very rare in the Mediterranean and sensitive to any changes in land use (Gómez Bustillo, 1981).

Traditionally these habitats have been considered useless, and certainly steppe-like habitats are unattractive to the general public. Overgrazing, fire, urbanization or changes in land use can be considered the main threats to these grasslands. The extremely peculiar butterfly fauna of one locality in central Spain (El Regajal in Aranjuez, Madrid Province) is the reason for the creation of the first nature reserve in Spain devoted to butterfly conservation (Viedma *et al.*, 1985). This reserve has a mixture of Mediterranean chaparral and dry grasslands with many of the typical species mentioned above. Its protection has been one of the milestones for invertebrate conservation in Spain, and certainly a good starting point for the other Mediterranean countries.

19.4.5   Mediterranean woodlands

Well-conserved forests are not particularly rich in butterfly species (Balletto, 1992), because they lack the open habitats in which these insects are abundant. Nevertheless as many Mediterranean woodlands are managed in some way, species typical to other habitats can be found in clearings. In this region, species such as *Callophrys avis* (Chapman's green hairstreak) and *Charaxes jasius* (two-tailed pasha) are restricted to these woodlands. *Hipparchia balletoi*, found in the beech woods of Mount Faito in Naples, is suffering from inappropriate woodland management.

Some of these woodlands are adversely affected by domestic pigs in Italy (Balletto, 1992) or even wild boar in Doñana National Park and other nature reserves in Spain. Other causes of concern are the extremely damaging fires during the very dry summers, and plantation with alien species after the removal of natural vegetation. This second activity favours fires, which then frequently spread to adjoining natural woodlands destroying huge areas (for example in the 1980s an average 91 000 ha of woodland, on average, were burnt in Spain per year; García, 1992).

19.4.6   Wetlands

In all Europe wetland drainage has endangered all the species exclusive to these habitats. The Mediterranean countries share this problem with northern countries, and the same species are involved. Both in Italy and Spain the following wetland or wet grassland species are considered vulnerable or endangered: *Maculinea alcon* (alcon blue), *M. nausithous* (dusky large blue), *M. teleius* (scarce large blue), *Boloria eunomia* (bog fritillary), *Lycaena dispar* (large copper) and *Coenonympha oedippus* (false ringlet). The last four are only present in Italy.

In the Po Valley most of the wetland habitats have given way to rice fields, and wetland species such as *L. dispar* that used canals as secondary habitats are now disappearing due to the use of pesticides (Balletto, 1992; Pullin, McLean and

Webb, Chapter 11 this volume). In Spain, wet meadows will probably suffer from a reduction in cattle grazing as a result of EC agricultural policy. The meadows where the few populations of *M. nausithous* live are threatened in this way.

19.4.7 Coastal habitats

The best-known Mediterranean areas are also the most thoroughly destroyed. In some places they are 'covered in an almost continuous ribbon of concrete' (Balletto and Casale, 1991), a common sight in most tourist areas, although Greece is probably less affected. This massive destruction of natural habitats has not yet resulted in any documented butterfly extinction, but the habitat loss is probably one of the greatest to have taken place in Europe after the industrial revolution.

The species involved are mainly those living in sclerophilous forests and their accompanying shrublands. Butterflies such as *C. jasius* or *Nordmannia esculi* (false ilex hairstreak) have certainly lost a good deal of their habitat to urbanization. None of these species is endangered as a whole, but at the edges of their distribution ranges any impact can make them disappear from large areas. *Tarucus theophrastus* (common tiger blue) is an African species established in Europe in a few coastal areas of south-eastern Spain (Munguira, Martín and Rey, 1991); the same situation is true for an established population of *Danaus plexippus* (monarch) in Málaga (Martín and Gurrea, 1988). These species need very specific climatic conditions and are unable to colonize localities inland, being vulnerable to changes in land use, and particularly to urbanization. *Tarucus theophrastus* is protected in a nature reserve (Cabo de Gata, Almería), but the rest of the populations of the species are indeed in vulnerable areas, and the whole of the species will certainly be at risk in Europe if steps are not taken to protect its habitat.

19.5 ADVANCES IN CONSERVATION PRACTICE

Two main steps have been taken in Mediterranean Europe and are shared with other European countries: the creation of nature reserves and the legislation to protect butterflies. A third measure, the EC habitats directive, could be grouped with the legislation measures, although it is somewhat different in its approach. The latter is still being implemented and it is probably too soon to comment on its effectiveness.

19.5.1 National parks and nature reserves

The national park network is mainly devoted in Mediterranean Europe to outstanding landscapes, relevant forested areas or rare vertebrate species. In some areas rare or endemic plants are also taken into consideration. Nevertheless, as they are meant to be sanctuaries for wildlife they can also protect butterflies wherever this is compatible with the protection of other natural values. In Spain, for example, national parks and nature reserves are habitats for 45% of the endangered butterfly species (Viejo *et al.*, 1992). The only drawbacks of relying on nature reserves for butterfly conservation are that, in most cases, the most interesting sites are not included in protected areas and some rare or endangered species do not live in protected areas. For example, in Spain only two of the nine national parks have endangered butterflies in their boundaries. In Greece the three most important areas for endemic or endangered butterflies are not represented in the 10 national parks (65 000 ha) that have been declared up to now. The situation is similar in Italy, where the most relevant sites are not included among the five existing national parks (271 400 ha), although some of the endangered species may be represented in some of the parks, as in the Spanish case.

The conclusion is that it is not enough to rely on national parks to conserve butterflies, and nature reserves or other protection measures should be used to protect relevant butterfly habitats.

19.5.2 Legislation

European legislation protecting butterflies was reviewed by Heath (1981a). In this survey the only Mediterranean country with legislation protecting butterflies was France, with 17 protected species. In those days legislation was mainly centred on protecting butterflies against collecting activities, but the situation has changed substantially in recent times. In 1988 the Bern Convention legislated for the protection of invertebrates, among which were several endangered Mediterranean butterflies mentioned in this paper: *P. hospiton*, *P. apollo*, *M. arge*, *C. oedippus*, *L. dispar*, *M. teleius*, *M. nausithous* and *L. golgus* (Fernández-Galiano, 1992). Cyprus, Greece, Italy, Portugal, Spain and Turkey are Mediterranean countries acting as Parties to the Convention. While previous legislation has been strongly criticized because of its emphasis on collecting (Balletto, 1992), the advantage of the Bern Convention is that it stresses habitat protection. The real effectiveness of this legislation to protect declining or endangered species is yet to be proved in the Mediterranean, but no doubt it is going to be an important weapon for conservationists in the foreseeable future.

Spanish legislation to protect butterfly species has been reviewed by Viejo and co-workers (Viejo *et al.*, 1992). Local or national legislation undoubtedly exists in more Mediterranean countries, but its effectiveness relies mainly in its capacity to save relevant butterfly habitats against the impact of human activities.

19.6 CONSERVATION PROPOSALS

Developed industrialized countries have certainly gone further in awareness for insect conservation and have taken more action to conserve species or habitats. Among Mediterranean countries this awareness has developed mainly in Italy, and it is evident that poorer countries are less interested in conservation, although in fact their natural habitats are still in good condition.

Any effective conservation action must include the following three guidelines:

1. Information: politicians and land managers need to perceive a general concern before any action is taken. The general public is certainly aware of the problems with pollution, the greenhouse effect or extinction of mammals and birds. Therefore the first step to be taken is to let people know about the problems that butterflies face in the Mediterranean, and which species are involved. In most countries this action has already begun, in part, with the publication of red data books, posters and leaflets, but we are still far from the level of knowledge concerning butterflies that the general public has in northern countries.

2. Research: the ecology of most Mediterranean butterflies is very poorly understood, and even for some endangered butterflies basic biological knowledge is still wanting. Studies of butterfly ecology should be centred on endangered species, particularly those restricted to Mediterranean countries. This would certainly produce some insight into the specific problems faced by these species, and their possible solution. Large-scale research programmes, such as international mapping schemes, are certainly necessary, but the past years have shown how difficult it is to receive the proper funding, even for programmes on a national scale like the one suggested by Balletto and Kudrna (1985) for Italy.

3. Active conservation: this involves first the creation of nature reserves to enclose relevant butterfly habitats and, secondly, the proper management of these reserves to maintain suitable population levels. The first action is the easiest, except when it conflicts with economic interests. The actual network of nature reserves and national parks in Mediterranean countries can certainly be used as a basis, and would conserve a good percentage of the Mediterranean fauna, endangered species included.

Adequate management for butterflies in nature reserves, to prevent changes in land use, is more difficult to implement, and here a good information campaign is necessary. In most places the necessary actions would need proper funding, and this is the main reason why prospects are not very good. Subsidies are nowadays the only way to encourage traditional land uses where they are dramatically disappearing, or no longer profitable. Subsidies are also the best way to prevent overgrazing in the places where this practice poses a problem for butterfly populations.

Wetland drainage and the transformation of wetlands into agricultural land should be stopped, particularly in the Mediterranean where wetlands are very scarce. One possible action is to declare relevant wetlands for butterflies as RAMSAR sites, so far this has only been used for habitats rich in bird species. The creation of specific nature reserves is necessary, always bearing in mind that proper management is required. Information is also necessary, particularly in places where the use of wetlands by tourists can produce more profit than devoting the land to agriculture, particularly considering that Europe has a surplus of agricultural products.

Actions against fires include the implementation of efficient prevention programmes, afforestation with autochthonous species that are more resistant against fire and having specialized and well-equipped fire brigades in all the vulnerable areas.

The negative effects of urbanization are indeed the most difficult to fight. First of all, reversal of urbanized areas to the previous state is almost impossible. But above all, developers have strong arguments to carry on with their proposals, because according to public opinion the richness of a region depends more on the activities of developers than on those of conservationists. As a result the conservation of butterfly habitats is a battle for every relevant biotope. The battle is not won even after the declaration of a site as nature reserve, as shown by the situation in Sierra Nevada (Spain), where new developments are taking place in the very centre of a natural park to fit the ski station for the needs of the 1995 World Ski Championships.

The only hope to reverse this situation lies in the increasing demand for 'ecological tourism', because if it becomes profitable to keep the environment in a good state of conservation, this will encourage developers to be more concerned with nature preservation. A new understanding of tourism and development is certainly beginning, and the best action we can possibly take as conservationists is to support it, showing the potential natural values and beauties of the habitats we want to protect.

19.7   CONSERVATION AREAS IN MEDITERRANEAN EUROPE

In Mediterranean countries some of the most endangered species of butterfly are those that are also present in other European countries (i.e. *M. nausithous* in Spain, or *C. oedippus* in Italy). Nevertheless, when considering the Mediterranean as a whole, the butterflies most urgently in need of conservation are those exclusive to restricted areas. For this reason the following areas suggested as priorities for conservation in Mediterranean Europe are centred on these restricted endemics. Although the selection may be subjective, the chosen areas cover 18 (35%) true endemic species (see Table 19.1) present in the Mediterranean.

19.7.1   Sierra Nevada (south-eastern Spain)

This area has two exclusive species (*L. golgus* and *A. zullichi*) and seven more species listed in the Spanish red data book (Viedma and Gómez Bustillo, 1985). Some habitats for these species are threatened by tourist development. The area is a nature reserve, but this has not protected it against urbanization.

19.7.2   Massiccio del Pollino (southern Italy)

This area is the habitat of *A. galloi* and is also very rich in species. Parenzan (1975) recorded 104 butterflies, including two more Italian endemics (*M. arge* and *Agrodiaetus virgilius*). Overgrazing is one of the threats to this valuable mountain range (Balletto, 1993). This area should be protected against damaging impacts to conserve its rich butterfly fauna.

19.7.3 Vall d'Aosta (Italian Alps)

An extension of the Gran Paradiso National Park would protect *A. humedasae* (Balletto, 1992) and other exclusive insects and plants (Balletto, 1993). Threats to the area are mainly represented by changes in land use.

19.7.4 Mountains of Corsica and Sardinia (France and Italy)

As with the island of Crete, several species have evolved since the isolation of these islands from the mainland. Particularly attractive species such as *P. hospiton* may be endangered by collecting in some areas (Balletto, 1992) and by tourist developments such as ski resorts in Corsica (Heath, 1981a). Creation of nature reserves in well-conserved areas would be a first step towards the conservation of at least six endemic species.

19.7.5 Idhi and Dhíkti Mountains (Crete)

The creation of a nature reserve in these mountains, together with the already existing Samaria National Park, would protect the endemic Cretan fauna (*Z. cretica, K. psylorita, C. thyrsis* and *H. cretica*). The development of ski stations and other activities related to tourism can be considered the main threat to the area.

19.7.6 Taígetos Mountains and Mont Chelmos (Peloponnesus, Greece)

These mountains, together with Sierra Nevada in Spain, are very important for the conservation of endemic species. The species involved are: *P. menelaos, Agrodiaetus aroaniensis* (Grecian anomalous blue) and *Pseudochazara graeca* (Grecian grayling). Other eastern species limited in Europe to the mentioned ranges are *Agrodiaetus iphigenia* (Chelmos blue), *A. coelestinus* (Pontic blue), *Turanana panagea* (odd-spot blue) and *Colias aurorina*. The creation of a national park in each range is highly recommended to protect their wildlife.

19.8 ACKNOWLEDGEMENTS

I am grateful to José Martín and José Luis Viejo who made valuable suggestions to a previous version of the manuscript. Enrique García-Barros provided relevant information on the distribution of several Mediterranean species.

CHAPTER 20

Butterfly biodiversity and conservation in the Afrotropical region

T.B. Larsen

20.1 OVERVIEW OF THE AFRICAN BUTTERFLIES

20.1.1 Number of species

About 3600 butterfly species are now recognized from the Afrotropical region (Ackery, Smith and Vane-Wright, 1994), constituting 20% of the 18 000 known species. This places Africa as the second richest of the major biogeographical regions as far as butterflies are concerned (Table 20.1), although it is noteworthy that the Neotropical fauna is more than twice as rich.

Of these 3600 species, more than a third have been described during the past 30 years. When Aurivillius (1898) compiled the first ever catalogue of African butterflies, the total came to about 1600. By the time of the volume of Seitz (1925),

Table 20.1 Approximate number of butterfly species of the major biogeographical regions of origin

Biogeographical region	Species	%
Neotropical	8000	44
Afrotropical	3600	20
Oriental	2700	15
Palaearctic	2000	11
Papuan/Australian	1000	6
Nearctic	700	4
Total	18 000	100

Sources: D'Abrera (1982–86), Shields (1989), Heppner (1991), Larsen (1991), Ackery, Smith and Vane-Wright (1994).

Ecology and Conservation of Butterflies Edited by Andrew S. Pullin.
Published in 1995 by Chapman & Hall. ISBN 0 412 56970 1

the total was just under 2000. Next came the checklist of Peters (1952) with a total of 2615 or so. Carcasson (1964) dealt with 2700 species, which he thought might represent 85% of the true total. By the time he published his checklist for the first time, the number already exceeded his earlier estimate, reaching 3200 (Carcasson, 1981). In the decade since then, the total has grown by another 400 species. The true total is probably 15% higher than the 3600 species currently recognized.

20.1.2 Biogeographical affinities

The Afrotropical butterfly fauna is most closely related to that of the Oriental region, with which it shares all butterfly families, most subfamilies, many tribes and about 40 genera. Affinities with the Neotropical and Palaearctic regions are very weak. The number of species shared with other regions is negligible (Larsen, 1991). There are two subfamilies limited to Africa: Pseudopontiinae with the single species *Pseudopontia paradoxa* Felder (the ghost) and the Lipteninae with more than 500 species. The latter has affinities with the purely Oriental Poritiinae (Eliot, 1973); some authors consider them a single subfamily (e.g. Heppner, 1991). Conversely, the small subfamily Amathusiinae is the only other subfamily proper to the Oriental Region (see also Ackery, 1984).

The genus *Neptis* (Sailers) is a good example of the relationship between the Afrotropical and Oriental regions. There are more than 50 species in each region, but it is quite impossible to say that any one Oriental species is more closely related to one of the Afrotropical species than to another.

Clearly, the two regions once constituted a joint evolutionary unit which later split into two. Most evidence suggests that a cooling of the world climate during the Miocene, when the climate in what are now London and Tokyo was tropical, pushed the fauna southwards into two areas which have since remained more or less disjunct.

20.2 ECOLOGY AND BIOGEOGRAPHY OF AFRICAN BUTTERFLIES

While working at the Coryndon Museum in Nairobi (now the National Museums of Kenya), Carcasson (1964) published a concise biogeographical analysis of the African butterflies. His paper remains one of the finest biogeographical accounts of any large group of animals in Africa. The remainder of this section is based on Carcasson's paper as well as on my own work on the fauna of Kenya (Larsen, 1991), Botswana (Larsen, in preparation), and West Africa.

20.2.1 Main ecological and biogeographical groupings

Virtually all African butterflies are restricted to one of a limited number of ecological zones (Figure 20.1).

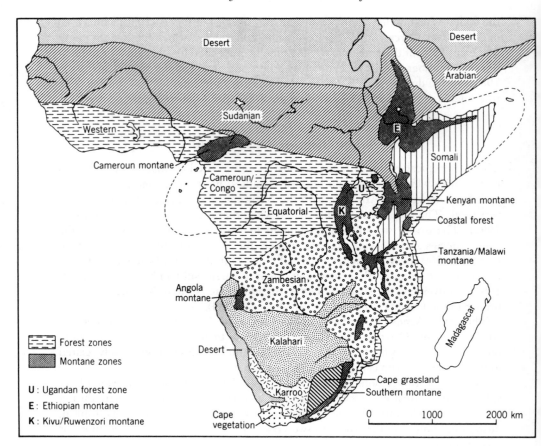

Figure 20.1 Biogeographical map of Africa, adapted after Carcasson (1964); the 'Dahomey Gap' between the western and Cameroun–Congo forest zone is not shown (by kind permission of Oxford University Press).

The number of species that can be said to be ubiquitous is very small, less than 1% of the total. However, many genera have species that are adapted to different ecological zones. The *Neptis* are, again, illustrative. Each of the major zones has its own set of *Neptis* species, the distributions of which hardly overlap.

20.2.2 Lowland forest and its subdivision

Lowland forest (rainforest) is by far the most important ecological zone for butterflies and most other organisms, and may be amazingly rich in species. One small hill in Cameroun has so far yielded almost 800 species, and a small area of seriously degraded forest/bush area near Lagos, 385 species (Larsen, Riley and Cornes, 1980). The forest stretches from the Basse Casamance in Senegal, along the West African coast, briefly interrupted in Togo and Benin by the 'Dahomey Gap'. It then stretches continuously from western Nigeria, through the Congo Basin to Uganda, western Tanzania and western Kenya. Additionally, there are small isolated forest areas in southern Sudan, Ethiopia and south-central Africa, as well as a narrow strip of coastal forest from southern Somalia to Mozambique and Natal.

Many large and important butterfly genera are hardly found outside the lowland forest zone, although some have montane derivatives: *Mylothris* in the Pieridae, virtually all the Lipteninae (excepting a few genera such as *Baliochila*, *Cnodontes* and *Alaena*), many genera of the Theclinae (*Oxylides*, *Syrmoptera*, *Actis*, *Diopetes*, *Hypokopelates*) and a few of the Polyommatinae (*Oboronia*, *Athysanota*). The Nymphalidae have numerous genera restricted to forest, mostly in the Limenitinae (e.g. *Euphaedra*, *Euriphene*, *Bebearia*, *Catuna*, *Euptera*, *Pseudathyma*) and the Nymphalinae (*Kallimoides*, *Kamilla*). Many genera of the Hesperiidae are restricted to forests, in both the Pyrginae (*Pyrrhochalcia*, *Pyrrhiades*, *Katreus*) and the Hesperiinae (*Ceratrichia*, *Mopala*, *Caenides*, *Leona*).

There are five main subdivisions in the lowland forest zone. Western West Africa, from Senegal to the Dahomey Gap, probably has about 800 forest species, of which perhaps 120 are endemic to the subdivision. At present the Dahomey Gap almost blocks contact between the western fauna and that of Nigeria, but there has been contact at various times in the past, and the Nigerian fauna is intermediate. The second large subdivision covers eastern Nigeria (especially east of the Cross River), much of Gabon, Congo, Central African Republic and the Mayumbe area of Zaïre. This is one of the richest areas of butterflies, with a total probably surpassing 1700 forest species, since there is significant overlap with neighbouring zones (West Africa and Zaïre) – Cameroun alone has 1500 species (Libert, 1992). The third subdivision is the main forests of the Congo Basin in Zaïre, the fauna of which is not well known, but probably almost as rich as the previous zone. The Ugandan zone, covering parts of eastern Zaïre, Uganda and western Kenya, has seen a fair amount of speciation, partly because the forest habitat was always rather fragmented, and contains about 900 forest species, of which some 120 are endemic. Finally, there is the coastal forest zone of eastern Africa, with about 150 forest species, many of which are specifically or subspecifically endemic. However, it is worth noting that in both West Africa and in the coastal forests there is virtually no endemism at the level of genus, while many genera of the Cameroun and Zaïre forests do not extend to the outlying zones.

20.2.3 Montane forests and their subdivisions

Montane forests occur mainly in those parts of East Africa bordering the great Rift Valley with its relatively recent tectonic and volcanic activity, as well as the Cameroun and Adamawa mountains further west on the Cameroun/Nigeria border. There is also a montane zone in Angola. However, some of these mountains, such as the Ulugurus and Usambaras in Tanzania, are much older. At the Equator, montane conditions begin at a level of 1800 m, decreasing as one moves south, to the point where the lowland forest of the Cape is biogeographi-

Table 20.2 Endemic butterflies in selected mountains of Tanzania, still little researched

Mountain	Endemic spp.	Endemic ssp.
Uluguru	8	11
Nguru	4	6
Rubeho	4	5
Uzungwa/Iringa	7	12
Usambara/Pare	11	7

Source: Kielland (1990).

cally montane. The montane ecological zone is of special interest since it consti-tutes a vast archipelago, with many isolated mountains, often of small geographical extent. Each of these has its own species composition, normally with a significant level of endemism at both species and subspecies level (e.g. Table 20.2).

The number of species is always considerably lower than in lowland forest (250–300 species). The fauna is mainly derived from genera also prominent in the lowland forest, but there are a number of essentially montane genera. In the Pieridae, *Pieris* and *Colias** are montane, as are *Lycaena** (Lycaeninae), *Capys* (Theclinae) and the Polyommatinae (*Uranothauma, Harpendyreus*). In the Nymphalidae we find *Argyreus**, *Issoria, Antanartia* (with one lowland species) (Nymphalinae), *Lasiommata**, *Aphysoneura* (Satyrinae) (genera marked * are mainly Palaearctic). In the skippers most of the true *Metisella* and *Chondrolepis* are montane. Many of the largest African butterfly genera have montane sections which are characteristic of all the montane zones (e.g. *Mylothris, Lepidochrysops, Charaxes, Acraea, Celaenorrhinus* and many more).

The montane forests exist under two conditions. They either gradually emerge from lowland forest as altitude increases, or they emerge from the savannahs where mountains gather sufficient rain for forest to develop. The rocky ground often assists the development of forest, and its extent may be quite limited. As mentioned, the montane forest forms an archipelago from Ethiopia and southern Sudan via Kenya, Uganda, Kivu, Rwanda, Burundi, Tanzania, Zambia, Zimbabwe and Malawi to South Africa. The mountains of Angola and Cameroun are very isolated from the remainder of the montane forests. The natural extent of forested area (much degraded by human activity) varies from large expanses in Ethiopia, Central Kenya, Kivu and Rwanda to very limited areas such as the small massifs of Tanzania, Malawi and Zimbabwe.

Six rough groupings can be discerned:

1. Ethiopia, southern Sudan, Central Kenya;
2. Uganda, Kivu, Rwanda;
3. Tanzania, Malawi, Zimbabwe;
4. South Africa;
5. Angola; and
6. Cameroun.

Each zone has its characteristic montane butterflies, but there are many anomalies in both the absence and the presence of various species.

20.2.4 Savannah formations and their subdivisions

Savannah formations of various types cover most of Africa. They include the tall '*miombo*' woodlands of Zambia, Zimbabwe, Angola and parts of Tanzania, as well as the dense Guinea savannah bordering the West African rainforest. Most savannah formations are more open, like the Kalahari, the South African low-veldt, the Kenyan and Tanzanian savannahs and the Sudan savannah of West and Central Africa. North of these areas, the dry Sahel savannahs merge with the desert; Sahel-type conditions also occur in Namibia and southern Angola as transitions to the Namib Desert and the Skeleton Coast. Although each type of savannah does have its special species, the structure of the savannah fauna is generally rather similar, and many of the savannah species have vast ranges, from the Kalahari to Somalia, Sudan and Senegal. The savannahs of southern and eastern Africa have many more species than the vast West African savannahs. The number of species in a limited area ranges between 100 and 250.

Colotis and *Eronia* (Pieridae) are the savannah butterflies *par excellence*, often being found in vast numbers, and hardly penetrating the forests. In the Lycaenidae there are *Alaena*, *Cnodontes* (Lipteninae), *Chloroselas*, *Desmolycaena* (Theclinae), *Lepidochrysops*, *Azanus*, *Euchrysops*, and *Tarucus* (Polyommatinae). In the Nymphalidae the small genera *Hamanumida*, *Catacroptera* and *Byblia* (Nymphalinae) are limited to savannah habitats, as are some genera of the Satyrinae (*Physcaeneura*, *Coenyra*, *Coenyropsis*). In the Hesperiidae, the *Spialia* (with one exception), *Leucochitonea*, *Netrobalane* and *Caprona* (Pyrginae) are savannah species, as are some Hesperiinae (*Zenonia*, *Kedestes*, *Gegenes*).

Starting from the south, the Kalahari Savannah of Botswana and western Transvaal is relatively impoverished, with only 170 species of butterflies, only a handful of which are endemic. The Zambesian Savannah subdivision (Angola, Zambia, Zimbabwe, Tanzania, south-eastern Kenya) is much more ecologically varied than the Kalahari. It supports a rich and diverse fauna of up to 400 savannah species, many of which do not occur in other savannah subdivisions, in addition to the widespread savannah species. There is a proliferation of *Charaxes* (Charaxinae), *Acraea* (Acraeinae), *Bicyclus* (Satyrinae) and *Lepidochrysops* (Polyommatinae). The Somali subdivision (eastern Ethiopia, Djibouti, Somalia, Kenya and dry parts of north-central Tanzania) is much more arid than the Zambesian. None the less, it has a rich and diverse butterfly fauna of perhaps 300 species in all, many endemic, although some penetrate to Arabia, Sudan and Lake Chad. Among the endemics are several *Colotis* (Pieridae), *Chloroselas*, *Iolaus* (Theclinae), *Anthene* (Polyommatinae), *Acraea* (Acraeinae) and *Spialia* (Pyrginae). The relative richness, diversity and endemism of the Somali zone is in marked contrast to the impoverished fauna of the vast Western Savannah subdivision, stretching from Sudan to Senegal. West of northern Nigeria it is largely limited to the most widespread and common savannah butterflies found in all the subdivisions. There are no more than 200 species in all.

The less specialized savannah species are increasingly invading agricultural lands and other disturbed areas within the forest zone.

20.2.5   South African special zones

The South African special zones are technically not in the tropics, and are not closely matched elsewhere in Africa. They contain a large number of butterfly genera which are effectively limited to South Africa. The three main zones are the Cape, the Karroo and the Cape Grasslands, each of which has many distinctive butterflies. Curiously, few of these seem to feed on the multitude of endemic plant genera. A few of these genera do extend as far north as Kenya and Tanzania, and rather more to the Zimbabwe highlands, but most do not. They probably evolved at a time when a more southerly positioning of Africa made the temperate and subtropical zones much larger than they are today.

The special South African fauna is too complex to be dealt with further in this chapter, but it must be noted that many of the species have very limited ranges and need direct conservation. *Aloeides dentatis dentatis* may occur in a single colony, for which the Rumisig Nature Reserve has been created.

20.2.6   Desert

Deserts occur where rainfall decreases to 250 mm/yr or less. Generally there is a gradual transition between Sahel savannah and desert, following a rainfall gradient. Deserts are inimical to butterflies, and there are a very few desert specialists. Three desert areas occur in Africa: the Namib, parts of Somalia and the vast Sahara desert. The only butterflies that can exist on a permanent basis are those that have developed exceptional diapause capabilities or have made themselves independent of fresh plants. In the Sahara, the Pierid *Pontia glauconome* (desert bath white) can spend 5 years in its pupal stage waiting for rain. Its close relative, *Pontia daplidice* (bath white), is a powerful opportunistic migrant, providing an interesting example of radically different evolutionary solutions to the same problem. Members of the genus *Apharitis* (Lycaenidae) live commensally with ants and in periods of stress can feed on the early stages of ants, as do the ants themselves (Larsen and Pittaway, 1982). The few desert species are under no threat and will not be discussed further.

20.2.7 Islands

All the African islands, except Cabo Verde and Annobon, have endemic forest species that are threatened by encroachment on the forest. Chief of these is Madagascar, where nearly 250 of the 300 or so species are endemic. Fernando Po (Bioko) also has several endemic species. The Comores, Principe and Sao Thomé each have 50 or so species, with several endemics. Réunion and Mauritius have very few butterflies, some of which are endemic. *Libythea cinyras*, which was on Mauritius, is the only definitely extinct African butterfly. On all the islands, forest conservation is a priority; the issues are too specific to be discussed here.

20.3   THREATS TO THE BUTTERFLY FAUNA

The only real threat to African butterflies is the rapid loss of habitat due to human interference. This has been accelerated by rapid population growth, which has seen the population of sub-Saharan Africa increase at least fourfold to 500 million since the beginning of the twentieth century, and which is still growing at 3% a year. Agriculture and the need for firewood have drastically decreased tree cover in all the main ecological zones.

Habitat loss has been speeded up by the extensive growing of plantation and large-scale crops in some areas: coffee, rubber, cocoa and palm-oil in the forest areas; coffee and tea in the montane zone; and sisal, ground nuts and cashew in the savannahs. In the driest zones, such as the Sahel and the Kalahari, overgrazing by cattle has led to drastic alterations in the composition of the plant cover.

This is in addition to the high value of tropical hardwoods, which makes their overexploitation very tempting in times of economic and political stress. Under stable political conditions, their value is an argument for careful long-term management of forest resources, and this is increasingly being understood in Africa. Unfortunately, stability cannot be guaranteed; the Liberian civil war is largely being financed by indiscriminate, non-sustainable logging.

Butterflies are used for 'decorative' mosaic pictures sold in major cities throughout Africa. It is sometimes suggested that this may constitute a threat to the fauna. Irrespective of where they are sold, they all seem to come from the same group of manufacturers in the Central African Republic. Their representatives actually travel as far as Botswana to peddle them to the souvenir shops. Each involves the dismemberment of not more than six or eight butterflies, mainly the green *Graphium*, *Belenois*, *Appias*, and other Pierinae, *Eurema*, black *Charaxes* and a sprinkling of other common species. The species composition clearly shows that they have been collected or trapped at damp patches in disturbed areas and not from intact forest, and that each specimen is used judiciously and not profligately. Nothing is known about the scale of this industry (the three major hotels in Gaborone sell no more than one or two a week each), but 20 000–30 000 sales a year seems an outside maximum (there are probably 100 sales outlets in all of Africa, selling two or four a week). Thus, at the very most, a quarter of a million common butterflies are involved per year, although it would be interesting to study the matter more closely. Assessments in Papua New Guinea also support the view that overcollecting is unlikely to pose a real threat, unless combined with habitat destruction (Parsons, 1992a).

20.3.1   Relative levels of threat to habitat

(a)   Montane forests

The smaller montane forests are probably the most threatened African habitats, and many will disappear unless they are protected or existing protection is continued. Each of these has a special faunal composition and endemic species, but few have been adequately surveyed by either botanists or zoologists. Rwanda, which used to be wholly clad in trees, now has the highest population density of any mainly agricultural society, and essentially only the Nyungwe Forest is left. The remaining forests of most of the mountains in Tanzania have been reduced in size; several are rated very high on a list of key forests for bird preservation (Collar and Stuart, 1988). Most remaining forest of the Uzungwa Mountains is conserved as part of the management of the Brooke Bond Tea Estate at Mufindi. Forests of the Nguu, Nguru, Uluguru, Rubeho and other mountains have been greatly encroached upon in recent times.

(b)   Coastal forests

The coastal forests of East Africa probably now exist only where they are managed as parks or forests reserves. They were never very extensive, and are few in number. The main ones are Lower Tana River, Arabuko-Sokoke, Simba Hills (Kenya), Pugu Hills (Tanzania) and the Sofala coastal forests (Mozambique). Their continued conservation is of the utmost importance.

(c)   Lowland forests

The lowland forests of western Africa have shrunk dramatically in size and number, and are continuing to shrink. Quite a few butterflies are only known from forests that no longer exist. The relatively large and very distinctive *Euriphene kiki* is known from a single male, which I collected in an isolated forest near Abeokuta in western Nigeria in 1967; the forest has now been destroyed and the butterfly may well be extinct. Proper management and conservation of most of the remaining forest is essential. Most of the forests are poorly researched, but they seem to differ in species composition. Thus, the westernmost forests in Senegal's Basse Casamance and in Guinea (Conakry), have species and subspecies which seem to have evolved in response to the greater seasonality and dryness of the climate. The few remaining forests on the rocky ridge that forms the Ghana/Togo border seem to support a number of interesting endemic species (although we cannot be sure, because collecting has been too patchy in the neighbouring areas). The Kagoro Forest in northern Nigeria is of exceptional interest, yet it is so little known that it was not included in the key forest survey of Collar and Stuart (1988), although this was rectified by Wilkinson and Beecroft (1988). Its entomological interest has long been known (St. Leger, personal communication), though not published. It has several undescribed species and subspecies of butterflies. Most interestingly, the fauna is a mixture of western and equatorial elements, thus providing a potential base for study of the biogeographical dynamics of the area.

The Ugandan rainforests have always been fragmented, and not much of this type of forest remains. Its westernmost point is the Kakamega Forest in western Kenya, with more than 400 species (Larsen, unpublished). The conservation of forests in this zone is imperative, and most currently appear well managed.

The forests of the Cameroun and Zaïre zones are much more extensive than the other lowland forests, and thus under less potential threat in the short term. Each has special areas of particular interest, but there is time to study this, and to get conservation plans into place.

(d) Savannah formations

These are less fragile than the forests, and while they may be degraded, they rarely disappear completely as an ecosystem as forests may do, although the savannah areas are shrinking because of desertification in the Sahel zone. Savannahs are also usually very large expanses of relatively homogeneous habitat, which even when degraded allow for continued gene flow. An isolated lowland forest is genuinely isolated, and its gene-flow cut off. Also, the bulk of the very large parts of Africa which have been designated as national parks and game reserves are in savannah areas.

20.4 BUTTERFLIES AND CONSERVATION

20.4.1 Species conservation

Although most butterflies will survive or die with their habitats, there may be the occasional case for measures to conserve specific threatened species. As mentioned, the only colony of *A. dentatis* in the Transvaal is in a fenced nature reserve, and has received much positive media coverage. At present only one colony is known of the monobasic *Erikssonia acraeina*. Since it is a unique and fascinating butterfly, the colony is a candidate for active protection, despite the fact that there must be many undiscovered colonies elsewhere in Transvaal, Botswana, Zambia and Namibia/Angola. Species that can easily be trapped may also rate higher than species that cannot, and probably a case could be made for protecting certain very local, montane species of *Charaxes*.

However, if their views are to be taken seriously, it is important that naturalists do not cry wolf too often. Governments and conservation agencies have to make critical decisions with little information, less time and uncertain political support. Their time and resources should not be wasted on spurious species conservation. Thus, in 1983, the Transvaal government was persuaded to ban the collecting of all *Charaxes* except by special permit; yet *Charaxes saturnus* (foxy charaxes) is extremely common and several others only slightly less so. Such a blanket ban simply looks silly.

Many of the species placed on red data lists may be scarce in collections and have ranges that appear small on the map, while on the ground the species may be so common and widespread at the right time of year, that even busloads of single-minded collectors could make no dent in the population.

Generally speaking, we know too little about African butterflies to identify legitimate candidates for species conservation. In the lowland forest, scores of species are known only from the type(s). *Ornipholidotos issia*, for instance, was only known from the type locality in Côte d'Ivoire until earlier this year, when I studied a Nigerian specimen that I collected 12 years ago; its true range may be even wider.

The inclusion of phantom species on red data lists also detracts from credibility. Thus, none of the three *Neptis* listed for Madagascar (International Union for the Conservation of Nature, 1990) ever actually occurred on the island.

20.4.2 Habitat conservation

The immediate needs of butterfly conservation in Africa are the preservation of:

1. most of the montane forests;
2. all of the coastal forests of East Africa;
3. much of the remaining lowland forests in West Africa; and
4. the lowland forests of Uganda, Kenya and Tanzania.

Each time one of these forests disappears, some butterflies will be among the many organisms lost forever.

Many models are under development to assist in conservation choices: hotspots, centres of endemism, centres of origin, indices of biological value and so on. Where the most threatened habitats of Africa are concerned, we, unfortunately, do not have the luxury of strategic choice. Conservation of what is left is the overriding priority. More sophisticated methods may be used to plan the conservation of less threatened habitats and to formulate research strategies.

20.4.3 The role of butterflies in conservation

In Africa, we should not be thinking too much about butterfly conservation in isolation but about how to use butterflies as arguments and tools for habitat preservation. Butterflies are peculiarly suited for this purpose. Despite the many gaps, they are the best-known major groups of arthropods in Africa. They are biogeographically diverse, and often strongly habitat specific. Most can be readily identified, at least where identification guides are available, and many can be identified even without capture. They can be marked for population and ecological studies. Probably only birds and certain groups of plants are as suitable.

Small-scale conservation (for hunting, religious and even sentimental or aesthetic reasons) has always been carried out by the local population. Patches of *ju-ju* woods (imbued with magic) were largely responsible for a small area of secondary bush near Lagos containing no fewer than 385 species of butterflies (Larsen, Riley and Cornes, 1980). Game parks have an obvious, immediate economic rationale, and some have been in existence for 50 years or more. But large-scale conservation of entire 'uneconomic' ecosystems and centres of endemicity is relatively new, very largely inspired from outside the continent, and often undertaken by central authorities with little consultation with the local population affected (for some interesting cases see Adams and McShane, 1992). The fact that it is possible to watch hours of television footage from African nature reserves without ever seeing an African is one clear reflection of this outside bias.

Conservation efforts in Africa demand a level of political will, administrative skill and long-term persistence that is not easily mobilized. This depends on a complex matrix of continued motivation, including pride in the local flora and fauna, knowledge about it, national and international recognition, research activities, field stations, economic viability and the support of the local population in and near conservation areas, who, even today, often view the forests as inherently hostile.

With increasing interest and concern for biodiversity, the activities of international organizations, and the mounting evidence of irreversible environmental damage, there is growing understanding for conservation among the African élite. There has been less effort to involve the local population in conservation, although this, too, is beginning. And local villagers do have pride and interest in their environment. They may not have thought about butterflies before, but when you tell them that there are 400 species, and that this one has never been found anywhere else than in their little patch of forest, the reaction is normally very positive.

20.4.4 Butterfly project potential

A priority need is butterfly inventories from key forests all over Africa. Hardly any exist. They can be used to argue for habitat conservation and be integrated with general ecological studies. They will also be useful for assessing the minimum size of nature reserves and rates of extinction in small, isolated forests. These might be organized through a system whereby entomological societies 'adopt' key forests and run the scheme in collaboration with local conservation bodies. With the recent books by D'Abrera (1980), Carcasson (1981), Berger (1981), Kielland (1990), Larsen (1991) and Pennington (1994) most butterflies in much of Africa can be identified without the need for referral to collections abroad. A book entitled *Butterflies of West Africa – Origins, Natural History, Diversity, and Conservation* will be forthcoming (Larsen, in preparation).

Butterflies have a useful role to play in ecological impact assessments, and in continued monitoring of ecological health. Butterflies would also be very suitable organisms for studying the effects of forest management. Even slight disturbance may cause the loss of some species (Africa's two largest swallowtails seem among the most susceptible), but as shown by the study from Lagos previously referred to, even strongly degraded forest may retain significant diversity.

African butterflies are extremely suitable for research into behaviour, evolution, population dynamics, community structure, ecology, prey–predator relationships, chemical communication, etc., but very little is currently being carried out. This gap is particularly glaring when seen in relation to the stream of high-grade research now coming out of the Neotropics, and when we take into account that during the first 40 years of the twentieth century, Africa accounted for much fruitful activity by butterfly giants such as Eltringham, Poulton and Hale Carpenter. The book by Owen (1971), and his own wide-ranging research, is an excellent illustration of what can be done even with relatively modest resources.

Ecological tourism is a sector of the tourist market that is growing fast. When suitable accommodation is available close to good habitats, the potential is considerable. Danger of overcollecting is slight, and a liberal policy of issuing collecting permits would carry little risk. Some special management to make nature trails, clearings, observation towers, etc. may be needed. In the Amazon area of Ecuador private hoteliers now conserve patches of forest.

Commercial exploitation of butterflies comes in four main types:

1. the sale of specimens to collectors;
2. the breeding or ranching of selected butterflies for sale;
3. the sale of decorative butterfly collections as souvenirs; and
4. the use of butterflies in decorative art work (as in the Central African Republic).

Someone managing to set up captive breeding of Africa's largest butterflies, *Papilio antimachus* and *P. zalmoxis*, would be assured of a steady demand at good prices, as would those breeding many species of *Charaxes*.

Any type of butterfly research not only provides data that can help in conservation, it also brings funding, foreign exchange, potential prestige and a greater understanding that nature (and butterflies) is important. It also brings researchers, and eventually field stations, into the forests, and this raises the physical security of the forest. Research should use workers recruited from the area studied as much as possible, and involve the local population in the activities. I suspect that ecological field stations might be commercially viable.

Finally, if research findings are to be useful for, and give prestige to, the country where the research was conducted, the findings must be disseminated not just in scientific circles, but to the general public and the people in the area researched. In most of Africa, literature is not easy to get hold of, and any projects should build dissemination of their results into the budget and plan of operations. Every naturalist working in an African country should see it as an obligation to publish at least one interesting article in local newspapers or magazines.

20.5 CONCLUSIONS

The butterfly fauna of the Afrotropical region is diverse and of great interest. Our knowledge about it is growing fast, but not fast enough. The most urgent need in Africa is the immediate conservation of the most threatened habitats, which are the montane forests of East Africa, the lowland rainforests of West Africa proper and the coastal rainforest of East Africa. Butterflies can be used as powerful arguments for the preservation of habitats in Africa, and butterfly research can assist in strengthening the motivation for conservation work in general.

The philosophy of conservation in Africa, important as it is, was to a large extent inspired from aboard; often by people and groups with little understanding of the needs of Africans in Africa. Botswana, for instance, has many more elephants than it knows what to do with. It is hardly surprising that people there get somewhat cynical about campaigns for elephants as a 'threatened' species; by no stretch of imagination can the elephant be called threatened, although individual populations may be.

There is growing understanding among the élite in Africa that conservation is important, partly due to international conservation efforts. This is promising, and deserves the support it is receiving.

However, ultimately there is probably no substitute for ensuring that local people have a vested interest in conservation and proper management of the forests where they live. This will demand the creation of forest-related jobs, the employment of local people in Forestry Departments, ecological tourism, research activities and so on. Conservation of biodiversity will have to be paid for; it will not come for free.

20.6 ACKNOWLEDGEMENTS

This is the first in a series of papers under the auspices of the project *Butterflies of West Africa - Origins, Natural History, Diversity, and Conservation (1993–1997)*. I am grateful to the Carlsberg Foundation of Denmark for their support. Nancy Fee, Denis F. Owen, Andrew Pullin, Bob St. Leger and an anonymous referee all kindly reviewed the manuscript at various stages.

CHAPTER 21

Butterfly conservation in Australasia – an emerging awareness and an increasing need

T.R. New

21.1 INTRODUCTION

Development of the science of butterfly conservation in northern temperate regions has set a sound basis on which other parts of the world, less advanced in aspects of their conservation capability, awareness or planning, can draw. But equally, incorporation of experiences and perspectives from other zoogeographical regions can contribute to more effective general understanding of the protocols and practice of butterfly conservation and the development of unifying principles. Particularly, it may be important to avoid imposition of solely 'eurocentric' values on less-developed countries. This chapter indicates how this two-way process is occurring in parts of Australasia, so that the twin roles of butterflies in fostering awareness of invertebrate conservation (namely, as species-focused conservation programmes to save notable taxa *per se*, and employing butterflies as flagship or umbrella taxa to monitor the well-being of assemblages) are advanced.

'Australasia' is defined usually as 'Australia, New Zealand, and nearby islands of the Western Pacific', a large geographically, climatically and biologically complex area, which supports a range of faunal elements and has areas subject to markedly varied human pressures. The regional concerns are noted here by comments on the butterflies and conservation needs of Australia, New Zealand and Papua New Guinea.

Ecology and Conservation of Butterflies Edited by Andrew S. Pullin.
Published in 1995 by Chapman & Hall. ISBN 0 412 56970 1

Table 21.1 Comparative appraisal of knowledge and status of butterfly faunas of Australia, New Zealand and Papua New Guinea

	Australia	*New Zealand*	*Papua New Guinea*
No. species	*c.*400	23	>800
Taxonomic knowledge	Very good	Excellent	Moderate
Distributional knowledge	Very good	Very good	Poor
Ecological knowledge	Good	Good	Poor
Proportion of spp. described	>90%	?100%	?80%
Proportion with life history/ biology known	*c.*80%	100%	(very few)
Taxa of major conservation concern	Lycaenidae Hesperiidae Nymphalidae: Satyrinae Papilionidae	(Lycaenidae)	Papilionidae

Sources: Common and Waterhouse (1981), Dunn and Dunn (1991) (Australia); Gibbs (1980) (New Zealand); Parsons (1992a) (Papua New Guinea).

21.2   STATUS OF BUTTERFLY FAUNAS

Some data on the butterfly faunas of the three countries are summarized in Table 21.1 (see also Figure 21.1).

Major sources of information are handbooks by Common and Waterhouse (1981) and Dunn and Dunn (1991) (Australia), Gibbs (1980) (New Zealand) and Parsons (1992a) (Papua New Guinea).

21.2.1   Australia

Australia has around 400 butterfly species (an equivalent number to Europe), with major taxonomic radiations in the Hesperiidae: Trapezitinae, Lycaenidae and Nymphalidae: Satyrinae. Distributional and ecological knowledge of the fauna is reasonably good and, although new taxa, distribution records and biological information continue to accumulate, butterflies are one of the best-known invertebrate groups in the country and are a 'flagship group' for insect conservation activities (New, 1991, 1992). Endemism is high, especially in the temperate fauna. Many southern species and subspecies are restricted to that region. Many northern butterflies have strong affinities with those of New Guinea (see Kitching, 1981) and in Australia represent attenuation from the richer faunas of the Indo-Malayan region. Many Australian butterflies have very restricted distributions, some being known from single colonies or sites. Diversity is generally highest in the east of Australia, including major concentrations of diversity in northernmost Queensland (around 140 species on Cape York: Monteith and

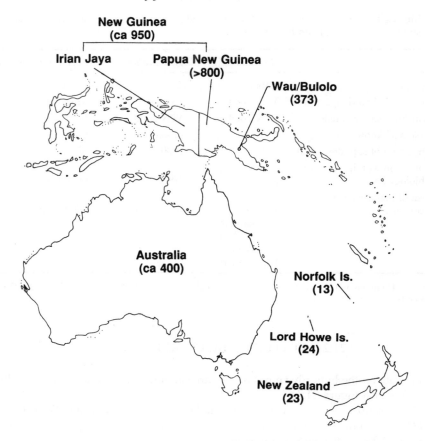

Figure 21.1   Butterfly faunas of the Australian region. Numbers of species indicated; sources of information as in Table 21.1; figures for Norfolk Island and Lord Howe Island are also included.

Hancock, 1977), in southern Queensland (where the northern and southern faunas overlap) and around Cairns between these two. In contrast, butterfly diversity is low in the arid interior of the continent.

21.2.2   New Zealand

The small New Zealand fauna, of only 23 species, includes 11 endemic taxa, and it is likely that no further endemic species will be found there. The endemic taxa include alpine Satyrinae and more widespread Lycaenidae and, highly unusually, none is of major conservation concern. The common role of butterflies as invertebrate 'flagships' for conservation has been assumed in New Zealand by the spectacular giant wetas (Orthoptera), to which a great deal of attention has been paid (see Meads, 1990; Howarth and Ramsay, 1991).

Table 21.2 Status evaluation of the butterflies of the Bulolo-Wau Valley, Morobe Province, Papua New Guinea (Parsons, 1992a)

| | DACOR rating (no. spp.) | | | | | |
	D	A	C	O	R	Total spp.
Hesperiidae	–	–	15	20	39	74
Papilionidae	–	2	9	3	6	20
Pieridae	1	2	12	16	13	44
Lycaenidae	–	–	25	55	48	128
Nymphalidae	1	5	39	40	21	106
Total	2	9	100	134	127	372

The data are based on DACOR scale of abundance: dominant (D), 21–41+; abundant (A), 11–20; common (C), 6–10; occasional (O), 2–5; rare (R), 1. Ratings are on an arbitrary time period, here over 4 years study, so that 'rare' species were seen only once or twice during that period.

21.2.3 Papua New Guinea

Knowledge of the Papua New Guinea butterflies is far less adequate than for the foregoing countries, but the detailed recent appraisal of the fauna of the Bulolo-Wau Valley, Morobe Province, by Parsons (1992a) may be taken to reflect the difficulties of appraising the complex fauna adequately. That small area, of approximately $350 \, km^2$, has yielded 373 species (i.e. nearly as many as the whole of Australia or Europe) and approximately half the number from the whole of Papua New Guinea. With few exceptions, biological knowledge is fragmentary or non-existent, although there are marked changes in taxa at different altitudes, and many species are 'rare' (Table 21.2).

However, against this background of tropical diversity, the spectacular *Ornithoptera* spp., birdwings, have long been flagship taxa (Parsons, 1980; Collins and Morris, 1985; New and Collins, 1991) and early measures to protect them, and to promote butterflies as a sustainable economic resource, were pioneering and innovative (Parsons, 1992b).

21.3 CONSERVATION PROBLEMS AND NEEDS

21.3.1 Status and threats

Collectively, these areas represent a wide range of butterfly conservation concerns. Papua New Guinea exemplifies the difficulties of assessing status of butterflies in a diverse tropical fauna and, with the exception of some Papilionidae, the focus is not generally on single species, but on the more widespread concern of habitat preservation for assemblages (cf. Larsen, Chapter 20 this volume). Australia, with a larger (but still small) number of resident collectors and concerned entomologists, has followed the northern hemisphere trend of species focusing, although the exercise of butterfly conservation is rather recent. The massive degradation of natural vegetation in Australia during only 200 years of European settlement has undoubtedly had major influences on native invertebrates (see Greenslade and New, 1991), but knowledge of the butterfly fauna is now sufficient to enable very constructive focus on species and habitat conservation in parts of the continent, with ability to enumerate the specific threats to many taxa.

Status evaluation and distribution mapping has commenced in both countries, through grid-mapping. In Papua New Guinea, the Insect Farming and Trading Agency (IFTA, below) commenced a distribution-recording scheme based on 10×10 km square units. Distribution maps produced recently for Australian butterflies by Dunn and Dunn (1991) were based on more than 88 000 individual records. There is no nationally organized monitoring or mapping scheme, and this would be impracticable because of the small number of collectors and the vast areas that need to be surveyed. Likewise, there is no equivalent to the British Butterfly Conservation Society or the USA's Xerces Society, and most progress in Australia comes from the efforts of a few tens of enthusiasts. Distribution of New Zealand butterflies is also well documented (maps in Gibbs, 1980). However, there are few detailed autecological studies of particular taxa in any country, and historical trends in distribution are poorly known, apart from a few records of decline or loss of local colonies caused directly by site alienation. Thus even the detailed effects on butterflies of rainforest fragmentation in Australia have not been quantified, although they may reasonably be inferred to have been adverse (Nadolny, 1987).

Despite lack of major conservation concern for New Zealand butterflies, those concerns which have been raised there exemplify the broad range of threats apparent elsewhere:

1. the endemic copper *Lycaena raupahara*, has declined in range through habitat alienation;
2. pupae of the endemic *Bassaris* admirals are attacked by parasitoid wasps introduced for biological control;
3. the endemic *Zizina oxleyi* may be at risk from hybridization with the introduced Australian *Z. labradus*, in one of very few cases of such an occurrence (Gibbs, 1980); and
4. the alpine taxa (as in south-east Australia) might be at long-term risk from global climatic change, that great imponderable which we need to incorporate urgently into conservation management planning.

Exotic species introductions, notably the widespread replacement of native vegetation with other plants, is associated commonly with other changes in land management, such as ploughing, wetland drainage, addition of agricultural chemicals and of exotic stock. Over much of Australia, the preservation of remnant habitats for butterflies, some of which are presumed to have been more widespread formerly, is a major concern. This concern is shared with New Zealand, where much endemic vegetation is under threat from grass invasion, and some butterflies are suffering because of the decline of native grasses. In Australia many areas of introduced vegetation (such as softwood plantations and introduced pasture grasses and crops) are managed intensively to exclude native species. This contrasts with aspects of plantation management in parts of Papua New Guinea. Parsons (1992a) noted that the open plantations of 15–20-year-old *Araucaria* pines support a great diversity of butterfly food-plants. These comprise saplings of forest trees, which may be cut back regularly (thus promoting flush growth) and the food-plant vines of many species growing on the pines themselves. Indeed, Parsons implied that a high proportion of the Bulolo-Wau butterflies might thrive in such plantations. By contrast to this, the conversion of forests to oil-palm plantations is regarded as a serious threat to *Ornithoptera alexandrae* (Queen Alexandra's birdwing) and pressure for agricultural land has led to clearing of large areas of forest and creation of fire-prone grasslands. As elsewhere in the world, reversion of such areas to any near-pristine condition is unlikely to occur over any moderate time scale.

21.3.2 Species focusing

(a) Protective legislation

No butterflies are listed in the New Zealand Protected Invertebrate schedules (Ramsay *et al.*, 1988), where wetas and beetles gain priority. However, they (and all other insects) are protected within Department of Conservation estate, and can thus be collected in about a quarter of New Zealand's land area only after obtaining permits to do so. By contrast, the listing of seven species of *Ornithoptera* under the Papua New Guinea 1966 Fauna Protection Ordinance gave those species an unusually high status as 'national butterflies' and undoubtedly influenced the later inclusion of birdwings in CITES appendices (*O. alexandrae* is now in Appendix 1, all others in Appendix 2).

As it is elsewhere, such legislation is controversial and an emotive topic. In Australia, several butterflies are accorded protected status under various State legislations (Queensland: *Papilio ulysses joessa*, *Ornithoptera* spp. (since 1974, although the new Nature Conservation Act 1992 may result in changes to this status), *Acrodipsas illidgei* (Permanently Protected Fauna, 1990); Victoria: *A. brisbanensis*, *A. myrmecophila*, *Paralucia pyrodiscus lucida* (Flora and Fauna Guarantee Act, listed 1991)) but no species has national-level protection. Other states do not list butterflies in this way so that, for example, no species in Western Australia is protected specifically. Preliminary lists of insects perceived to be 'threatened' or 'endangered' (Hill and Michaelis, 1988; International Union for the Conservation of Nature, 1990) include a number of butterfly species, some of highly local occurrence. Recent establishment of the new federal 'Endangered Species Protection Act 1992' provides a vehicle for potential consolidation and rationalization of protective legislation in the country.

The Victorian Flora and Fauna Guarantee Act 1988 provides also for protection of 'threatened communities', and a novel focus of the act is the designation of 'Butterfly Community No.1.', comprising the possibly unique joint occurrence of two rare myrmecophilous lycaenids, *A. brisbanensis* and *A. myrmecophila*, at Mt. Piper, some 80 km north of Melbourne. The legislation thus provides for effective preservation of the site from intrusion, as a vehicle for conservation of the co-occurring species. This Act (as with the new federal legislation) is also unusual in Australia in providing for preparation of an 'Action Statement' detailing the conservation needs of taxa and its later implementation. Any species nominated for inclusion under the Act is subject to evaluation of its status so that, unlike much 'protective legislation', there is a strong practical orientation to conservation rather than merely prevention/prohibition of collecting.

(b) Other priority taxa

Several other butterflies have been proposed for protective legislation in Australia. Thus, in New South Wales, *Ornithoptera richmondia*, *Euschemon rafflesia*, *Argyreus hyperbius inconstans* and *Tisiphone abeona joanna* were listed for coverage under the State's National Parks and Wildlife Act in 1980: the proposals were later withdrawn, together with those for other insects.

However, despite lack of any legal status, the documents noted above (Hill and Michaelis, 1988; International Union for the Conservation of Nature, 1990) imply the need for priority evaluation of conservation status. The former list includes 61 butterfly taxa (28 Hesperiidae, 24 Lycaenidae, 7 Nymphalidae, 2 Pieridae; no Papilionidae were listed by the 54 respondents to the questionnaire survey initiated by Hill and Michaelis, cf. the Queensland legislation noted earlier).

A number of these species have been, or are currently, the subjects of more intensive study to evaluate their status and needs more effectively.

(c) Ecological studies for status evaluation

A number of species noted above as having conservation interest have been the subject of ecological studies to help clarify their status. In some cases this has led to their removal from any further conservation priority listing at present; in others it has confirmed their parlous existence, or suggested that they might not be as vulnerable as considered earlier. The following examples are selected to demonstrate a range of outcomes and the increasing range of capability and opportunity in the region. The first three represent the major Australian endemic butterfly radiations in which many taxonomically complex forms occur over very small areas of the country.

The subspecies *Hesperilla flavescens flavescens* (Altona skipper or yellowish skipper) is closely allied to the very variable *H. donnysa* and is one of the more significant endemic Australian Trapezitinae, notable for the extensive yellowish markings and, until recently, known only from colonies near Melbourne. One colony was eliminated by industrial expansion in the 1960s, and population decline at the type locality, Altona, caused by drainage and alienation of swampland, prompted concern from the mid-1980s, and nomination for listing under the Flora and Fauna Guarantee Act. A detailed survey and review of likely habitats of the subspecies, and of its biology, was undertaken by Crosby (1990).

The butterfly was regarded as threatened at Altona, to the west of Melbourne. It was restricted to colonies at Cherry Lake and Truganina Swamp, where larvae feed on the saw-sedge *Gahnia filum*, restricted to mildly saline, exposed swampy conditions. Young, tender foliage is preferred. In 1989, Crosby (1990) estimated the size of the two Cherry Lake colonies at 250 and about 100 individuals, and the combined populations at Truganina Swamp to include about 600 individuals. Butterflies with the *flavescens* phenotype also occurred at another Altona site (200–300 individuals) and three other sites in the area (100, 250, 200–300), but those at another 21 sites in Western Victoria intergraded towards the appearance of *H. donnysa*. The Altona sites were thus regarded as of prime importance to conserve the extreme yellow phenotype of this variable species complex/cline. Crosby suggested that, ideally, all sites should be reserved to conserve the range of variability present, but four sites should be preserved as 'reference sites' and managed actively for the butterfly. Detailed protocols for this were suggested. The butterfly is of considerable evolutionary interest, but listing under the Act has not proceeded, because of the larger number of sites on which some '*flavescens*-type' populations have now been found.

The local Victorian subspecies *Paralucia pyrodiscus lucida* (Eltham copper), more than any other butterfly, has become a flagship for the development of butterfly conservation in Australia, although its New South Wales congenor, *P. spinifera* (Bathurst copper), has been described variously as 'Australia's rarest butterfly' (Kitching and Baker, 1990) or 'the most vulnerable species in N.S.W.' (Nadolny, 1987) (see Dexter and Kitching, 1993). Accounts of both species are included in New (1993), together with several other Australian Lycaenidae.

Briefly, the rediscovery of *P. p. lucida*, believed to have become extinct because of urbanization near Melbourne, in 1987 prompted a massive rescue effort and status evaluation with Australia's first 'dedicated butterfly reserve' being established in Eltham, an outer suburb of Melbourne. Monitoring and habitat restoration continues, together with investigation of captive breeding potential or translocations from one of the more distant Victoria colonies, themselves also very isolated. No genetic information on the similarities or differences between the colonies is yet available, so that options for these approaches are limited.

The endemic Tasmanian satyrine *Oreixenica ptunarra* (Ptunarra brown) occurs only in native grasslands across central Tasmania. The total area of suitable habitat is estimated at around 4000 ha (Neyland, 1992) and is being reduced: 4 of 33 colonies known in 1988 were extinct by 1992. Although Neyland's (1992) detailed survey revealed around 150 extant colonies, all three subspecies of *O. ptunarra* were assessed as 'endangered', because the small size of most colonies (only 20 were greater than 50 ha) rendered them very susceptible to disturbance. Most individual colonies were 'vulnerable'. *Oreixenica ptunarra* is restricted to areas with high ground cover (more than 25%) of *Poa* grasses, and disappears from areas which are heavily grazed by introduced stock, or cleared. Its continued survival will depend on sympathetic land management. Neyland's management recommendations included the need to fence as many 'core areas' (of 2–5 ha) as possible to control stock grazing, and to reduce the frequency of habitat burning. It is likely that *O. ptunarra* has only low ability to recolonize former sites naturally, as it is a very weak flier.

Status reports and management plans for the above taxa (and others, such as the Tasmanian races of the Australian hairstreak *Pseudalmenus chlorinda*: Prince, 1988) have been sponsored by the various state bodies or the Australian National Parks and Wildlife Service (now the Australian National Conservation Agency). This level of 'species focus' has not been attained for Papua New Guinea species, although one spectacular species has indeed been the target of much recent effort.

Ornithoptera alexandrae, the world's largest butterfly, listed in CITES Appendix 1, is the major flagship species for Papua New Guinea's magnificent birdwing fauna and has been assessed as 'endangered' because of habitat loss and its desirability to collectors (Parsons, 1984). It has been known until recently only from small areas near Popondetta, where expansion of the oil-palm industry led to concern for its well-being and recommendation for areas to be protected from such alienation, as wildlife management areas.

However, habitat destruction has continued on the Popondetta Plain, and several recent surveys for *O. alexandrae* have been sponsored by international agencies. It was highlighted in the recent Swallowtail Butterfly Action Plan (New and Collins, 1991), but surveys by Mercer in 1992 strongly suggested that it may be more widespread than previously supposed, and Mercer is attempting to form an integrated conservation and development project, to include a range of activities to foster the conservation of *O. alexandrae*. Its conservation may necessitate considerable social change in the Oro Province.

21.3.3   A threatened butterfly community, Mt. Piper, Victoria

The putatively unique co-occurrence of *A. brisbanensis* and *A. myrmecophila* at Mt. Piper, together with other notable Lycaenidae and a total of 37 butterfly species (around a third of resident Victorian species), has led to investigations (Britton and New, 1992, 1993) to quantify and elucidate the biology and status of notable taxa there, together with a survey of ants and ant–lycaenid interactions. This investigation is part of the Action Statement plan designed to evaluate the status of the butterfly community associated, in some species only by hill-topping (since their breeding sites are unknown), with this isolated volcanic plug. Mt. Piper, declared as an education reserve in 1980, extends to 456 m (from a base plain of around 230 m) and is threatened by possible subdivision for urban development, and mining exploration. Conservation steps taken so far include exclusion of grazing stock, closing of the summit area to vehicle access and provision of information signs in the base car park (Jelinek, Britton and New, 1994).

21.3.4   Ranching and farming

This topic is in its infancy in Australia for conservation purposes, although feasibility of captive breeding for *P. p. lucida* is at present being considered by the Eltham copper's 'Committee of Management', utilizing the expertise gained by the Royal Melbourne Zoological Gardens in developing its notable butterfly house, furnished largely by captive-bred Australian butterfly stocks.

In contrast, the extensive ranching operation in Papua New Guinea, co-ordinated through IFTA at Bulolo broke new ground in this approach to conservation, promoting the rearing of butterflies as a means of exploiting rationally a fully sustainable resource in a manner that provides financial benefits to the participants and fosters reduced intrusion into natural ecosystems and less intensive field collecting. This operation now involves more than 500 people throughout the country and, since it is now allied formally with the technical and commercial branch of the PNG University of Technology at Lae, may be poised for even greater expansion. Economic and logistic aspects of this pioneering operation are discussed by Mercer and Clark (1992). Other butterfly farming operations attempt to emulate the IFTA: in the Arfak Mountains of Irian Jaya, for example (Parsons, 1992b). Parsons (1992a) also pointed out the commonly supposed synonymy between 'rarity' and 'low availability' to collectors, so that ranching of commercially desirable species for sale may prevent them from being so highly coveted and avoid the increased demand/prices that unavailability tends to create.

21.3.5 Translocation or reintroduction

Neither of these strategies has yet played a major role in butterfly conservation in the region, although there is clear potential for them to do so, and the need to develop local protocols, perhaps using JCCBI guidelines (see New, 1994). Translocation, following habitat restoration, has been undertaken to establish a colony of *Trapezites symmomus* (symmomus skipper) within its former range in Victoria (Braby, 1991), whence it has spread naturally to a nearby area. The potential for range expansion by natural dispersal is also being investigated in northern New South Wales, by planting food-plants (*Aristolochia praevenosa*) of *Ornithoptera richmondia* (Richmond birdwing), which now has only around 1% of its original suitable forest habitat remaining, and where the butterfly has consequently become rare. Nearly 40 schools are involved in an educational project (Sands, personal communication), and a nursery owned by an Aboriginal co-operative will produce the food-plants for commercial sale.

21.3.6 Habitat focusing

Nature reserves and protected areas in Australia have been declared predominantly for criteria other than butterflies, although several commentators have drawn attention to the importance of habitat security for butterflies, for example in major centres of butterfly diversity such as the Iron Range (northern Queensland: Monteith and Hancock, 1977), or the value of particular habitat associations, such as rainforests, in New South Wales (Nadolny, 1987). Other than for *P. p. lucida* and *Acrodipsas illidgei* (Illidge's ant-blue) no butterflies in Australia have come near to the prominence given to New Guinea *Ornithoptera* in recommending habitat preservation. The two situations contrast markedly. *Paralucia pyrodiscus lucida* represents a situation possible only in some affluent environments, where several hundred thousand dollars could be raised to purchase and dedicate a small area of land on the outskirts of Melbourne especially for the butterfly. The recommendations for *Ornithoptera* spp. in Papua New Guinea, which have remained unheeded, have included those for declaration of wildlife management areas (which are of rather limited conservation value because they can still be exploited by landowners). Australia has a large network of national parks, in which collecting or taking of any wildlife without a permit is prohibited, and these collectively encompass many vegetation associations and climatic regimes. Papua New Guinea has only three national parks, totalling slightly over 7000 ha (Parsons, 1992a). Parsons also noted that none of the various reserves in Papua New Guinea is of especial significance for butterfly conservation, and inventory data are not available. By contrast, butterfly species lists, of varying degrees of completeness, are available for a wide range of Australian protected areas, and are likely to become of increasing value in conservation assessment as an ever-increasing suite of values comes to be appreciated in designating priorities.

21.4   DISCUSSION

Butterflies in Australia are a small but popular and highly significant fraction of a vast invertebrate biodiversity, which has been estimated as high as 300 000 species. As in Papua New Guinea, their role as flagship or umbrella taxa for promoting invertebrate awareness has gained significant impetus recently. This is being translated, albeit gradually, into active conservation action. Parsons (1992a) noted that 'The Constitution of Papua New Guinea is the only national constitution in the world that specifically encompasses the rationale behind insect conservation', with a stated goal of conserving natural resources and a call for 'all necessary steps to be taken to give adequate protection to all our valued birds, animals, fish, insects, plants and trees'. However, the capability to translate such laudable ideals into practice is far more advanced in Australia and New Zealand, reflecting the greater logistic commitment, number of informed personnel, chances for effective habitat preservation and/or restoration and a greater capability to determine a wide spectrum of species- and other priorities. The greater regional awareness of insect (and other invertebrate) conservation needs, resulting from recent studies, bodes well for the future, and the climate of sympathy for them seems likely to continue. Education at all levels, through media such as butterfly houses, and increased numbers of popular publications, has a role in Australia, and the social benefits induced and projected through the activities of IFTA are reflected in habitat protection and enhancement for birdwings, in particular.

However, there is indeed no room for complacency. It is one thing to acquire knowledge for which the impetus is perceived threats to species or habitats, and quite another to control or regulate the threats themselves. These may be highly localized, but totally devastating in impact. A number of Australian butterflies are known only from single sites, some of them very small. *Hypochrysops piceatus*, for example, is known only from a small length of roadside near Leyburn (Queensland) which would be susceptible to 'casual' roadworks or roadside vegetation clearing. The main concerns voiced by respondents to a questionnaire by Hill and Michaelis (1988) (where 61 butterfly taxa in Australia were listed as of conservation concern) included fire, urbanization, agricultural clearing and roadworks.

In principle, the last three of these could be controlled sufficiently to maintain particular sites known to support high-priority species, but may lead to additional intrusion or threats from nearby disturbance.

Habitat destruction, ranging from such local effects to more widespread loss of forests over a greater (and less predictable) time, remains the prime concern of butterfly conservationists in Australasia. Rainforest butterflies in tropical Australia are now restricted to small remnant habitats. In Papua New Guinea, much of the rainforest area is largely intact, much of it protected by terrain which is steep and difficult to access. Hoyle (1977) noted that three-quarters of the 40 million ha of forest in Papua New Guinea is not likely to be violated at present. However, pressures on rainforests in Papua New Guinea are likely to increase, and current pressures on lower montane forests are significant because these areas contain most of the endemic species.

By contrast, other perceived threats may not be amenable to regulation but are likely to be no less devastating in their impact on butterflies and other biota. Many Satyrinae endemic to Australia and New Zealand are alpine, and occur only in areas of grassland covered by snow in winter and at characteristic ranges of higher altitude. Moderate projections for global warming suggest that, over the next 40–50 years in Australia, a rise of 2 °C and decrease of precipitation by 20% would reduce the area amenable to cross-country skiing in New South Wales from around $1400 \, km^2$ to $270 \, km^2$ with elevation increase from $1500 \, m$ to $1760 \, m$ (Galloway, 1988). Such a change over Australia would eliminate the current habitats or encompass the whole range of some colonies or species (such as that of *Oreixenica latialis theddora*, known only on the Mt. Buffalo Plateau, Victoria) as well as have severe effects on the winter sports industries. In the meantime, in the interests of assured short-term gain being preferred to long-term doubt, there is substantial commercial pressure to develop alpine areas as rapidly and intensively as possible. A number of alpine butterfly colonies may, therefore, be at more proximal risk. The effects of similar changes on the montane butterflies of New Guinea, where a substantial number of taxa are confined to isolated mountains and high altitudes, are also imponderable, but a significant proportion of species of some major radiations is involved. Nine of the 32 species of *Delias* on Mt. Kaindi (Wau) occur predominantly over 1800 m, for example, and very few butterflies occur at both the highest altitude of the mountain and on the valley floor. Important though it may be to incorporate consideration of such distributions and projections into any long-term satisfactory conservation plan for Australasian butterflies (and other biota), this will be extraordinarily difficult to achieve.

21.5 ACKNOWLEDGEMENTS

I am very grateful to Dr G.W. Gibbs for his comments on a draft of this paper, and to Dr D.P.A. Sands for permission to refer to his innovative project on the Richmond birdwing. My participation in Butterfly Conservation's Silver Jubilee Symposium was facilitated by financial support from La Trobe University and Butterfly Conservation.

CHAPTER 22

Conservation and management of butterfly diversity in North America

P.A. Opler

22.1 INTRODUCTION

Butterfly conservation in North America began in 1971 with the formation of the Xerces Society, a group devoted to terrestrial arthropods and their conservation. In the 22 years since its founding, it has played a major role in the education of biologists, naturalists and decision-makers about the importance of invertebrates and the need for their conservation. The Society founded the North American butterfly counts, one of the first continent-wide efforts to monitor a wide array of Lepidoptera. Currently it publishes *Wings*, a quarterly magazine highlighting invertebrate conservation needs, and sponsors projects on behalf of Madagascar butterfly conservation and conservation of *Papilio homerus* (homerus swallowtail) in Jamaica.

In 1973, the passage of the US Endangered Species Act provided for conservation programmes on behalf of endangered and threatened insects. Since then, 28 insects have been added to the US List of Endangered and Threatened Wildlife. Nineteen of the listed species are butterflies, mostly endemic to habitats or ecosystems which are themselves endangered. Under provisions of the Act, listed species benefit by being protected from taking, commerce and import; by being the subject of recovery efforts (including land acquisition); by being considered in planning efforts by all federal agencies; and by being eligible for co-operative conservation programmes in the states where they occur.

More recently, several individual states have included conservation of butterflies as features of their own endangered species programmes. This allows regionally important butterfly populations to be recognized and protected even though they may not rank high on the national scale.

North America north of Mexico is moderately rich in butterfly species, with

Ecology and Conservation of Butterflies Edited by Andrew S. Pullin.
Published in 1995 by Chapman & Hall. ISBN 0 412 56970 1

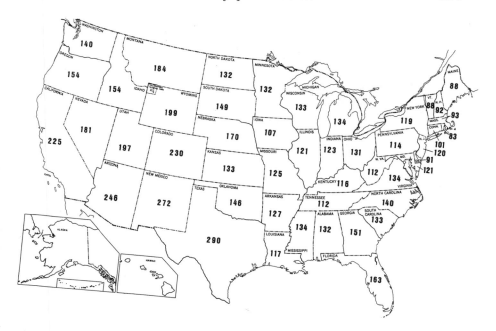

Figure 22.1 The number of resident and regular colonist butterfly species in each of the contiguous states. Note that the numbers do not include strays or infrequent breeders.

more than 700 recorded. I present information on butterfly diversity patterns that demonstrates that some areas of the continent are relatively depauperate, while others are especially species rich. Local or regional butterfly richness generally decreases as latitude increases, but there are areas of local richness at relatively high latitudes. Highest species richness occurs near the Mexican border and in the western mountains. These areas include the mountains of south-eastern Arizona and adjacent New Mexico, South Texas (including the subtropical lower Rio Grande Valley) and the southern Rocky Mountains of Colorado and northern New Mexico. Ecosystems with regional endemicity that should have high priority for conservation action include the above mentioned areas and southern California.

As in Europe, most North American butterflies are creatures of open, sunny places. Such open areas may be weedy, disturbed early successional habitats, or they may be disclimax sites related to soil type, aridity, slope, or regular disturbance. Some endangered or declining species have habitats that are seral stages in a succession that requires periodic fire, grazing or other disturbance.

22.2 PATTERNS OF SPECIES RICHNESS

I have determined the number of butterfly species in each of the 48 contiguous US states (Figure 22.1) and several biogeographical regions of the US (Figure 22.2).

Figure 22.2 Four regions of the contiguous United States that are especially rich in butterfly species: southern California, south-eastern Arizona and adjacent south-western New Mexico, the Rocky Mountain Front Range of Colorado and northern New Mexico, and south Texas. The larger number is that for resident species and regular colonists, while the smaller is the number of species with the majority of their range in the US restricted to that region.

The number of resident (present in area all year round) and colonist (breed in area but not present all year round) species are counted as separate from vagrants (recorded but not as an established breeding population) recorded in recent butterfly county atlases (Opler, 1982; Stanford and Opler, 1993) to give a truer picture of species richness.

Species richness varies from as low as 83 species in Rhode Island to as high as 290 in Texas. There is naturally a sampling problem, because large states tend to have more recorded species than small states. States or regions that have high species richness are southern or western. A wide range of topographic relief and complexity in these states appears to enhance their species richness. Additionally, states or regions adjacent to Mexico, with its very rich Neotropical fauna, share resident butterflies or receive many more vagrants and colonists than other states. Eastern states that include coastal areas as well as mountain areas have more species than states that do not. However, Florida, the richest eastern state with 163 species, is exceeded by all mountainous western states except Idaho (154), Oregon (154) and Washington (140).

22.3 PRIORITY AREAS

22.3.1 South Texas

The lower Rio Grande Valley of south Texas holds the richest tropical forests on the mainland US. Only a few thousand hectares of natural habitat remain in several private, state and federal parks and preserves. Notable habitats include sabal palmetto forest, thorn scrub and riparian forest. Of the 115 recorded species, 21 are found nowhere else in the US. All of these species range south into the Neotropics, although some are limited in their occurrence to a few regions of Mexico. Well over 100 other Mexican species have occurred in the valley as strays, and some have established temporary breeding populations. It is likely that more species would become permanent residents if there were more habitat available for the formation of metapopulations.

Within the valley there is considerable variation in physical and biotic factors. Where the Rio Grande River enters the Gulf of Mexico, conditions are more humid, and the temperatures are moderated by the Gulf, whereas conditions become progressively drier as one proceeds upstream, and the daily temperatures are more disparate. As a result, each park or preserve is unique and so are their butterfly faunas.

The Audubon Society's Sabal Palmetto Preserve lies closest to the Gulf in Cameron County. The preserve features a 15 ha grove of Texas sabal palmettos, the largest remaining grove of this tree. Several butterflies are resident in this preserve and nowhere else in the US; *Phycoides frisia* (a crescentspot); and *Timochares ruptifasciatus, Cabares potrillo* and *Decinea percosius*, all hesperids (skippers). There is no other natural habitat adjoining the preserve, but the Audubon Society has purchased neighbouring agricultural land and is restoring it to a natural state.

The Santa Ana National Wildlife Refuge and the Bentsen–Rio Grande Valley State Park are located upstream in Hidalgo County. Together the two comprise more than 2000 ha of thorn forest, palo verde swamp and river forest. Recently the Fish and Wildlife Service has purchased large units of natural habitat that serve as stepping stones between the refuge and the park. At present, *Myscelia ethusa* (blue wing) is found in the US only at Santa Ana. Since both areas are so large, they serve as important core areas for many of the valley's resident butterfly populations.

The Santa Margarita Ranch is a large area of natural habitat in Starr County. Most of the area is mesquite forest, but there is a significant area of tropical riparian forest adjacent to the river. This holds the only US resident population of *Emesis tenedia* (falcate emesia), although upstream water releases from the Falcon Dam during 1992 inundated the habitat and may have caused its extirpation.

22.3.2 Southern California

This is an area from Death Valley and the Tehachapi Mountains southward, having 159 species with 11 that occur principally in the area (Emmel and Emmel, 1973). Concern for this area comes mainly because of the large human population and extensive habitat destruction. Habitats with endemic species include salt marsh, coastal chapparal, mountain pine forest and desert mountains.

An example species is *Panoquina errans* (wandering skipper), which occurs at the upper edges of salt marshes in coastal bays and lagoons along the Pacific coast from Santa Barbara County, south to Baja California in Mexico. Most of its habitat in California has been destroyed by drainage and filling of marshes for development.

Several types of coastal chapparal occur in southern California and most are good habitats for butterflies. *Lycaena hermes* (Hermes copper) is found in a limited area of coastal San Diego County and adjacent Baja California Norte. The larval host is *Rhamnus crocea* (redberry), which ranges much further north than the butterfly. A significant proportion of its habitat has been destroyed by housing development in San Diego County.

Plebejus neurona (veined blue) is found only at high elevations in the Transverse Ranges of Kern, Los Angeles and Ventura counties, while *Plebulina emigdionis* (San Emigdio blue) is found in the vicinity of its larval host *Atriplex* spp. (saltbrush) primarily along dry stream beds in desert areas. There are many butterfly sub-species endemic to southern California, including the endangered *Euphilotes baltoides allyni* (El Segundo blue) and the extinct *Speyeria adiaste atossa* (Atossa fritillary).

22.3.3 South-eastern Arizona and south-western New Mexico

This area holds 182 species and, astonishingly, 32 species that reside nowhere else in the US (Bailowitz and Brock, 1991). It has several more or less isolated mountain ranges that have been referred to as the 'Sky Islands', the best known of which are the Chiricahuas and Huachucas. The fauna and flora of these are dominated by many species that range south into the Sierra Madre Occidentale of western Mexico. The habitats in these ranges include desert scrub at their bases with juniper woodland, oak–juniper, pine–oak and fir forest further up.

Most of the restricted local butterflies are hesperids, but there are a few true butterflies as well. *Neophasia terlootii* (Chiricahua pine white), with its sexual dimorphism (orange female and white male), is best known, but others include *Calephelis arizonensis* (Arizona metalmark) and *Pyrrhopyge araxes* (araxes skipper).

22.3.4 Rocky Mountain Front Range

This is the first range of mountains bordering the western edge of the Great Plains. This ecotone generally lies between 1500 and 2000 m elevation and holds 176 species, but only three endemics. There is more or less continuous habitat extending from northern New Mexico to northern Canada, but the richest butterfly habitats are from Colorado south. Habitats range from prairie grassland and sandy sage scrub at lowest elevations, through chapparal dominated by *Cercocarpus* sp. (mountain mahogany), juniper, yellow pine forest, Douglas-fir forest, spruce-fir forest to alpine rock fields and tundra at the highest. Each of these has its own species; those at highest elevations have affinities to the north (e.g. *Colias, Boloria, Oeneis, Erebia*), while the lower elevations include western endemics and a few species that range south to the mountains of central Mexico and beyond.

The highest species richness occurs in the low foothills and canyons. Typically, these habitats are privately owned or at the lowest elevations of National Forests. Several hesperids (*Polites rhesus, Stinga morrisoni*) are restricted to this area.

22.4 PATTERNS OF PROTECTION

Although there are vast uninhabited regions, areas of the US that may be considered secure for butterfly species are rare. Most of the land in eastern and midwestern states is privately owned, whereas extensive areas of the west are in public ownership, primarily by the Bureau of Land Management and the US Forest Service. In an extreme instance, 87% of Nevada is owned and managed by public agencies.

Protected areas include public lands administered by agencies in the US Department of the Interior (USDI), the US Department of Agriculture (USDA) and the US Department of Defense. In the USDI, the land administering bureaus are the Bureau of Land Management, the National Park Service, the Bureau of Indian Affairs and the Fish and Wildlife Service. The Bureau of Land Management is probably responsible for most lands in the contiguous 48 states. Most of this is in the arid intermountain west, and most are subject to grazing and mining. Protected lands overseen by the National Park Service include national parks, national monuments and national seashores. Public interpretation is the major mission of the Park Service and preservation is secondary, although no artificial manipulation of natural ecosystems is usually permitted. The Fish and Wildlife Service manages hundreds of national wildlife refuges, but, with a few exceptions noted below, the vast majority are managed primarily for wildfowl through the construction of water impounds or wild food crops. Livestock grazing and hay-making, etc. may be permitted on selected refuges.

The Forest Service in the USDA administers the national forests that compose the most extensive system of protected forested habitat in the US. They are managed for many purposes, but the production of timber is primary. The cutting practices, cutting schedules and building of access roads into previous roadless areas during the 1980s and early 1990s caused serious concern among conservationists.

The Department of Defense administers many military bases, proving grounds and test sites. Some of these, particularly in coastal areas and in the arid intermountain west, form important *de facto* preserves. The cessation of cold war tensions and the concomitant reduction in the US military establishment will result in the transfer of some of these lands to other public agencies, both federal and state. It is hoped that conservation of native species and ecosystems will be considered in these new arrangements.

Most states have park and preserve systems that can be considered protected, although human use is usually a high priority. In some states, some kinds of habitats have been devastated by development and heavy human use in state parks. A good example is the beach dune ecosystem in many California State beach parks.

Private lands with specific protected status are rare. The Nature Conservancy has acquired an extensive system of preserves where conservation of native ecosystems and species is the highest priority, and appropriate management is practised. Other private agencies such as Ducks Unlimited and private hunting organizations, own extensive lands for hunting of vertebrates. Such areas form extensive *de facto* preserves for butterflies and other invertebrates.

22.5 STATUS OF BUTTERFLY MANAGEMENT

The habitats of many rare butterflies are seral stages in a succession that is initiated by some type of disturbance. Often human influence on the environment, through the introduction of exotic plants, fire suppression or development, interrupts normal succession and causes restriction or loss of butterfly populations. In these cases management of the habitats may be necessary to restore declining populations.

In the US, although management of butterfly populations is uncommon, it is more frequent than in most other countries and there are some notable success stories.

22.5.1 Endangered and threatened butterflies

As mentioned earlier, 28 insect species are listed under the Endangered Species Act, and 15 of these are native butterflies. Active habitat management designed to recover the butterfly populations is under way for four of these: *E. b. allyni*, *Plebejus icarioides missionensis* (mission blue), *Apodemia mormo langei* (Lange's metalmark) and *Speyeria zerene hippolyta* (Oregon silverspot). Efforts for all four have been successful to a degree, but those for the last two have led to dramatic increases in population numbers.

Habitat management for these species has first meant securing land, then management to enhance larval food-plant abundance and nectar sources.

In some cases, securing land simply meant designating habitat areas without changing land ownership, while in others it was more involved. Habitats for *P. i. missionensis* were unmanaged private lands that were converted to managed county park land as part of the San Bruno Mountain Habitat Conservation Plan, while the remaining habitats of *A. m. langei* were acquired by the US Fish and Wildlife Service and now form the Antioch Dunes unit of the San Francisco Bay National Wildlife Refuge.

Habitat management for most of these species has included control or removal of weedy exotics, but has occasionally meant removal of some native woody plants. Physical removal, herbicides and fire have all been used successfully. In the case of *S. z. hippolyta* efforts by the Forest Service and Nature Conservancy resulted in unassisted recovery of the larval food-plant *Viola adunca*, while in the case of the other three species propagation of food-plants has been necessary.

Most states have their own endangered species programmes, which may be partially funded through co-operative agreements with the US Fish and Wildlife Service. The state programmes may or may not include insects under their purview. Under state endangered species law, species not listed under the federal act may be listed and protected, and several states list and protect butterflies that are endangered at state level. Additionally, almost all states have natural heritage programmes, many of which include butterflies on their lists of species of special concern. The Natural Heritage Programme, initiated by The Nature Conservancy, still maintains a nationwide natural heritage database that includes most North American butterflies.

22.5.2 Management of non-listed butterflies

Management efforts for habitats of non-listed butterflies are usually not aimed at specific species, but at communities. For example, recovery of native prairies in Illinois has been assisted by the use of fire, and several prairie butterflies were listed among species that benefited. Management of water impoundments and mowing of levee vegetation on the Santa Ana National Wildlife Refuge is now being programmed to benefit local butterfly populations (D. Pritchett, personal communication).

22.6 MONITORING

Monitoring of North American butterflies is in its infancy. Arnold's unpublished mark–release–recapture studies of several endangered Californian butterflies provide early population estimates and valuable information on their population structure and habitat needs. More non-invasive monitoring methods are now used for endangered species because of uncertainty in the confidence levels of estimates obtained from mark–recapture, and because endangered butterflies, especially Lycaenidae, may be injured by the handling required.

The transect method pioneered in the UK (see Pollard and Yates, 1993a; Harding, Asher and Yates, Chapter 1, and Pollard and Eversham, Chapter 2 this volume) is gaining favour in North America. It has been used in studies of the endangered *Boloria acrocnema* (Uncompahgre fritillary) and a modification is being considered for a continent-wide monitoring programme by the North American Butterfly Association.

The Xerces Society initiated the annual butterfly count that has just completed its nineteenth field season. More than 200 one-day counts were conducted in more than 30 states, five Canadian provinces and Mexico. The counts can provide regional trend information and document the occurrence and frequency of invasions of migratory and emigratory butterflies. A one-day count each year cannot provide year-to-year population estimates or indices for any single site for any species.

22.7   CONSERVATION NEEDS

The potential for butterfly conservation and management is almost unlimited. For endangered species protection there is a need to begin habitat acquisition and management for listed species that have not yet benefited from such attention. Unlisted candidate species should be added to the list promptly and habitat protection and management initiated. One of North America's most attractive butterflies, *Speyeria idalia* (regal fritillary), is a listing candidate that stands out as a species whose populations are declining and becoming extirpated.

One of the highest priorities for conservation in North American butterflies should be the increase and management of protected lands in the lower Rio Grande Valley. The Sabal Palmetto Sanctuary has a core of only 40 ha, but has additional adjacent land where propagated native woody plants and secondary succession will eventually provide a much larger reserve. Even more acreage should be added to this sanctuary, since it is the wettest tropical remnant in the valley.

Recently the concept of a wildlife corridor along the river was formalized with the creation of the Rio Grande Valley National Wildlife Refuge. This comprises more than 40 separate pieces of property, some of which lie adjacent to and increase the long-term security of pre-existing parks and refuges.

The conservation status of south-eastern Arizona and adjacent New Mexico consists of a patchwork of private land, national forests and preserves. Most notable is the Gray Ranch, a Nature Conservancy property that includes the Las Animas mountains, and another Nature Conservancy Preserve in the Huachuca mountains. The richest areas still remain in the national forests and in Fort Huachuca, a military reserve. It is of urgent importance that the US Forest Service includes management and maintenance of butterfly habitats in its plan for the forests.

References

Ackery, P.R. (1984) Systematic and faunistic studies on butterflies, in *Biology of Butterflies*, (eds R.I. Vane-Wright and P.R. Ackery), Symposium of the Royal Entomological Society, London, no. 11, Academic Press, London, pp. 9–21.

Ackery, P., Smith, C.R. and Vane-Wright, R.I. (1994) *Carcasson's African Butterflies*, The Natural History Museum and CSIRO, Canberra.

Adams, J. and McShane, T. (1992) *The myth of wild Africa*, Norton.

Andrewartha, H.G. and Birch, L.C. (1954) *The distribution and abundance of animals*, University of Chicago Press, Chicago.

Anon. (1981) Cover. *Radio Times*, 24 January 1988.

Anon. (1984) *Nature Conservation in Britain*, Nature Conservancy Council, Peterborough.

Anon. (1985) Survey of environmental topics on farms. England and Wales: 1985. *MAFF Statistical Note No 244/85*, Ministry of Agriculture, Fisheries and Food, London.

Anon. (1986) Insect re-establishment – a code for conservation practice. *Antenna*, **10**, 13–18.

Anon. (1988) *Set-aside*. Booklet SA1, Ministry of Agriculture, Fisheries and Food, Welsh Office Agriculture Department.

Anon. (1990) *1990 IUCN Red List of Threatened Animals*, IUCN, World Conservation Monitoring Centre, Cambridge, UK.

Anon. (1992a) *Action for the Countryside*, Department of the Environment, London.

Anon. (1992b) *Area Arable Payments: Explanatory Booklet*, Ministry of Agriculture, Fisheries and Food, London.

Anon. (1993a) The large blue is back in business. *English Nature*, **6**, 4–5.

Anon. (1993b) Four winged stamps. *PFC-Journalen*, **2**, 2–11.

Anon. (1993c) *Agriculture and England's Environment*. A consultation document on Environmental Schemes under the Common Agricultural Policy, MAFF.

Archer-Lock, A. (1982) The chequered skipper: *Carterocephalus palaemon* Pallas in England, 1976. *Entomologist's Record*, **94**, 123.

Asher, J. (1992) A programme for the coordination of butterfly recording in Britain and Ireland. Unpublished report to Butterfly Conservation and Biological Records Centre.

Aurivillius, C.A. (1898) *Rhopalocera Aethiopica*, Stockholm.

Bailowitz, R.A. and Brock, J.P. (1991) *Butterflies of Southeastern Arizona*, Sonoran Arthropod Studies, Inc., Tucson, Arizona, USA.

Baker, H.G. and Baker, I. (1983) Floral nectar sugar constituents in relation to pollinator type, in *Handbook of Experimental Pollination Biology*, (eds C.E. Jones and R.J. Little), Van Nostrand Reinhold, New York, pp. 117–40.

Baker, R.R. (1969) The evolution of the migratory habit in butterflies. *Journal of Animal Ecology*, **38**, 703–46.

Baker, R. R. (1984) The dilemma, when and how to go or stay, in *The Biology of Butterflies*, (eds R.I.Vane-Wright and P.R.Ackery), Symposium of the Royal Entomological Society of London, no. 11, Academic Press, London, pp. 279–96.

Balfour-Browne, F. (1958) The origin of our British swallowtail and our large copper butterflies. *Entomologists Record*, **70**, 33–6.

Balint, Z. (1991) Conservation of butterflies in Hungary. *Oedippus*, **3**, 5–36.

Balint, Z. and Fielder, K. (1992) Plebeius sephirus (Frivaldszky, 1835) in Pannonia, with special reference to its status and ecology in Hungary. *Oedippus*, **4**, 1–24.

Ball, S.G. (1994) The Invertebrate Site Register – objectives and achievements, in *Invertebrates in the Landscape*, (ed. P.T. Harding), Supplement to the British Journal of Entomology and Natural History (in press).

Balletto, E. (1992) Butterflies in Italy: status, problems and prospects, in *Future of Butterflies in Europe: Strategies for Survival*, (eds T. Pavlicek-van Beek, A.H. Ovaa and J.G. van der Made), Agricultural University, Wageningen, pp. 53–64.

Balletto, E. (1993) Species accounts on *Polyommatus humedasae* and *P. galloi*, in *Conservation Biology of Lycaenidae*, (ed. T.R. New), IUCN, Gland.

Balletto, E. and Casale, A. (1991) Mediterranean insect conservation, in *The Conservation of Insects and their Habitats*, (eds N.M. Collins and J.A. Thomas), Academic Press, London, pp. 121–42.

Balletto, E. and Kudrna, O. (1985) Some aspects of the conservation of butterflies in Italy, with recommendations for a future strategy (Lepidoptera Hesperiidae and Papilionoidea). *Bolletino Societa Entomologica. Italiana*, **117**, 39–59.

Balletto, E., Barberis, G. and Toso, G.G. (1982) Aspetti dell' ecologia dei Lepidotteri Ropaloceri nei consorzi erbacei delle alpi italiane. *Quaderni sulla Struttura delle Zoocenosi Terrestri*, **II. 2. I** *Pascoli Altomontani*, 11–96

Balletto, E., Toso, G.G. and Lattes, A. (1989) Studi sulle comunità di Lepidotteri ropaloceri dell littorale tirrenico. *Bolletino Musei Istituto Biologia Università Genova*, **53**, 141–86.

Balletto, E., Toso, G., Barberis, G. and Rossaro, B. (1977) Aspetti dell'ecologia dei Lepidotteri ropaloceri nei consorci erbacei alto Apeninninici. *Animalia*, **4**, 277–343.

Balletto, E., Lattes, A., Cassulo, L. and Toso, G.G. (1988) Studi sull'ecologia di lepidotteri ropaloceri in alcuni ambienti delle Dolomiti. *Studi Trentini Di Scienza Naturali*, **64**, 87–123.

Barbour, D.A. (1986a) Why are there so few butterflies in Liverpool? An answer. *Antenna*, **10**, 72–5.

Barbour, D.A. (1986b) Expansion of the range of the speckled wood *Pararge aegeria* (L) in N.E. Scotland. *Entomologist's Record and Journal of Variation*, **81**, 98–105.

Barnham, M. and Foggitt, G.T. (1987) *Butterflies in the Harrogate district*, Barnham and Foggitt, Harrogate.

Bartlein, P.J. and Prentice, I.C. (1989) Orbital variation, climate and paleoecology. *Trends in Ecology and Evolution*, **4**, 195–9.

Baudry, J. (1988) Hedgerows and hedgerow networks as wildlife habitat in agricultural landscapes, in *Environmental Management in Agriculture*, (ed. J.R. Park), Belhaven Press, London and New York, pp. 111–24.

Baz, A. (1991) Ranking species and sites for butterfly conservation using presence–absence data in Central Spain. *Nota Lepidopterologica*, suppl. 2, 4–12.

Bennett, K.D., Tzedakis, P.C. and Willis, K.J. (1991) Quaternery refugia of north European trees. *Journal of Biogeography*, **18**, 103–15.

Berger, A. and Loutre, M.F. (1991) Insolation values for the climate of the last 10 million years. *Quaternary Science Reviews*, **10**, 297–317.

Berger, L. (1946) *Maculinea rebeli* Hirschke, bona species. *Lambillionea*, **46**, 95.

Berger, L.A. (1981) *Les papillons de Zaïre*. Prèsidence de la Rèpublique, Kinshasa.

Bergmann, A. (1952) *Die Grossschmetterlinge Mitteldeutschlands*, Vol. 2, Urania Verlag, Jena.

Bernardi, G. (1947) Notes apropos d'un article de L. Berger: *Maculinea rebeli* etc. *Bulletin de la Sociéte Entomologique, Mulhouse*, **7–8**, 61.

Bink, F.A. (1962) De grote vuurvlinder van het Fries-Overijsselse moerasgebied (*Thersamonia dispar batavus* Oberth.). *Linneana belgica*, **1**, 2–13.

Bink, F.A. (1970) A review of the introductions of *Thersamonia dispar* Haw. (Lep. Lycaenidae) and the speciation problem. *Entomologische Berichten*, **30**, 179–83.

Bink, F.A. (1972) Het onderzoek naar de grote vuurvlinder (*Lycaena dispar batava* (Ober-thür)) in Nederland (Lep. Lycaenidae). *Entomologische Berichten*, **32**, 225–39.

Bink, F.A. (1992) The butterflies of the future, their strategy, in *Future of butterflies in Europe*; *Strategies for Survival*, (eds T. Pavlicek-van Beek, A.H. Ovaa and J.G. van der Made), Agricultural University, Wageningen, pp. 134–8.

Bischof, N. (1980) Pflanzensoziologische Untersuchungen von Sukzessionen aus gemähten Magerrasen in der subalpinen Stufe der Zentralalpen. Ph.D. thesis, Universität Basel.

Bischof, N. (1984) Pflanzensoziologische Untersuchungen von Sukzessionen aus gemähten Magerrasen in der subalpinen Stufe der Zentralalpen. *Beiträge zur Geobotanischen Landesaufnahme der Schweiz*, **60**, 1–128.

Blab, J. and Kudrna, O. (1982) Hilfsprogramm für Schmetterlinge. *Naturschutz aktuell*, **6**, 1–135.

Blab, J., Ruckstuhl, Th., Esche, T. and Holzberger, R. (eds) (1987) *Aktion Schmetterling: so können wir sie retten*, Otto Maier, Ravensburg.

Boggs, C.L. (1987) Ecology of nectar and pollen feeding in Lepidoptera, in *Nutritional Ecology of Insects, Mites, Spiders and Related Invertebrates*, (eds F. Slansky and J.G. Rodriguez), John Wiley, New York, pp. 369–91.

Braby, M.F. (1991) City butterflies go bush. *Wildlife Australia*, (winter 1991), 20–1.

Brakefield, P.M. (1984) The ecological genetics of quantitative characteristics of *Maniola jurtina* and other butterflies, in *The Biology of Butterflies*, (eds R.I. Vane-Wright and P.R. Ackery), Symposium of the Royal Entomological Society, no. 11, Academic Press, London, pp. 167–90.

Bretherton, R.F. (1951) The early history of the swallow-tail butterfly (*Papilio machaon* L.) in England. *Entomologists Record*, **63**, 206–11.

Bretherton, R.F. (1966) A distribution list of the butterflies (Rhopalocera) of western and southern Europe. *Transactions of the Society for British Entomology*, **17**(1), 1–94.

Bretherton, R.F. (1981) *Carterocephalus palaemon* (Pallas) (Lepidoptera: Hesperiidae) at Woodwalton Fen and elsewhere. *Entomologist's Gazette*, **32**, 167–8.

Bretherton, R.F. (1983) The incidence of migrant Lepidoptera in the British Isles, in *Moths and Butterflies of Great Britain and Ireland*, Vol. 10, (eds J. Heath and A.M. Emmet), Harley Books, Colchester, pp. 9–34.

Brian, M.V. (1979) Caste differentiation and division of labor, in *Social Insects*, (ed. H. Hermann), Academic Press, New York, Ch. 5.

Bristow, R., Mitchell, S.H. and Bolton, D.E. (1993) *Devon Butterflies*, Devon Books, Tiverton.

Britton, D.R. and New, T.R. (1992) *Ecology of the butterfly and ant community at Mt. Piper, Victoria*, Report to Department of Conservation and Environment, Victoria.

Britton, D.R. and New, T.R. (1993) Communities of diurnal Lepidoptera in central Victoria, with emphasis on the Mt. Piper region, Broadford. Report to Department of Conservation and Natural Resources.

Brouwer, F. (1989) Determination of broad-scale landuse changes by climate and soils. *Journal of Environmental Management*, **29**, 1–15.

Brown, A.H.F. and Warr, S.J. (1992) The effects of changing management on seed banks in ancient coppices, in *Ecology and Management of Coppice Woodlands*, (ed. G.P. Buckley), Chapman & Hall, London, pp. 147–66.

Bryner, R. (1987) Dokumentation über den Rückgang der Schmetterlingsfauna in der Region Biel-Seeland-Chasseral. *Beiträge zum Naturschutz in der Schweiz*, **9**, 1–92.

Bullock, D.J. (1991) The feral goat, in *Handbook of British Mammals*, (eds S. Harris and G.B. Corbet), Blackwell, Oxford.

Burn, A.J. (1988) Assessment of the impact of pesticides on invertebrate predation in cereal crops. *Aspects of Applied Biology*, **17**, 279–88.

Buszko, J. (1992) A long-term research project on Polish butterflies, in *Future of butterflies in Europe*, (eds T. Pavlicek-van Beek, A.H. Ovaa and J.G. van der Made), Agricultural

University, Wageningen, pp. 83–8.

Butterflies Under Threat Team (1986) *The Management of Chalk Grassland for Butterflies*, Nature Conservancy Council, Focus on Nature Conservation no. 17, Peterborough.

Butterfly Conservation (1992) *Introducing Butterfly Conservation*, British Butterfly Conservation Society, Dedham.

Cadbury, C.J. (1990) The Status and Management of Butterflies on RSPB Reserves, in *RSPB Conservation Review no. 4*, (eds C. J. Cadbury and M. Everett), Royal Society for the Protection of Birds, Sandy, pp. 40–6.

Carcasson, R.H. (1964) A preliminary survey of the zoogeography of African butterflies. *East African Wildlife Journal*, 2, 122–57.

Carcasson, R.H. (1981) *Collins Field Guide to the Butterflies of Africa*, Collins, London.

Carter, C.I. and Anderson, M.A. (1987) Enhancement of lowland forest ridesides and roadsides to benefit wild plants and animals. *Forestry Commission Research Information Note 126*, HMSO.

Chalmers-Hunt, J.M. and Owen, D.F. (1952) The history and status of *Pararge aegeria* (Lep. Satyridae) in Kent. *Entomologist*, 85, 145–54.

Chapman, A. (ed.) (1987) *RSPB Reserves Visiting*. Royal Society for Protection of Birds, Sandy.

Chapman, T.A. (1916) What the larva of *Lycaena arion* does during its last instar. *Transactions of the Entomological Society of London 1915*, 291–7.

Chapman, T.A. (1919) Notes on *Lycaena alcon* F., as reared in 1918–1919. *Transactions of the Entomological Society of London 1919*, 443–9.

Clarke, C.A. and Larsen, T.B. (1986) Speciation problems in the *Papilio-machaon* group of butterflies, Lepidoptera, Papilionidae. *Systematic Entomology*, 11, 175–82.

Clarke, J.H. and Cooper, F.B. (1992) Vegetation changes and weed levels in set-aside and subsequent crops, in *BCPC Monograph No. 50: Set-aside*, (ed. J.H. Clarke), BCPC, Farnham, Surrey, pp. 103–10.

Clarke, S.A. and Robertson, P.A. (1993a) The relative effects of woodland management and pheasant predation on the survival of pearl-bordered and small pearl-bordered fritillaries (*B. euphrosyne* and *B. selene*) in the South of England. *Biological Conservation*, 65, 199–203.

Clarke, S.A. and Robertson, P.A. (1993b) Pheasants and butterflies. *The Game Conservancy Annual Review*, 24, 83–4.

Cleveland, W.A. (ed.) (1987) *Britannica Atlas*, Encyclopaedia Britannica, Chicago.

Coleman, W.S. (1860) *British Butterflies*, Routledge, London.

Collar, N.J. and Stuart, S.(1988) Key forests for threatened birds in Africa. *International Council for Bird Preservation, Monograph No. 3*, Cambridge, UK.

Collier, R.V. (1966) Status of butterflies on Castor Hanglands, N.N.R. 1961–1965 inclusive. *Northamptonshire Natural History Society and Field Club Journal*, 35, 451–6.

Collier, R.V. (1972) Chequered skipper (*Carterocephalus palaemon*). Unpublished Report to Nature Conservancy Council.

Collier, R.V. (1978) The status and decline of butterflies on Castor Hanglands NNR 1919–1977. Unpublished report to Nature Conservancy Council.

Collier, R.V. (1986) The conservation of the chequered skipper in Britain. *Focus on Nature Conservation*, Vol. 16, Nature Conservancy Council.

Collins, N. M. and Morris, M.G. (1985) *Threatened Swallowtail Butterflies of the World*, IUCN, Gland.

COM (1984) *The Regions of Europe*. Second periodic report on the social and economic situation and development of the regions of the community, Commission of the European Communities, Vol. 40, 4 April, 1984.

COM. (1985) *European Regional Development Fund*. Tenth report of the Commission, Commission of the European Communities, Vol. 516 (final), 4 October, 1985.

Common, I.F.B. and Waterhouse, D.F. (1981) *Butterflies of Australia*, 2nd edn, Angus and Robertson, Sydney.

Corbet, S.A. (1978) Bee visits and the nectar of *Echium vulgare* L. and *Sinapsis alba* L. *Ecological Entomology*, **3**, 25–37.

Corbet, S.A., Unwin, D.M. and Prys-Jones, O.E. (1979) Humidity, nectar and insect visits to flowers, with special reference to *Crataegus*, *Tilia* and *Echium*. *Ecological Entomology*, **4**, 9–22.

Corke, D. (1989) Of pheasants and fritillaries: is predation by pheasants (*Phasianus colchicus*) a cause of the decline in some British butterfly species? *British Journal of Entomology and Natural History*, **2**, 1–14.

Corke, D. (1991) Stinging nettle butterflies. *British Wildlife*, **2**(6), 325–34.

Corke, D. and Davis, S. (1992) Essex garden butterflies. *Essex Wildlife*, no. 27, p. 9.

Corke, D. and Harding, P.T. (1990) A decade of county butterfly lists. *Butterfly Conservation News*, **45**, 33–7.

Courtney, S.P. (1980) Studies on the biology of the butterflies *Anthocharis cardamines* L. and *Pieris napi* L. in relation to speciation in Pierinae. Ph.D. Thesis, University of Durham.

Courtney, S.P. (1981) Coevolution of pierid butterflies and their cruciferous foodplants. III. *Anthocaris cardamines* survival, development and oviposition on different hostplants. *Oecologia*, **51**, 91–6.

Courtney, S.P. (1986) The ecology of pierid butterflies: dynamics and interactions. *Advances in Ecological Research*, **15**, 51–131.

Courtney, S.P. and Duggan, A.E. (1983) The population biology of the orange-tip butterfly *Anthocharis cardamines* in Britain. *Ecological Entomology*, **8**, 271–81.

Cribb, P.W. (1982) *How to Encourage Butterflies to Live in Your Garden*, Amateur Entomologists Society, Feltham.

Cronin, T.M. and Schneider, C.E. (1990) Climatic influences on species: evidence from the fossil record. *Trends in Ecology and Evolution*, **5**, 275–9.

Crosby, D.F. (1990) A management plan for the Altona Skipper butterfly *Hesperilla flavescens flavescens* Waterhouse (Lepidoptera: Hesperiidae). *Technical Report Series, No. 98, Arthur Rylah Institute for Environmental Research*, Melbourne.

D'Abrera, B. (1980) *Butterflies of the Afrotropical Region*, Lansdown Press, Melbourne.

D'Abrera, B. (1982–86) *Butterflies of the Oriental Region*, (3 vols), Hill House, Victoria, Australia.

Dabrowski, J.S. and Krzywicki, M. (1982) Ginace i zagrozone gatunki motyli w faunie Polski. *Studia Naturae, (B)*, **31**, 1–171.

Davies, M. (1992) *The White-letter Hairstreak Butterfly*, Booklet No. 12, Butterfly Conservation, Colchester.

Davis, B.N.K., Lakhani, K.H. and Yates, T.J. (1991) The hazards of insecticides to butterflies of field margins. *Agriculture, Ecosystems and Environment*, **36**, 151–61.

Davis, S. (1990) An investigation of the nature of perch based territories adopted by male *Carterocephalus palaemon*, and an analysis of the behaviour exhibited by males in such territories. Unpublished M.Sc. thesis, University of Aberdeen.

Dempster, J.P. (1968) The control of *Pieris rapae* with DDT. II. Survival of the young stages of *Pieris* after spraying. *Journal of Applied Ecology*, **5**, 451–62.

Dempster, J.P. (1989) Insect introductions: natural dispersal and population persistence in insects. *The Entomologist*, **108**, 5–13.

Dempster, J.P. (1991) Fragmentation, isolation and mobility of insect populations, in *The Conservation of Insects and their Habitats*, (eds N. M. Collins and J.A. Thomas), Academic Press, London, pp. 143–53.

Dempster, J.P. (1992) Evidence of an oviposition-deterring pheromone in the orange-tip butterfly, *Anthocharis cardamines* (L). *Ecological Entomology*, **17**, 83–5.

Dempster, J.P. and Hall, M.L. (1980) At attempt at re-establishing the swallowtail butterfly at Wicken Fen. *Ecological Entomology*, **5**, 327–34.

Dempster, J.P., King, M.L. and Lakhani, K.H. (1976) The status of the swallowtail butterfly in Britain. *Ecological Entomology*, **1**, 71–84.

Dennis, R.L.H. (1977) *The British Butterflies: Their Origin and Establishment*, Classey, Faringdon.

Dennis, R.L.H. (1982–83) Mate location strategies in the Wall Brown butterfly: Wait or seek? *Entomologist's Record and Journal of Variation*, **94**, 209–14: **95**, 7–10.

Dennis, R.L.H. (1982a) Observations on habitats and dispersion made from oviposition markers in north Cheshire *Anthocharis cardamines* (L.) (Lepidoptera: Pieridae). *Entomologist's Gazette*, **33**, 151–9.

Dennis, R.L.H. (1982b) Patrolling behaviour in orange-tip butterflies within the Bollin valley in north Cheshire, and a comparison with other pierids. *Vasculum*, **67**, 17–25.

Dennis, R.L.H. (1983) Egg laying cues in the Wall Brown butterfly. *Entomologists' Gazette*, **34**, 89–95.

Dennis, R.L.H. (1986) Motorways and cross-movements. An insect's mental map of the M56 in Cheshire. *The Bulletin of the Amateur Entomologists' Society*, **45**, 228–43.

Dennis, R.L.H. (1992) An evolutionary history of British butterflies, in *The Ecology of Butterflies in Britain*, (ed. R.L.H. Dennis),Oxford University Press, Oxford, pp. 217–45.

Dennis, R.L.H. (1993) *Butterflies and Climate Change*, University Press, Manchester.

Dennis, R.L.H. and Shreeve, T.G. (1989) Butterfly morphology variation in the British Isles: the influence of climate, behavioural posture and the hostplant-habitat. *Biological Journal of the Linnean Society*, **38**, 323–48.

Dennis, R.H.L. and Shreeve, T.G. (1991) Climate change and the British butterfly fauna: opportunities and constraints. *Biological Conservation*, **55**, 1–16.

Dennis, R.L.H., Williams, W.R. and Shreeve, T.G. (1991) A multivariate approach to the determination of faunal structure among European butterfly species (Lepidoptera: Rhopalocera). *Zoological Journal of the Linnean Society*, **101**, 1–49.

DeVries, P.J., Cocroft, R.B. and Thomas, J.A. (1993) Comparison of acoustical signals in *Maculinea* butterfly caterpillars and their obligate host *Myrmica* ants. *Biological Journal of the Linnean Society*, **49**, 229–38

Dexter, E.M. and Kitching, R.L. (1993) The Bathurst Copper, *Paralucia spinifera* in *Conservation Biology of the Lycaenidae*, (ed. T.R. New), IUCN, Gland, pp. 168–70.

Dorp, D. van and Opdam, P.F.M (1987) Effects of patch size, isolation and regional abundance on forest bird communities. *Landscape Ecology*, **1**, 59–73.

Douwes, P. (1975) Distribution of a population of *Heodes virgaureae*. *Oikos*, **26**, 332–40.

Dover, J.W. (1989) The use of flowers by butterflies foraging in cereal field margins. *Entomologist's Gazette*, **40**, 283–91.

Dover, J.W. (1991) The conservation of insects on arable farmland, in *The Conservation of Insects and their Habitats*, (eds N.M. Collins and J.A. Thomas), Academic Press, London, pp. 293–317.

Dover, J.W., Sotherton, N.W. and Gobbett, K. (1990) Reduced pesticide inputs on cereal field margins: the effects on butterfly abundance. *Ecological Entomology*, **15**, 17–24.

Downes, J.A. (1948) The history of the speckled wood butterfly (*Pararge aegeria*) in Scotland, with a discussion of recent changes of range of other British butterflies, *Journal of Animal Ecology*, **17**, 131–8.

Duffey, E. (1968) Ecological studies on the large copper butterfly *Lycaena dispar* (Haw.) *batavus* (Obth.) at Woodwalton Fen National Nature Reserve, Huntingdonshire. *Journal of Applied Ecology*, **5**, 69–96.

Duffey, E. (1977) The reestablishment of the large copper butterfly *Lycaena dispar batavus* Obth. on Woodwalton Fen National Nature Reserve, Cambridgeshire, England, 1969–73. *Biological Conservation*, **12**, 143–58.

Duffey, E. (1982) *Naturparks in Europa*, Christian Verlag, München.

Duffey, E. and Mason, G. (1970) Some effects of summer floods on Woodwalton Fen in 1968/69. *Entomologist's Gazette*, **21**, 23–6.

Duffey, E., Morris, M.G., Ward, L.K., Wells, D.A. and Wells, T.C. (1974) *Grassland Ecology and Wildlife Management*, Chapman & Hall, London.

Dunlap-Pianka, H., Boggs, C.L. and Gilbert. S.E. (1977) Ovarian dynamics in heliconiine butterflies: programmed senescence versus eternal youth. *Science*, **197**, 487–90.

Dunn, K.L. and Dunn, L. E. (1991) *Review of the Australian Butterflies* (4 vols), Melbourne, privately published.

Dunn, T.C. and Parrack, J.D. (1986) The moths and butterflies of Northumberland and Durham. *Vasculum*, suppl. 2.

Dutreix, C. (1992) Mapping and conservation of butterflies: remarks on *Lycaena dispar*, in *Future of Butterflies in Europe: Strategies for Survival*, (eds T. Pavlicek-van Beek, A.H. Ovaa and J.G. van der Made), Agricultural University, Wageningen, p. 279.

Ebert, G. and Falkner, F. (1978) Rote Liste der in Baden-Württemberg gefährdeten Schmetterlingsarten (Macrolepidoptera) 1. Fassung, Stand 1. 11. 77. *Beihefte der Veröffentlichungen für Naturschutz und Landschaftspflege, Baden-Württemberg*, **11**, 323–65.

Ebert, G. and Rennwald, E. (1991) *Die Schmetterlinge Baden-Württembergs*, Ulmer Verlag, Stuttgart.

Ehrlich, P.R. (1984) The structure and dynamics of butterfly populations, in *Biology of Butterflies*, (eds R.I. Vane-Wright and P.R. Ackery), Symposium of the Royal Entomological Society, no. 11, Academic Press, London, pp. 25–40.

Elfferich, N.W. (1963) Kweekervaringen met *Maculinea alcon* Schiff. *Entomologische Berichten*, **23**, 46–52.

Eliot, J.N. (1973) The higher classification of the Lycaenidae (Lepidoptera): a tentative arrangement. *Bulletin of the British Museum, Natural History (Entomology)*, *28*(6), 373–505.

Ellis, E.A. (1965) *The Broads*, The New Naturalist Series 46, Collins, London.

Elmes, G.W. and Thomas, J.A. (1987) a Die Gattung Maculinea, in *Tagfalter und ihr Lebensraum*, (ed. W. Geiger), Schweizerisches Bund für Naturschutz, Basel, pp. 354–68.

Elmes, G.W. and Thomas, J.A. (1992) The complexity of species conservation: interactions between *Maculinea* butterflies and their ant hosts. *Biodiversity and Conservation*, **1**, 155–69.

Elmes, G.W., Thomas, J.A. and Wardlaw, J.C. (1991) Larvae of *Maculinea rebeli*, a Large Blue butterfly, and their *Myrmica* host ants: wild adoption and behaviour in ant nests. *Journal of Zoology, London*, **223**, 447–60.

Elmes, G.W., Wardlaw, J.C. and Thomas, J.A. (1991) Larvae of *Maculinea rebeli*, a Large Blue butterfly and their *Myrmica* ant hosts: patterns of caterpillar growth and survival. *Journal of Zoology, London*, **224**, 79–92.

Elmes, G.W., Thomas, J.A., Hammarstedt, O., *et al.* (1994) Differences in host-ant specificity between Spanish, Dutch and Swedish populations of the endangered butterfly *Maculinea alcon* (Schiff.) (Lepidoptera). *Zoologica memorabilia*, **48**, 55–68.

Emmel, T.C. and Emmell, J.F. (1973) *The Butterflies of Southern California*. Natural History Museum of Los Angeles County, Sciences Series, no. 26.

Emmet, A.M. (1989) The vernacular names and early history of British butterflies, in *The Moths and Butterflies of Great Britain and Ireland*, Vol. 7, part 1, (eds A.M. Emmet and J. Heath), Harley Books, Colchester, pp. 7–21.

Emmet, A.M. and Heath J. (eds) (1989) *The Moths and Butterflies of Great Britain and Ireland*, Vol. 7, part 1, Harley Books, Colchester.

Emmet, A.M. and Pyman, G.A. (1985) *The Larger Moths and Butterflies of Essex*, The Essex Field Club, London.

Entomologischer Verein Alpstein, St. Gallen, Schweiz (1989) *Inventar der Tagfalter-Fauna (Lepidoptera) der Nordostschweiz und Veränderungen seit der Jahrhundertwende, Beobachtungen des Entomologischen Vereins Alpstein, St. Gallen 1960–1978*, Entomologischer Verein Alpstein, St. Gallen.

Erhardt, A. (1985a) *Lepidoptera Fauna in Cultivated and Abandoned Grassland in the Subalpine Region of Central Switzerland*. Proceedings of the 3rd Congress of European Lepidopterology, Cambridge 1982, pp. 63–73.

Erhardt, A. (1985b) Wiesen und Brachland als Lebensraum für Schmetterlinge. Eine Feldstudie im Tavetsch (GR). *Denkschriften der Schweizerischen Naturforschenden Gesellschaft*, **98**, 1–154.

Erhardt, A. (1985c) Day-active Lepidoptera: Sensitive indicators of cultivated and abandoned grassland. *Journal of Applied Ecology*, **22**, 841–61.

Erhardt, A. (1990) Chloridea ononis D. and S.: Evidence for an autochthonous population in the Swiss Alps (Lepidoptera, Noctuidae). *Nota lepidopterologica*, **13**, 207–12.

Erhardt, A. (1991) Nectar sugar and amino acid preferences of *Battus philenor* (Lepidoptera, Papilionidae). *Ecological Entomology*, **16**, 425–34.

Erhardt, A. (1992a) Preferences and non-preferences for nectar constituents in *Ornithoptera priamus* Poseidon (Lepidoptera, Papilionidae). *Oecologia*, **90**, 581–5.

Erhardt, A. (1992b) Impact of grassland management on diurnal Lepidoptera in the Swiss Central Alps, in *Future of Butterflies in Europe: Strategies for Survival*, (eds T. Pavlicek-van Beek, A.H. Ovaa, and J.G. van der Made), Agricultural University Wageningen, pp. 146–55.

Erhardt, A. and Thomas, J.A. (1991) Lepidoptera as indicators of change in the semi-natural grasslands of lowland and upland Europe, in *The Conservation of Insects and Their Habitats*, (eds N.M. Collins and J.A. Thomas), Academic Press, London, pp. 213–36.

Evans, C.E. (1949) The chequered skipper *C. palaemon* in West Inverness. *Scottish Naturalist*, **61**, 176.

Evers, W.M.J., Maaren, N.G.J. van and Made, J.G. van der (1987) De grote vuurvlinder in der Wieden (Overijssel). *De Levende Natuur*, **88**, 82–8.

Eversham, B.C., Harding, P.T., Loder, N., *et al.* (1993) Research applications using data from species surveys in Britain, in *Faunal Inventories of Sites for Cartography and Nature Conservation*, (eds J.L. van Goethem and P. Grootaert), Bulletin de l'Institut des Sciences naturelles de Belgique, pp. 29–40.

Farrell, L. (1973) A preliminary report on the status of the chequered skipper (*Carterocephalus palaemon* (Pall.)). Unpublished Report to Joint Committee for the Conservation of British Insects.

Farrell, L. (1974) Chequered skipper survey 1974. Unpublished Report to Institute of Terrestrial Ecology.

Feber, R.E. (1993) The ecology and conservation of butterflies on lowland arable farmland. D. Phil. thesis, University of Oxford.

Fernández-Galiano, E. (1992) Conservation of butterflies in the Bern Convention, in *Future of Butterflies in Europe: Strategies for Survival*, (eds T. Pavlicek van Beek, A.H. Ovaa and J.G. van der Made), Agricultural University, Wageningen, pp. 244–5.

Fiedler, K. (1990) New information on the biology of *M. nausithous* and *M. teleius* (Lepidoptera: Lycaenidae). *Nota Lepidopterologica*, **12**, 246–56.

Floyd, W.D. (1992) Political aspects of set-aside as a policy instrument in the European Community, in *BCPC Monograph No. 50: Set-aside*, (ed. J.H. Clarke), British Crop Protection Council, Farnham, Surrey, pp. 13–20.

Ford, E.B. (1945) *Butterflies*, Collins, London.

Forestry Commission (1984) *Census of woodland trees 1979–82: Great Britain*, Forestry Commission, Edinburgh.

Forster, W. and Wohlfahrt, T.A. (1955) *Die Schmetterlinge Mitteleuropas. II. Tagfalter*, 1st edn, Franckh'sche Verlagshandlung, Stuttgart.

Fowles, A.P. (1985) Population studies of the Marsh Fritillary colony at Rhos Llawr-cwrt NNR, Dyfed. Unpublished report, Dyfed Powys Region, Nature Conservancy Council, Aberystwyth.

Frazer, J.F.D. and Hyde, G.E. (1965) The decline of the chalk grassland butterflies. *Wildlife* (formerly *Animals*), **7**, 427–35.

Frohawk, F.W. (1892) Life-history of *Carterocephalus palaemon*. *Entomologist*, **25**, 225–8; 254–6.

Frohawk, F.W. (1916) Further observations on the last stage of the larva *Lycaena arion*. *Transactions of the Entomological Society of London 1915*, 313–16.

Frohawk, F.W. (1924) *Natural History of British butterflies*, London.

Frohawk, F.W. (1934) *A Complete Book of British Butterflies*, Ward Lock.

Frost, M.P. and Madge, S.C. (1991) *Butterflies in South-East Cornwall*, The Caradon Field and Natural History Club, Torpoint.

Fuller, R. J. (1992) Effects of coppice management on woodland breeding birds, in *Ecology and Management of Coppice Woodlands*, (ed. G.P. Buckley), Chapman & Hall, London, pp. 169–92.

Fuller, R.J. and Warren, M.S. (1990) *Coppiced Woodlands: Their Management for Wildlife*, Nature Conservancy Council, Peterborough.

Fuller, R.M. (1987) The changing extent and conservation interest of lowland grasslands in England and Wales: A review of grassland surveys 1930–84. *Biological Conservation*, **40**, 281–300.

Fussell, M. and Corbet, S.A. (1991) Forage for bumble-bees and honey bees in farmland: a case study. *Journal of Apicultural Research*, **30**, 87–97.

Fussell, M. and Corbet, S.A. (1992) Flower usage by bumble-bee: a basis for forage plant management. *Journal of Applied Ecology*, **29**, 451–65.

Galloway, R.W. (1988) The potential impact of climate changes on Australian ski fields, in *Greenhouse Planning for Climate Change*, (ed. G.I. Pearman), CSIRO, Melbourne, pp. 428–37.

Garciá, M.A. (1992) La política forestal en el Estado de las Autonomías. *Quercus*, **75**, 13–16.

Garciá-Barros, E. (1988) *Estudio comparativo de la biología y morfología de cuatro especies de* Hipparchia fabricius *(Lepidoptera, Satyridae)*, Ediciones Universidad Autónoma de Madrid, Madrid.

Garland, S.P. (1981) *Butterflies of the Sheffield area*, Sorby Record Special Series No. 5.

Geiger, H.J. and Scholl, E. (1981) Wiederfangversuche an markierten Alpengelblingen (*Colias phicomone* Esp., Lep. Pieridae). *Mitteilungen der Naturforschenden Gesellschaft Bern*, **38**, 145–56.

Geiger, W. and Gonseth Y. (1992) Conservation of butterflies in Switzerland, in *Future of Butterflies in Europe: Strategies for Survival*, (eds T. Pavlicek-van Beek, A.H. Ovaa, and J.G. van der Made), Agricultural University Wageningen, pp. 36–44.

Gepp, J. (1991) Jahr der Schmetterlinge. *Natur und Land*, **77**, 22–44.

Gibbs, G.W. (1980) *New Zealand Butterflies*, Collins, Auckland.

Gilbert, L.E. (1972) Pollen feeding and reproductive biology of Heliconius butterflies. *Proceedings of the National Academy of Sciences of the USA*, **69**, 1403–7.

Gilpin, M. and Hanski, I. (eds) (1991) *Metapopulation dynamics: empirical and theoretical investigations*, Academic Press, London.

Godwin, H. (1956) *The History of the British Flora*, Cambridge University Press, Cambridge.

Godwin, H., Clowes, D.R. and Huntley, B. (1974) Studies in the ecology of Wicken Fen. V. Development of fen carr. *Journal of Ecology*, **62**, 197–214.

Gómez Bustillo, M.R. (1981) Protection of lepidoptera in Spain. *Beih. Veröff. Naturschutz Landschaftspflege Baden-Würtemberg*, **21**, 67–72.

Gómez Bustillo, M.R. and Fernández-Rubio, F. (1974) *Mariposas de la Península Ibérica*, Vol. II, ICONA, Madrid.

Gonseth, Y. (1987) Verbreitungsatlas der Tagfalter der Schweiz (Lepidoptera, Rhopalocera). *Documenta Faunistica Helvetica*, **6**, 1–242.

Goodden, R. (1978) *British Butterflies*, David and Charles, Newton Abbot.

Gower, J.C. (1967) Multivariate analysis and multidimensional geometry. *The Statistician*, **17**, 13–28.

Greatorex-Davies, J.N. and Marrs, R.H. (1992) The quality of coppice woods as habitats for invertebrates, in *Ecology and Management of Coppice Woodlands* (ed. G.P. Buckley), Chapman & Hall, London, pp. 271–96.

Greatorex-Davies, J.N., Hall, M. L. and Marrs, R.H. (1992) The conservation of the pearl-bordered fritillary (*B. euphrosyne*): preliminary studies on the creation and management of glades in conifer plantations. *Forest Ecology and Management*, **53**, 1–14.

Greatorex-Davies, J.N., Sparks, T.H. and Moy, I.L. (1993) The influence of shade on butterflies in rides of coniferous lowland woods in southern England and its implications for conservation management. *Biological Conservation*, **63**, 31–41.

Greenslade, P. and New, T.R. (1991) Australia: conservation of a continental insect fauna, in *Conservation of Insects and Their Habitats*, (eds N.M. Collins and J.A. Thomas), Academic Press, London, pp. 33–70.

Hall, M.L. (1981) *Butterfly Monitoring Scheme – Instructions for Independent Recorders*, Institute of Terrestrial Ecology, Cambridge.

Hall, M.R. (1991) *An Atlas of Norfolk butterflies, 1984–1988*, British Butterfly Conservation Society, Norfolk Branch, Norwich.

Hanski, I. (1994) A practical model of metapopulation dynamics. *Journal of Animal Ecology* (in press).

Hanski, I. and Thomas, C.D. (1994) Metapopulation dynamics and conservation: a spatially explicit model applied to butterflies. *Biological Conservation* (in press).

Hanski, I., Kuussaari, M. and Nieminen, M. (1994) Metapopulation structure and migration in the butterfly *Melitaea cinxia*. *Ecology* (in press).

Harding, P.T. (1986) Do you watch butterflies – if so read on. *BBCS News*, **36**, 26–7.

Harding, P.T. and Green S.V. (1989) A review of butterfly recording and survey in the United Kingdom, unpublished report.

Harding, P.T. and Green S.V. (1991) *Recent Surveys and Research on Butterflies in Britain and Ireland: A Species Index and Bibliography*, Biological Records Centre, Huntingdon.

Harding, P.T. and Greene, D.M. (1984) Butterflies in the British Isles : a new data base. *Annual Report of the Institute of Terrestrial Ecology, 1983*, pp. 48–9.

Harding, P.T. and Sheail, J. (1992) The Biological Records Centre – a pioneer of data gathering and retrieval, in *Biological Recording of Changes in British Wildlife*, (ed. P.T. Harding), HMSO, London, pp. 5–19.

Harrison, F. and Sterling, M.J. (1985) *Butterflies and Moths of Derbyshire*, Derbyshire Entomological Society, Derbyshire.

Harrison, S. (1991) Local extinction in a metapopulation context: an empirical evaluation. *Biological Journal of the Linnean Society*, **42**, 73–88.

Harrison, S. (1994) Metapopulations and conservation, in Large Scale Ecology and Conservation Biology (eds R.M. May, P.J. Edwards and N.R. Webb) Symposium of the British Ecological Society.

Harrison, S., Murphy, D.D. and Ehrlich, P.R. (1988) Distribution of the Bay checkerspot butterfly, *Euphydryas editha bayensis*: evidence for a metapopulation model. *American Naturalist*, **132**, 360–82.

Harvey, H.J. and Meredith, T.C. (1981) The biology and conservation of milk-parsley, *Peucedanum palustre*, at Wicken Fen. *Nature in Cambridgeshire*, **24**, 38–42.

Haworth, A.H. (1803–28) *Lepidoptera Britannica*, London.

Heath, J. (1974) A century of change in the Lepidoptera, in *The Changing Flora and Fauna of Britain*, (ed. D.L. Hawksworth), Academic Press, London, pp. 275–92.

Heath, J. (1981a) *Threatened Rhopalocera (Butterflies) in Europe*, Nature and Environment Series 23, Council for Europe, Strasbourg.

Heath, J. (1981b) Chequered skipper (*Carterocephalus palaemon*) survey, 1980. Unpublished Report to Joint Committee for the Conservation of British Insects.

Heath, J. (1983) Is this the earliest record of *Lycaena dispar* (Haworth) (Lepidoptera: Lycaenidae)? *Entomologist's Gazette*, **34**, 228.

Heath, J., Pollard, E. and Thomas, J.A. (1984) *Atlas of Butterflies in Britain and Ireland*, Viking, Harmondsworth.

Hellawell, J.M. (1991) Development of a rationale for monitoring, in *Monitoring for Conservation and Ecology*, (ed. F.B. Goldsmith), Chapman & Hall, London, pp. 1–14.

Hengeveld, R. (1988) Mechanisms of biological invasions. *Journal of Biogeography*, **15**, 819–28.

Hengeveld, R. (1990) *Dynamic Biogeography*, Cambridge University Press, Cambridge.

Heppner, J. B. (1991) Faunal regions and the diversity of Lepidoptera. *Tropical Lepidoptera*, **2**(suppl. 1), 1–85.

Hickin, N. (1992) *The Butterflies of Ireland – A Field Guide*, Roberts, Rinehart and Schull, Dublin.

Higgins, L.G. (1975) *The Classification of European butterflies*, Collins, London.

Higgins, L.G. and Hargreaves, B. (1983) *The Butterflies of Britain and Europe*, Collins, London.

Higgins, L.G.and Riley, N.D. (1970) *A Field Guide to the Butterflies of Britain and Europe*, Collins, London.

Higgins, L.G. and Riley, N.D. (1983) *A Field Guide to the Butterflies of Britain and Europe*, 5th edn, Collins, London.

Hill, C.J. (1989) The effect of adult diet on the biology of butterflies. 2. The common crow butterfly, *Euploea core corinne*. *Oecologia*, 81, 258–66.

Hill, C.J. and Pierce, N.E. (1989) The effect of adult diet on the biology of butterflies. 1. The common imperial blue, *Jalmenus evagoras*. *Oecologia*, 81, 249–57.

Hill, L. and Michaelis, F.M. (1988) *Conservation of Insects and Related Wildlife*, Australian National Parks and Wildlife Service, Occasional Paper No. 13, Canberra.

Hill, M.O. (1979) *TWINSPAN – A FORTRAN Program for Arranging Multivariate Data in an Ordered Two-way Table by Classification of the Individuals and Attributes*, Cornell University, Ithaca, New York.

Hill, M.O, Bunce, R.G.H. and Shaw, M.W. (1975) Indicator species analysis. A divisive polythetic method of classification, and its application to a survey of native pinewoods in Scotland. *Journal of Ecology*, 63, 597–613.

Hochberg, M., Thomas, J.A. and Elmes, G.W. (1992) The population dynamics of a Large Blue Butterfly, *Maculinea rebeli*, a parasite of red ant nests. *Journal of Animal Ecology*, 61, 397–409.

Hochberg, M.E., Clarke, R.T., Elmes, G.W. and Thomas, J.A. (1994) Population dynamic consequences of direct and indirect interactions involving a large blue butterfly and its plant and red ant hosts. *Journal of Animal Ecology*, 63, 375–91.

Hockey, P.A.R. (1978) The ecology of the chequered skipper butterfly (*Carterocephalus palaemon* Pall.) at Doire Donn Wildlife Reserve. Unpublished report to Scottish Wildlife Trust.

Hodgson, J.G. (1993) Commonness and rarity in British butterflies. *Journal of Applied Ecology*, 30, 407–27.

Hope-Simpson, J. F. (1965) Conservation of chalk grassland: the need to preserve ancient grasslands. *Society for the Promotion of Nature Reserves, Handbook and Annual Report*, pp. 56–8.

Horton, P. (1977) Local migrations of Lepidoptera from Salisbury Plain in 1976. *Entomologist's Gazette*, 28, 281–83.

Houghton, J.T., Jenkins, G.J. and Ephraums, J.J. (1990) *Climate Change. The IPCC Scientific Assessment*, Academic Press, New York.

Houston, J. (1976) Chequered skipper survey. Unpublished Report to Scottish Wildlife Trust.

Howarth, F.G. and Ramsey, G.W. (1991) The conservation of island insects and their habitats, in *Conservation of Insects and Their Habitats*, (eds N.M. Collins and J.A. Thomas), Academic Press, London, pp. 71–107.

Hoyle, M.A. (1977) Forestry and conservation in Papua New Guinea. *Tiger Paper*, 4, 10–12.

Hruby, K. (1964) *Prodromus Lepidopterorum Slovaciae*, SAV, Bratislava.

Huemer, P. (1986) Neufunde von Kleinschmetterlingen aus Vorarlberg (Österreich). Insecta: Lepidoptera. *Berichte des naturwissenschaftlich-medizinischen Vereins in Innsbruck*, 73, 147–54.

Huemer, P. (1994) *Rote Liste der gefährdeten Schmetterlinge (Macrolepidoptera) in Südtirol* (in press).

Huemer, P. and Tarmann, G. (1989) Udea carniolica n. sp. – eine neue Pyraliden-Art aus den Süd- und Südostalpen (Lepidoptera: Pyralidae). *Zeitschrift der Arbeitsgemeinschaft österreichischer Entomologen*, 40, 83–90.

Huemer, P. and Tarmann G. (1992) Westpaläarktisch Gespinnstmotten der Gattung Kessleria (Nowicki): Taxonomie, Ökologie, Verbreitung (Lepidoptera, Yponomeutidae). *Mitteilungen der Münchner entomologischen Gesellschaft*, 81, 5–110.

Huemer, P. and Tarmann, G. (1994a) Wissenschaftliche Ergebnisse der lepidopterologischen Forschungen des Tiroler Landesmuseums Ferdinandeum im Alpenraum. – 1. Eine neue Catoptria – Art aus den Alpi Orobie. *Veröffentlichungen des Tiroler Landesmuseums Ferdinandeum* (in press).

Huemer, P. and Tarmann, G. (1994b) Wissenschaftliche Ergebnisse der lepidopterologischen Forschungen des Tiroler Landesmuseums Ferdinandeum im Alpenraum. – 2. Eine neue Kessleria – Art aus den Alpi Orobie (Insecta, Lepidoptera, Yponomeutidae). *Veröffentlichungen des Tiroler Landesmuseums Ferdinandeum* (in press).

Huemer, P., Reichl, E.R. and Wieser, C. (1994) Rote Liste der gefährdeten Grossschmetterlinge Österreichs (Macrolepidoptera), in *Rote Liste der gefährdeten Tiere Österreichs*, (ed. J. Gepp) (in press).

Huemer, P., Deutsch, H., Habeler, H. and Lichtenberger, F. (1992) Neue und bemerkenswerte Funde von Kleinschmetterlingen in Österreich. *Berichte des naturwissenschaftlich-medizinischen Vereins in Innsbruck*, **79**, 199–202.

Hywel-Davies, J., Thom, V. and Bennett, L. (1986) *The Macmillan Guide to Britain's Nature Reserves*, Macmillan, London.

International Union for the Conservation of Nature (1990) *The IUCN Red List of Threatened Animals*, IUCN, Gland.

Irwin, A.G. (1984) The large copper, *Lycaena dispar dispar* (Haworth) in the Norfolk Broads. *Entomologists Record and Journal of Variation*, **96**, 212–13.

Jaksic, P. (1988) *Privremene Karte Rastrostranjenosti Dnevnih Leptira Jugoslavije (Lepidoptera: Rhopalocera)*, Societas Entomologia Jugoslavica, Agreb.

Jeffcoate, G. (1992) Dark-green fritillary survey. *Butterfly Conservation News*, **51**, 37–9.

Jelinek, A., Britton, D.R. and New, T.R. (1994) Conservation of a threatened butterfly community at Mt. Piper, Victoria. *Memoirs of the Queensland Museum* (in press).

Johnson, W.C. and Adkisson, C.S. (1985) Dispersal of beech nuts by blue jays in fragmented landscapes. *American Midland Naturalist*, **113**, 319–24.

Joicey, J.J. and Noakes, A. (1907) Lepidoptera in Glenshian, Inverness-shire, in July, 1907. *Entomologist's Monthly Magazine*, **43**, 255–6.

Jones, P.D. (1990) The climate of the past 1000 years. *Endeavour*, **14**, 129–36.

Jordano, D., Fernández, J. and Rodríguez, J. (1990) The effect of seed predation by *Tomares Ballus* (Lep., Lycaenidae) on *Astragalus lusitanicus* (Fabaceae): determination of differences among patches. *Oikos*, **57**, 250–6.

Jordano, D., Retamosa, E.C. and Fernández Haeger, J. (1991) Factors facilitating the continued presence of *Colotis evagore* (Klug 1829) in Southern Spain. *Journal of Biogeography*, **18**, 637–46.

Kaaber, S. and Nielsen, O.F. (1988) 30-year changes of the butterfly fauna in an area of Central Jutland, Denmark. *Flora og Fauna*, **94**, 95–110.

Kappeler, M. (1987) Schmetterlinge. *Schweizer Naturschutz*, **3/87**, 1–23

Kelly, P.G. (1983) The ecology of the chequered skipper (*Carterocephalus palaemon* Pallas) in Scotland. Unpublished Report to Nature Conservancy Council.

Kielland, J. (1990) *The Butterflies of Tanzania*, Hill House, Melbourne and London.

Killingbeck, J. (1985) *Creating and maintaining a garden to attract butterflies*, National Association for Environmental Education, Walsall.

Kinkler, H. (1991) Der Segelfalter (*Iphiclides podalirius* L.) in Rheinland-Pfalz – ein Artenschutzprojekt. *Beiträge zur Landschaftspfl. Rheinland-Pfalz*, **14**, 7–94.

Kirby, K.J. and Patterson, G. (1991) Ecology and management of semi-natural tree species mixtures, in *The Ecology of Mixed Species Stands of Trees*, (eds M.G.R. Cannell, D.C. Malcolm and P.A. Robertson), Blackwell Scientific Publications, London, pp. 189–209.

Kitching, R.L. (1981) The geography of the Australian papilionoidea, in *Ecological Biogeography of Australia*, (ed. A. Keast), W. Junk, The Hague, pp. 979–1005.

Kitching, R.L. and Baker, E.J. (1990) Hello, Goodbye? *Geo*, **12**, 92–5.

Knapp, R. and Casey, T.M. (1986) Thermal ecology, behavior, and growth of gypsy moth and eastern tent caterpillars. *Ecology*, **67**, 598–608.

Knight, R. and Campbell, J.M. (1986) *An Atlas of Oxfordshire Butterflies*, Oxfordshire County Council, Occasional Paper no. 10. Woodstock.

KNMI (1985) *Klimatologische gegevens van Nederlandse stations no. 25: Het klimaat van het hoofdobservatorium De Bilt in de jaren 1901–1985*, KNMI, De Bilt.

Kostrowicki, A. S. (1969) *Geography of the Palaeartic Papilionoidea (Lepidoptera)*, Panstwowe Wydawnictwo Naukowe, Krakow.

Kralicek, M. and Povolny, D. (1980) K soucasnemu stavu faunistiky moravskych dennich motylu. *Entomologiché Problemy*, **16**, 107–31.

Krzywicki, M. (1959–70) [Butterflies], in *Klucze Oznaczania Owadow Polski.*, **27**(60–6), Panstwowe Wydawnictwo Naukowe, Warszawa.

Krzywicki, M. (1982) Der gegenwärtige Stand der Tagfalterfauna Polens unter besonderer Berücksichtigung ihrer Bedrohung. *Nota lepidopterologica*, **5**, 3–16.

Kudrna, O. (1986) *Butterflies of Europe. 8. Aspects of the conservation of European butterflies*, Aula Verlag, Wiesbaden.

Kudrna, O. (ed.) (1991) Schutz der Tagfalterfauna im Osten Mitteleuropas. *Oedippus*, **3**, 1–102.

Kudrna, O. (1992) Ein Plan für die Wiederherstellung der Rhopalozönose des NSG Rotes Moor in der hessischen Rhön. *Oedippus*, **5**, 1–31.

Kudrna, O. (1993) Verbreitungsatlas der Tagfalter (Rhopalocera) der Rhön. *Oedippus*, **6**, 1–138.

Kudrna, O. and Kralicek, M. (1991) Schutz der Tagfalterfauna in Böhmen und Mähren (Tschechoslowakei). *Oedippus*, **3**, 37–74.

Kudrna, O. and Mayer, L. (1990) Grundlagen zu einem Artenhilfsprogramm für *Colias myrmidone* (Esper, 1780). *Oedippus*, **1**, 1–46.

Kudrna, O. and Seufert, W. (1991) Ökologie und Schutz von *Parnassius mnemosyne* Linnaeus, 1758) in der Rhön. *Oedippus*, **2**, 1–44.

Kühnelt, W. (1943) Die Leitformenmethode in der Ökologie der Landtiere. *Biologia generalis*, **17**, 106–46.

Kulfan, J. and Kulfan, M. (1991) Die Tagfalterfauna der Slowakei und ihr Schutz. *Oedippus*, **3**, 75–102.

Lamb, H.F. (1974) The chequered skipper *Carterocephalus palaemon* (Pallas) field survey 1974. Unpublished report to Institute of Terrestrial Ecology.

Lane, C. and Rothschild, M. (1957) Note on the habitat of the wood white (*Leptidea sinapis* L.) and chequered skipper (*Carterocephalus palaemon* Pallas). *Entomologist*, **90**, 271–2.

Larsen, T. B. (1991) *The Butterflies of Kenya and Their Natural History*, Oxford University Press, Oxford.

Larsen, T.B. (1994) *The Butterflies of Botswana and Their Natural History*, Biologiske Skrifter, Kongelige Danske Videnskabernes Selskab, Copenhagen (in press).

Larsen, T.B. and Pittaway, A.R. (1982) Notes on the ecology, biology and taxonomy of *Apharitis acamas*. Klug (Lep., Lycaenidae). *Entomologists' Gazette*, **33**, 163–8.

Larsen, T.B., Riley, J. and Cornes, M.A. (1980) The butterfly fauna of a secondary bush locality in Nigeria. *Journal of Research in Lepidoptera*, **18**, 4–23.

Latour, J. and van Swaay, C.A. M. (1992) Dagvlinders als indicatoren voor de regionale milieukwaliteit. *De Levende Natuur*, **93**(1), 19–22.

Leigh, E.G. (1981) The average lifetime of a population in a varying environment. *Journal of Theoretical Biology*, **90**, 213–39.

Leigheb, G., Riboni, E. and Cameron-Curry, V. (1990) *Kretania psylorita* Freyer (Lepidoptera, Lycaenidae). Discovery of a new locality in Crete. *Nota lepidopterologica*, **13**, 242–5.

Levins, R. (1970) Extinction. *American Maths Society*, **2**, 77–107.

Libert, M. (1992). Notes faunistiques sur les Lepidoptères Rhopalocères du Cameroun. *Lambillionea*, **92**, 21–34.

Lindsay, J.M. (1975) Charcoal iron smelting and its fuel supply; the example of Lorn furnace, Argyllshire, 1753–1876. *Journal of Historical Geography*, **1**, 283–98.

Loader, C. and Damman, H. (1991) Nitrogen content of foodplants and vulnerability of *Pieris rapae* to natural enemies. *Ecology*, **72**, 1586–90.

Loertscher, M. (1991) Population biology of two satyrine butterfies, *Erebia meolans* (de

Prunner 1798) and *Erebia aethiops* (Esper 1777) (Lepidoptera: Satyridae). *Nota Lepidopterologica*, suppl. 2, 22–31.

Loertscher, M., Erhardt, A. and Zettel, J. (1994) Microdistribution of butterflies in a mosaic-like habitat – the role of nectar sources (in preparation).

Lowday, J.E. and Marrs, R.H. (1992) Control of bracken and the restoration of heathland. I. Control of bracken. *Journal of Applied Ecology*, 29, 195–203.

Mabbett, R. and Williams, M. (eds) (1993) *1992 West Midlands and Gloucestershire Butterfly and Moth Report*, West Midlands Branch of the British Butterfly Conservation Society, Kenilworth.

McAllister, D.W. (1993) The spread of the speckled wood *Pararge aegeria* in Easter Ross and Sutherland. *Butterfly Conservation News*, 53, 44–8.

MacArthur, R.H. and Wilson, E.O. (1967) *The Theory of Island Biogeography*, Princeton University Press, Princeton, NJ.

Macdonald, D.W. and Smith, H. (1991) New perspectives on agro-ecology, in *The Ecology of Temperate Cereal Fields*, (eds L.G. Firbank, N. Carter, J.F. Darbyshire and G.R. Potts), Blackwell Scientific Publications, Oxford, pp. 413–48.

Mackworth-Praed, C.W. (1942) *Carterocephalus palaemon* in Western Inverness-shire. *Entomologist*, 75, 216.

Mackworth-Praed, C.W. (1945) Three seasons in the Western Highlands. *Entomologist's Monthly Magazine*, 81, 114–17.

Maelfait, J.-P. and De Keer, R. (1990) The border zone of an intensively grazed pasture as a corridor for spiders (Araneae). *Biological Conservation*, 54, 223–38.

Malicky, H. (1969) Versuch einer Analyse der okologischen Beziehungen zwischen Lycaeniden (Lepidoptera) und Formiciden (Hymenoptera). *Tijdschrift voor Entomol.*, 112, 213–98.

Manley, G. (1974) Central England temperatures: monthly means 1659–1973, *Quarterly Journal of the Royal Meteorological Society*, 100, 389–405.

Marchant, J.H., Hudson, R., Carter, S.P. and Whittington, P. (1990) *Population Trends in Breeding Birds*, British Trust for Ornithology/Nature Conservancy Council publication.

Margules, C.R. (1989) Introduction to some Australian developments in conservation evaluation. *Biological Conservation*, 50, 1–11.

Margules, C.R. and Stein, J.L. (1989) Patterns in the distribution of species and the selection of nature reserves: and example of *Eucalyptus* forests in south-eastern New South Wales. *Biological Conservation*, 50, 219–38.

Marrs, R.H. (1987) Studies on the conservation of lowland *Calluna* heaths. I. Control of birch and bracken and its effects on heath vegetation. *Journal of Applied Ecology*, 24, 163–75.

Marrs, R.H., Frost, A.J. and Plant, R.A. (1989) A preliminary report on the impact of herbicide drift on plant species of conservation interest. *Brighton Crop Protection Conference – Weeds 1989*, 2, 795–802.

Marrs, R.H., Lowday, J.E., Jarvis, L., Gough, M.W. and Rowland, A.P. (1992) Control of bracken and the restoration of heathland. *Journal of Applied Ecology*, 29, 218–25.

Marsden, H. (1884). On the probable extinction of *arion* in England. *Entomologist's Monthly Magazine*, 21, 186–9.

Marshall, E.J.P and Smith, B.D. (1987) Field margin flora and fauna; interaction with agriculture, in *BCPC Monograph No. 35: Field Margins*, (eds J.M. Way and P.W. Greig-Smith), British Crop Protection Council, Farnham, Surrey, pp. 23–33.

Martín, J. and Gurrea, P. (1988) Establishment of a population of *Danaus plexippus* (Linnaeus, 1758) (Lep.: Danaidae) in Southwest Europe. *Entomologist's Record and Journal Variation*, 100, 163–8.

Martín, J. and Gurrea, P. (1990) The peninsular effect in Iberian butterflies (Lepidoptera: Papilionoidea and Hesperioidea) *Journal of Biogeography*, 17, 85–96.

Mason, C. and Long, S. (1987) Management of lowland broadleaved woodland, Bovington Hall, Essex, in *Conservation Monitoring and Management*, (ed. R. Matthews), Countryside Commission, Cheltenham, pp. 37–42.

Meads, M.J. (1990) *Forgotten Fauna. The Rare, Endangered and Protected Invertebrates of New Zealand*, DSIR, Wellington.

Mellanby, K. (1981) *Farming and Wildlife*. Collins, London.

Mendel, H. and Piotrowski, S.H. (1986) *The Butterflies of Suffolk, an Atlas and History*, Suffolk Naturalists' Society, Ipswich.

Mercer, C.W.L. and Clark, P.B. (1992) *Organisation and Economics of Insect Farming in Papua New Guinea*. Proceedings of the 1st. Symposium of the PNG Society for Animal Production (1989), pp. 62–70.

Meredith, G.H.J. (1989) Butterfly distribution in Gloucestershire 1975–1988. *The Gloucestershire Naturalist*, no. 4.

Meredith, T.C. (1979) The ecology and conservation of *Peucedanum palustre* at Wicken Fen. Ph.D. thesis, University of Cambridge.

Merriam, H.G. (1984) Connectivity: a fundamental ecological characteristic of landscape pattern, in *Methodology in Landscape Ecological Research and Planning*. Vol. 1, (eds. J. Brandt and P.A. Agger), Roskilde, pp. 5–15.

Merxmüller H. (1952) Untersuchungen zur Sippengliederung und Arealbildung in den Alpen. *Jahrbuch des Vereins zum Schutze der Alpenpflanzen und -tiere*, **17**, 1–105.

Meteorological Office (1976–1992) *Monthly Weather Reports*, HMSO, London.

Meyer, M. and Pelles, A. (1981) *Atlas Provisoire des Insectes du Grand-Duche de Luxembourg. Lepidoptera: Rhopalocera (+Hesperiidae)*, Natural History Museum, Luxembourg.

Meyrick, E. (1895) *A Handbook of the British Lepidoptera*, MacMillan.

Mikkola, K. (1991) The conservation of insects and their habitats in northern and eastern Europe, in *The Conservation of Insects and their Habitats*, (eds N.M. Collins and J.A. Thomas), Academic Press, London, pp. 109–19.

Mills, L.S., Soulé, M.E. and Doak, D.F. (1993) The key-stone concept in ecology and conservation. *BioScience*, **43**, 219–24.

Mitchell, B. (1991) Observing tips. *Butterfly Conservation West Midlands Branch Newsletter*, no. 26 (winter), 19–20.

Mitchell, B. (1992) Butterflies in the garden: the 1991 season. *Butterfly Conservation West Midlands Branch Newsletter*, no. 27 (summer), 23–4.

Mitchell, J.F.B., Manabe, S., Tokioka, T. and Meleshko, V. (1990) Equilibrium climate change, in *Climate Change. The IPCC Scientific Assessment*, (eds J.T. Houghton, G.J. Jenkins and J.J. Ephraums), Cambridge University Press, Cambridge, pp. 131–72.

Mitchell, J.F.B. and Zeng, Q.-C. (1991) Climatic change prediction. *The Meterological Magazine*, **120**, 153–63.

Mitchell, P.L. and Kirby, K.J. (1989) *Ecological Effects of Forestry Practices in Long-established Woodland and their Implications for Nature Conservation*, Oxford Forestry Institute Occasional Paper no. 39, University of Oxford.

Moffat, G. (1975) The distribution of the Checkered Skipper *Carterocephalus palaemon* on the O.S. sheet 46 (south of Fort William). Unpublished report to the Scottish Wildlife Trust.

Monteith, G.B. and Hancock, D.L. (1977) Range extensions and notable records for butterflies of Cape York peninsula, Australia. *Australian Entomological Magazine*, **4**, 21–38.

Morgan, I.K. (1989) A provisional review of the butterflies of Carmarthenshire, Nature Conservancy Council, Aberystwyth (unpublished).

Moriarty, F. (1969) Butterflies and insecticides. *Entomologist's Record and Journal of Variation*, **81**, 276–8.

Morisita, M. (1962) I_d-index, a measure of dispersion of individuals. *Research in Population Ecology*, **4**, 1–7.

Morris, M.G. (1967) The representation of butterflies (Lep., Rhopalocera) on British statutory nature reserves. *Entomologist's Gazette*, **18**(2), 57–68.

Morris, M. G. (1989) Legislation for Lepidoptera conservation – towards rationale. *Nota Lepidopterologica*, suppl. 1, 15.

Morris, M.G. and Thomas, J.A. (1989) Re-establishment of insect populations, with special reference to butterflies, in *The Moths and Butterflies of Great Britain and Ireland*, Vol. 7, (eds J. Heath and A.M. Emmet), Harley Books, Colchester. pp. 22–36.

Morris, M.G. and Thomas, J.A. (1991) Progress in the conservation of butterflies. *Nota Lepidopterologica*, suppl. 2, 32–44.

Morton, A.C.G. (1985) The population biology of an insect with a restricted distribution: *Cupido minimus* Fuessly (Lepidoptera; Lycaenidae). Ph.D. thesis, University of Southampton.

Morton, A.J. (1977) Mineral pathways in a *Molinietum* in autumn and winter. *Journal of Ecology*, 65, 993–9.

Moucha, J. (1959) Neueste Forschungsergebnisse über unsere Lepidoptera-Fauna und deren Bedeutung für die Lösung zoogeographischer Fragen in der Tschechoslowakei. *Sbornik faunishilych Prací Entomologischeto Narodniho Musea Praze*, 4, 3–81.

Munguira, M.L. (1987) Biología y biogeographìa de los Licénidos Ibéricos en peligero de extinción (Lepidoptera, Lycaenidae). Tesis Doctoral Madrid.

Munguira, M.L. (1989) *Biología y biogeografía de los licéndios ibéricos en peligro de extinción (Lepidoptera, Lycaenidae)*, Ediciones Universidad Autónoma de Madrid, Madrid.

Munguira, M.L. and Martín, J. (1989) Biology and conservation of the endangered lycaenid species of Sierra Nevada, Spain. *Nota Lepidopterologica*, suppl. 1, 16–18.

Munguira, M.L. and Thomas, J.A. (1992) Use of road verges by butterfly and burnet populations, and the effect of roads on adult dispersal and mortality. *Journal of Applied Ecology*, 29, 316–29.

Munguira, M.L., Martín, J. and Balletto, E. (1993) Conservation biology of lycaenidae: a European overview, in *Conservation Biology of Lycaenidae*, (ed. T.R. New), IUCN, Gland, pp. 23–4.

Munguira, M.L., Martín, J. and Rey, J.M. (1991) Use of UTM maps to detect endangered lycaenid butterflies in Spain. *Nota Lepidopterologica*, suppl. 2, 45–55.

Munguira, M.L., Martín, J. and Viejo, J.L. (1988) Distribución geográfica y biología de *Eumedonia eumedon* (Esper, 1780) en la Península Ibèrica (Lepidoptera: Lycaenidae). *Shilap Revista de Lepidopterología*, 16, 217–29.

Murphy, D.D., Launer, A.E. and Ehrlich, P.R. (1983) The role of adult feeding in egg production and population dynamics of the checkerspot butterfly, *Euphydryas editha*. *Oecologia*, 56, 257–63.

Murphy, D.D., Menninger, M.S. and Ehrlich P.R. (1984) Nectar source distribution as a determinant of oviposition host species in Euphydryas chalcedona. *Oecologia (Berlin)*, 62, 269–71.

Myers, J.H. (1985) Effect of physiological condition of the host plant on the ovipositional choice of the cabbage white butterfly, *Pieris rapae*. *Journal of Animal Ecology*, 54, 193–204.

Nadolny, C. (1987) *Rainforest Butterflies in New South Wales: their Ecology, Distribution and Conservation*, NSW National Parks and Wildlife Service, Sydney.

Nature Conservancy (1965) *Report on Broadland*, London.

Nature Conservancy Council (1989) *Guidelines for the Selection of SSSI's*, Nature Conservancy Council, Peterborough.

New, T.R. (1991) *Butterfly Conservation*, Oxford University Press, Melbourne.

New, T.R. (1992) Conservation of butterflies in Australia. *Journal of Research on the Lepidoptera*, 29, 237–53.

New, T.R. (ed.) (1993) *Conservation Biology of the Lycaenidae*, IUCN, Gland.

New, T.R. (1994) Needs and prospects for insect re-introduction for conservation in Australia, in *Reintroduction Biology of Australasian Biota*, (ed. M. Serena) (in press).

New, T.R. and Collins, N.M. (1991) *Swallowtail Butterflies. An Action Plan for their Conservation*, IUCN, Gland.

Newbery, P. (compiler)(1983) *RSPB Nature Reserves*, RSPB, Sandy.

Newman, L.W. and Leeds, H.A. (1913) *Text Book of British Butterflies and Moths*, Gibbs and Bamforth, St. Albans.

Neyland, M. (1992) The Ptunarra Brown butterfly *Oreixenica ptunarra* conservation research statement. *Department of Parks, Wildlife and Heritage, Tasmania. Scientific Report 92/2.*

Nie, N.N., Hull, C.H., Jenkins, J.G., *et al.* (1975) *Statistical Package for the Social Sciences,* McGraw Hill, New York.

Ni Lamhna, E. (1980) *Distribution Atlas of Butterflies in Ireland,* 3rd edn, An Foras Forbartha, Dublin.

Novak, I. and Spitzer, K. (1982) *Ohrozeny svet hmyzu,* Academia, Praha.

Oates, M.R. (1985) *Garden Plants for Butterflies,* Brian Masterson, Fareham.

Oates, M.R. (1992a) The role of butterfly releases in Great Britain and Europe, in *Future of Butterflies in Europe: Strategies for Survival,* (eds P. Pavlicek-van Beek, A.H. Ovaa and J.G. van der Made), Agricultural University, Wageningen. pp. 204–12.

Oates, M.R. (1992b) The principles of grassland management. *National Trust Views,* **16,** 41–6.

Oates, M.R. (1993) The management of southern limestone grasslands. *British Wildlife,* **5,** 73–82.

Oates, M.R. and Emmet, A.M. (1990) *Hamearis lucina* (Linnaeus), in *The Butterflies of Great Britain and Ireland,* (eds A.M. Emmet and J. Heath), Harley Books, Colchester, pp. 177–9.

Oates, M.R. and Warren, M.S. (1990) *A Review of Butterfly Introductions in Britain and Ireland,* JCCBI/WWF, Godalming.

O'Connor, R.J. and Shrubb, M. (1986) *Farming and Birds,* Cambridge University Press, Cambridge.

Opler, P.A. (1982) Atlas of butterflies of the eastern United States. US Fish and Wildlife Service, Washington, DC, unpublished report.

Owen, D.F. (1971) *Tropical Butterflies,* Clarendon Press, Oxford, UK.

Owen, J. (1991) *The Ecology of a Garden: the First Fifteen Years,* Cambridge University Press, Cambridge.

Pakeman, R.J. and Marrs, R.H. (1992) The conservation value of bracken *Pteridium aquilinum* (L.) Kuhn-dominated communities in the UK, and an assessment of the ecological impact of bracken expansion or its removal. *Biological Conservation,* **62,** 101–14.

Palik, E. (1980) The protection and reintroduction in Poland of *Parnassius apollo* Linnaeus. *Nota Lepidopterologica,* **2,** (1979), 163–4.

Parenzan, P. (1975) Contributi alla conoscenza della lepidotterofauna dell'Italia meridionale. I. Rhopalocera de Puglia e Lucania. *Entomologica,* **11,** 87–154.

Parrish, J.A.D. and Bazzaz, F.A. (1979) Difference in pollination niche relationships in early and late successional plant communities. *Ecology,* **60,** 597–610.

Parsons, M.J. (1980) A conservation study of *Ornithoptera alexandrae* (Rothschild) (Lepidoptera: Papilionidae). Report to PNG Division of Wildlife.

Parsons, M.J. (1984) The biology and conservation of *Ornithoptera alexandrae,* in *The Biology of Butterflies,* (eds R.I. Vane-Wright and P.R. Ackery), Symposium of the Royal Entomological Society, London, no. 11, Academic Press, London, pp. 327–31.

Parsons, M.J. (1992a) *Butterflies of the Bulolo-Wau Valley,* Bishop Museum Press, Honolulu.

Parsons, M.J. (1992b) The butterfly farming and trading industry in the Indo-Australian Region and its role in tropical forest conservation. *Tropical Lepidoptera,* **3** (suppl. 1), 1–32.

Pavlicek-van Beek, T., Ovaa, A.H. and van der Made, J.G. (eds) (1992) *Future of Butterflies in Europe: Strategies for Survival,* Agricultural University, Wageningen.

Payne, M. (1987) *Gardening for Butterflies,* British Butterfly Conservation Society, Quorn, Loughborough.

Peachey, C. (1982) National Butterfly Review, part 1: the representation of butterflies on National Nature Reserves. Invertebrate Site Register Report 10(1), Nature Conservancy Council, London. (Confidential unpublished report.)

Pennington, K.M. (1994) *Pennington's Butterflies of Southern Africa.* (revised) (in press).

Peterken, G.F. (1981) *Woodland Conservation and Management,* Chapman & Hall, London.

Peters, W. (1952) *A Provisional Checklist of the Butterflies of the Ethiopian Region*, E.W. Classey, Feltham, UK.

Philp, E.G. (1993) The butterflies of Kent, *Transactions of the Kent Field Club*, Vol. 12, Kent Field Club, Sittingborne.

Pictet, A. (1942) Les macrolépidoptères du parc national suisse et des régions limitrophes. *Ergebnisse der wissenschaftlichen Untersuchung des schweizerichen Nationalparkes*, 1, 81–263.

Pilcher, R.E.M. (1961) The Lepidoptera of Castor Hanglands (1919–1960). Unpublished Report to Nature Conservancy Council.

Plachter, H. (1991) *Naturschutz, UTB 1563*, Fischer, Stuttgart.

Plant, C.W. (1987) *The Butterflies of the London Area*, London Natural History Society, London.

Pleasants, J.M. and Chaplin, S.J. (1983) Nectar production rates of *Asclepias quadrifolia*: causes and consequences of individual variation. *Oecologia*, 59, 232–8.

Pollard, E. (1977) A method for assessing changes in the abundance of butterflies. *Biological Conservation*, 12, 115–34.

Pollard, E. (1979) Population ecology and change in range of the White Admiral butterfly *Ladoga camilla* L. in England. *Ecological Entomology*, 4, 61–74.

Pollard, E. (1988) Temperature, rainfall and butterfly numbers, *Journal of Applied Ecology*, 25, 819–28.

Pollard, E. (1991) Changes in the flight period of the hedge brown butterfly *Pyronia tithonus* during range expansion. *Journal of Animal Ecology*, 60, 737–48.

Pollard, E. and Yates, T.J. (1992) Extinction and foundation of local butterfly populations in relation to population variability and other factors. *Ecological Entomology*, 17, 249–54.

Pollard, E. and Yates, T.J. (1993a) *Monitoring Butterflies for Ecology and Conservation*, Chapman & Hall, London.

Pollard, E. and Yates, T.J. (1993b) Population fluctuations of the holly blue butterfly, *Celastrina argiolus* (L.) (Lepidoptera: Lycaenidae), *Entomologist's Gazette*, 44, 3–9.

Pollard, E., Hall, M.L. and Bibby, T.J. (1986) *Monitoring the Abundance of Butterflies, 1976–1985*, Research and Survey in Nature Conservation, no. 2, Nature Conservancy Council.

Pollard, E., Hooper, M.D. and Moore, N.W. (1974) *Hedges*, Collins, London.

Pollard, E., Moss, D. and Yates, T.J. (1994) Population trends of common British butterflies at monitored sites. *Journal of Applied Ecology* (in press).

Pollard, E., van Swaay, C.A.M. and Yates, T.J. (1993) Changes in butterfly numbers in Britain and the Netherlands 1990–91. *Ecological Entomology*, 18, 93–4.

Pollard, E., Elias, D.O., Skelton, M.J. and Thomas, J.A. (1975) A method of assessing the abundance of butterflies in Monk's Wood National Nature Reserve in 1973. *Entomologist's Gazette*, 26, 79–88.

Polunin, O. and Smithies, B.E. (1973) *Flowers of South-west Europe a Field Guide*, Oxford University Press, Oxford.

Poore, M.E.D. (1956) The ecology of Woodwalton Fen. *Journal of Ecology*, 44, 455–92.

Porter, K. (1982) Basking behaviour in larvae of the butterfly *Eurodryas aurinia*. *Oikos*, 38, 308–12.

Porter, K. (1989) Mountain Ringlet, in *The Moths and Butterflies of Britain and Ireland*, Vol. 7, (eds A.M. Emmet and J. Heath), Harley Books, Colchester, pp. 253–5.

Porter, K. (1992) Eggs and egg-laying, in *The Ecology of Butterflies in Britain*, (ed. R.L.H. Dennis), Oxford University Press, Oxford, pp. 46–72.

Porter, K. and Nowakowski, M. (1991) The use of a graminicide in conservation management, in *British Crop Protection Conference – Weeds, 1991*, British Crop Protection Council, Farnham, pp. 647–54.

Porter, K., Steel, C.A. and Thomas J.A. (1992) Butterflies and communities, in *The Ecology of Butterflies in Britain*, (ed. R.L.H. Dennis), Oxford University Press, Oxford, pp. 139–77.

Potts, G.R. (1986) *The Partridge: Pesticides, Predation and Conservation*, Collins, London.

Pratt, C. (1986–7). A history and investigation into the fluctuations of *Polygonia c-album* L. *Entomologist's Record and Journal of Variation*, **98**, 197–203; 244–250; and **99**, 21–7; 69–80.

Prendergast, J.R., Quinn, R.M., Lawton, J.H., Eversham, B.C. and Gibbons, D.W. (1993) Rare species, the coincidence of diversity hotspots and conservation strategies. *Nature*, **365**, 335–7.

Pressey, R.L. and Nicholls, A.O. (1989a) Efficiency in conservation evaluation: scoring versus iterative approaches. *Biological Conservation*, **50**, 199–218.

Pressey, R.L. and Nicholls, A.O. (1989b) Application of a numerical algorithm to the selection of reserves in semi-arid New South Wales. *Biological Conservation*, **50**, 263–78.

Pretscher, P. (1984) Rote Liste der Grossschmetterlinge (Macrolepidoptera), in *Rote Liste der gefährdeten Tiere und Pflanzen in der BRD, Naturschutz Aktuell 1*, 4th edn, (eds J. Blab, E. Nowak and W. Trautmann), pp. 53–66.

Prince, G.B. (1988) The conservation status of the Hairstreak butterfly *Pseudalmenus chlorinda* Blanchard in Tasmania. Report to the Department of Lands, Parks and Wildlife, Tasmania.

Prola, G. and Prola, C. (1990) Libro rosso delle farfalle italiane. *WWF Quaderni*, **13**.

Pullin, A.S. (1986a) The influence of the food plant *Urtica dioica* on larval development, feeding efficiencies and voltinism of a specialist insect, *Inachis io*. *Holartic Ecology*, **9**, 72–8.

Pullin, A.S. (1986b) Unusual egg laying strategies of the Small Tortoiseshell butterfly, *Aglais urticae*. *Entomologist's Record*, **98**, 9–10.

Pullin, A.S. (1987a) Changes in leaf quality following clipping and regrowth of *Urtica dioica* and consequences for a specialist insect herbivore, *Aglais urticae*. *Oikos*, **49**, 39–45.

Pullin, A.S. (1987b) Adult feeding time, lipid accumulation, and overwintering in *Aglais urticae* and *Inachis io* (Lepidoptera: Nymphalidae). *Journal of Zoology*, **211**, 631–41.

Purefoy, E.B. (1953) An unpublished account of experiments carried out at East Farleigh, Kent, in 1915 and subsequent years on the life history of *Maculinea arion*, the large blue butterfly. *Proceedings of the Royal Entomological Society of London, A*, **28**, 160–2.

Rackham, O. (1980) *Ancient Woodland*, Edward Arnold, London.

Rackham, O. (1986) *The History of the British Countryside*, Dent, London.

Rakonczay, Z. (1989) *Vörös Könyv*, Akademiai Kaido, Budapest.

Ramsay, G.W., Meads, M. J., Sherley, G.H. and Gibbs, G.W. (1988) Research on terrestrial insects of New Zealand. *Wildlife Research Liaison Group Research Review*, **10**, 1–49.

Rands, M.R.W. (1985) Pesticide use on cereals and the survival of grey partridge chicks: a field experiment. *Journal of Applied Ecology*, **22**, 49–54.

Ratcliffe, D.A. (1977) *A Nature Conservation Review*, Cambridge University Press, Cambridge.

Ratcliffe, P.R. (1992) The interaction of deer and vegetation in coppice woods, in *Ecology and Management of Coppice Woodlands*, (ed. G.P. Buckley), Chapman & Hall, London, pp. 223–45.

Ravenscroft, N.O.M. (1986). *An Investigation into the Distribution and Ecology of the Silver-studded Blue Butterfly (*Plebejus argus *L.) in Suffolk – an Interim Report*, Suffolk Trust for Nature Conservation, Suffolk.

Ravenscroft, N.O.M. (1990) The ecology and conservation of the silver-studded blue butterfly *Plebejus argus* L. on the Sandlings of East Anglia, England. *Biological Conservation*, **53**, 21–36.

Ravenscroft, N.O.M. (1991) The Chequered Skipper in Scotland. *British Wildlife*, **2**, 269–75.

Ravenscroft, N.O.M. (1992) The ecology and conservation of the chequered skipper butterfly *Carterocephalus palaemon* (Pallas). Ph.D. thesis, University of Aberdeen.

Ravenscroft, N.O.M. (1994a) The ecology of the chequered skipper butterfly *Carterocephalus palaemon* Pallas in Scotland. I: Microhabitat selection. *Journal of Applied Ecology* (in press).

Ravenscroft, N.O.M. (1994b) The ecology of the chequered skipper butterfly *Carterocephalus palaemon* Pallas in Scotland. II: Foodplant quality and population range. *Journal of Applied Ecology* (in press).

Ravenscroft, N.O.M. (1994c) The influences of environment on the mate location system of male chequered skipper butterflies *Carterocephalus palaemon* (Lepidoptera: Hesperiidae). *Animal Behaviour*, 47, 1179–87.

Ravenscroft, N.O.M. (1994d) The feeding behaviour of *Carterocephalus palaemon* (Lepidoptera: Hesperiidae) caterpillars: does it avoid host defences or maximise nutrient intake? *Ecological Entomology*, 19, 26–30.

Ravenscroft, N.O.M. (1994e) *A Review of Chequered Skipper Sites in Scotland 1993*, Butterfly Conservation, Dedham.

Ravenscroft, N.O.M. and Davis, S.M. (1992) *The Chequered Skipper* Carterocephalus palaemon *at Fôret de Spincourt, France and Surrounds*, British Butterfly Conservation Society, Dedham.

Ravenscroft, N.O.M. and Warren, M.S. (1992) Habitat selection by larvae of the chequered skipper *Carterocephalus palaemon* in northern Europe. *Entomologist's Gazette*, 43, 237–42.

Read, M. (1985) The silver-studded blue conservation report. M.Sc. thesis, University of London.

Redman, M. (1992) *Organic Farming and the Countryside*, special report from British Organic Farmers/Organic Growers Association in conjunction with the Soil Association.

Reichl, E.R. (1992) *Verbreitungsatlas der Tierwelt Österreichs. 1. Lepidoptera – Diurna*, Forschungsinstitut für Umweltinformatik, Linz.

Reiprich, A. and Okali, I. (1989) *Dodatky k prodromu Lepidopter Slovenska*, 3, Veda, Bratislava.

Richarz, N. (1989) Untersuchungen zur Ökologie des Apollofalters (*Parnassius apollo vinningensis* Stichel, (1899) im Weinbaugebiet der unteren Mosel. *Mitteilingen der Arbeitsgemainschaft rhein.-westfälischer Lepidopterologugen*, 5, 101–264.

Riess, W. (1978) Bedrohte Tierarten der Alpen – Signal für den Verlust an Heimat. *Jahrbuch des Vereins zum Schutz der Bergwelt*, 43, 39–102.

Riley, A.M. (1991) *A Natural History of the Butterflies and Moths of Shropshire*, Swan Hill Press, Shrewsbury.

Rimington, E. (1992) *Butterflies of the Doncaster district*, Sorby Record Special Series No. 9.

Robertson, P.A. (1992) Woodland management for pheasants. *Forestry Commission Bulletin 106*, HMSO, London.

Robertson, P.A. and Sotherton, N.W. (1992) Arable energy coppice as a wildlife habitat, in *Wood-Energy and the Environment and Energy Workshop*, (ed. G.E. Richards), Harwell Laboratories, Oxfordshire, pp. 143–7.

Robertson, P.A., Woodburn, M.I.A. and Hill, D.A. (1988) The effect of woodland management for pheasants on the abundance of butterflies. *Biological Conservation*, 45, 1–9.

Rollason, W.A. (1908) Life-history of *Hesperia paniscus*, F., *palaemon*, Pallas; Staud. Cat. *Entomologist*, 41, 102–6.

Rothschild, M. and Farrell, C. (1983) *The Butterfly Gardener*, Michael Joseph, London.

Rowell, T.A. (1986) The history of drainage at Wicken Fen, Cambridgeshire, England, and its relevance to conservation. *Biological Conservation*, 35, 111–42.

Rowell, T.A. and Harvey, H.J. (1988) The recent history of Wicken Fen, Cambridgeshire, England: A guide to ecological development. *Journal of Ecology*, 76, 73–90.

Rudge, A.J.B. (ed.) (1983) *The Capture and Handling of Deer*, Nature Conservancy Council.

Ruehlmann, T.E., Matthews, R.W. and Matthews, J.R. (1988) Roles for structural and temporal shelter-changing by fern-feeding lepidopteran larvae. *Oecologia*, 75, 228–32.

Rutherford, C.I. (1983) *Butterflies in Cheshire 1961 to 1982*, The Lancashire and Cheshire Entomological Society, A supplement to the proceedings for the year 1981–82.

Salter, J.H. (1880) Summer notes. *Natural History Journal*, 4, 112.

Sassoon, S. (1938) *The Old Century and Seven More Years*, Faber and Faber.

Sawford, B. (1987) *The Butterflies of Hertfordshire*, Castlemead, Ware.

Schreiber, H. (1976) Lepidoptera: Fam. Papilionidae, Pieridae und Nymphalidae, in *Fundortkataster der Bundesrepublik Deutschland*, Universität des Saarlandes, Saarbrücken.

Schroth, M. and Maschwitz, U. (1984). Zur larvalbiologie und wirtfidung von *Maculinea teleius* (Lepidoptera, Lycaenidae), eines parasiten von *Myrmica laevinodis* (Hymenoptera, Formicidae). *Entomologica Generalis*, 9(4), 225–30.

Schwarz, R. (1948–49) *Motyli. 1 and 2*. Vesmir, Praha.

Schweizerischer Bund für Naturschutz (1987) *Tagfalter und ihre Lebensräume, Arten, Gefährdung, Schutz*, Schweizerischer Bund für Naturschutz, Basel.

Scott, J.A. (1986) *The Butterflies of North America*, Stanford.

Seitz, A. (1925) *Macrolepidoptera of the World*, Vol. 13. African Rhopalocera, (ed. C. Aurivillius), Alfred Kernen Verlag, Stuttgart.

Settele, J. von (1990) Zur hypothese des bestandsrückgangs von insekten in der Bundesrepublik Deutschland: untersuchungen zu tagfaltern in der pfalz und die darstellung der ergebnisse auf verbreitungskarten. *Landschaft + Stadt*, 22, 88–96.

Shapiro, A.M. (1981) The Pierid red-egg syndrome. *American Naturalist*, 117, 276–94.

Shaw, G. (1975) The distribution of the chequered skipper butterfly in the Western Highlands in 1975. Unpublished report to the Scottish Wildlife Trust.

Shaw, M.R. (1978) The status of *Trogus lapidator* (F.) (Hymenoptera: Ichneumonidae) in Britain, a parasite of *Papilio machaon* L. *Entomologists Gazette*, 29, 287–88.

Shield, I.F. and Godwin, R.J. (1992) Changes in the species composition of a natural regeneration sward during the five-year set-aside scheme, in *BCPC Monograph No. 50: Set-aside*, (ed. J.H. Clarke), BCPC, Farnham, Surrey, pp. 123–34.

Shields, O. (1989) World number of butterflies. *Journal of the Lepidopterists' Society*, 43, 178–83.

Shirt, D.B. (ed.) (1987) *British Red Data Books: 2. Insects*, Nature Conservancy Council, Peterborough.

Shreeve, T.G. (1985) The Population Biology of the Speckled Wood Butterfly *Pararge aegeria* (L.) (Lepidoptera: Satyridae). Ph.D. Thesis (CNAA), Oxford Polytechnic.

Shreeve, T.G. (1986) Egg-laying in the speckled wood butterfly (*P. aegeria*): the role of female behaviour host plant abundance and temperature. *Ecological Entomology*, 11, 229–36.

Shreeve, T.G. (1992) Adult behaviour, in *The Ecology of Butterflies in Britain*, (ed. R.L.H. Dennis), Oxford University Press, Oxford, pp. 22–45.

Shreeve, T.G. and Mason, C.F. (1980) The number of butterfly species in woodland. *Oecologia*, 45, 414–18.

Siegel, S. (1956) *Non-parametric statistics*. McGraw Hill Kogakusha Ltd, London.

Simcox, D.J. and Thomas, J. A. (1979) *The Glanville Fritillary. Survey 1979*, Report to the Joint Committee for the Conservation of British Insects, Furzebrook.

Sinha, S.N., Lakhani, K.H. and Davis, B.N.K. (1990) Studies on the toxicity of insecticidal drift to the first instar larvae of the Large White butterfly *Pieris brassicae* (Lepidoptera: Pieridae). *Annals of Applied Biology*, 116, 27–41.

Skala, H. (1912) Die Lepidopterenfauna Mährens. *Verhandlungen der naturforschendau Ver. Brünn*, 50 (1911), 63–241.

Skala, H. (1932) Zue Lepidopterenfauna Mährens und Schlesiens. *Casiopsis moravscého Musea*. 30 (1931/32), 1–193.

Skapec, L. (1992) *Cervena Kniha*, 3, Priroda, Bratislava.

Slansky, F. and Feeny, P. (1977) Stabilisation of the rate of nitrogen accumulation by larvae of the cabbage butterfly on wild and cultivated food plants. *Ecological Monographs*, 47, 209–28.

Sly, J.M.A. (1981) *Review of Usage of Pesticides in Agriculture and Horticulture in England and Wales 1975–1979*, MAFF, London.

Smith, A.E. (ed.)(1982) *A Nature Reserves Handbook*, Royal Society for Nature Conservation, Lincoln.

Smith, H. and Macdonald, D.W. (1989) Secondary succession on extended arable field margins: its manipulation for wildlife benefit and weed control. *Brighton Crop Protection Conference - Weeds 1989*, **3**, 1063–8.

Smith, H. and Macdonald, D.W. (1992) The impacts of mowing and sowing on weed populations and species richness in field margin set-aside, in *BCPC Monograph No. 50: Set-aside*, (ed. J.H. Clarke), British Crop Protection Council, Farnham, pp. 117–22.

Smith, H., Feber, R.E., Johnson, P., *et al.* (1993) *The Conservation Management of Arable Field Margins*, English Nature Science No. 18, Peterborough.

Smith, J.E. and Bullock, D.J. (1993) A note on the summer feeding behaviour and and habitat use of free-ranging goats in the Cheddar Gorge SSSI. *Journal of Zoology*, **231**, 683–8.

Smith, R. and Brown, D. (1987) *The Lepidoptera of Warwickshire*, Warwickshire Museum Service, Warwick.

Smith, R.T. and Taylor, J.A. (eds) (1986). *Bracken, Ecology, Land Use and Control Techniques*, Parthenon Publishing Group, Carnforth.

Sneath, P.H. and Sokal, R.R. (1973) *Numerical Taxonomy. The Principles and Practice of Numerical Classification*, W.H. Freeman, San Francisco.

Soldat, M. (1987) Cervena Kniha CSR. *Motyli. Zpr. csl. Spol. ent. CSAV*, **23**, 1–36.

Sommerville, A. (1977) The conservation of the chequered skipper butterfly. *Scottish Wildlife*, **13**, 24–5; 49.

Sotherton, N.W. (1991) Conservation headlands: a practical combination of intensive cereal farming and conservation, in *The Ecology of Temperate Cereal Fields* (eds L.G. Firbank, N. Carter, J.F. Darbyshire and G.R. Potts), Blackwell, Oxford, pp. 373–97.

Sotherton, N.W., Rands, M.R.W. and Moreby, S.J. (1985) Comparison of herbicide treated and untreated headlands on the survival of game and wildlife. *British Crop Protection Conference – Weeds 1985*, British Crop Protection Council, Farnham, pp. 991–8.

Southwood, T.R.E. (1962) Migration of terrestrial arthropods in relation to habitat. *Biological Reviews*, **37**, 171–214.

Southwood, T.R.E. (1977) Habitat, the templet for ecological strategies? *Journal of Animal Ecology*, **46**, 337–65.

Southwood, T.R.E. and Cross, D.J. (1969) The ecology of the partridge. III. Breeding success and the abundance of insects in natural habitats. *Journal of Animal Ecology*, **38**, 497–509.

Spooner, G.M. (1963). On causes of the decline of *Maculinea arion* L. (Lep., Lycaenidae) in Britain. *Entomologist*, **96**, 199–210.

Stanford, R.E. and Opler, P.A. (1993) *Atlas of Western USA Butterflies*, Published by authors, Denver, Colorado.

Stark, M.J.R. (1975) Report on the status of *C. palaemon* in England 1975. Unpublished Report to Institute of Terrestrial Ecology.

Steel, C. and Parsons, M. (1989) *Butterflies on County Trust Reserves*, draft Invertebrate Site Register Report, Nature Conservancy Council, Peterborough.

Steel, C. and Steel, D. (1985) *Butterflies of Berkshire, Buckinghamshire and Oxfordshire*, Pisces Publications, Oxford.

Stephens, D.E.A. and Warren, M.S. (1992) *The Importance of Garden Habitats to Butterfly Populations*, Joint Committee for the Conservation of British Insects, Wareham, Dorset.

Stern, V.M. and Smith, R.F. (1960) Factors affecting egg production and oviposition in populations of *Colias eurytheme. Hilgardia*, **29**, 411–54.

Sterneck, J. (1929) *Prodromus der Schmetterlingsfauna Böhmens*, Selbstverlag, Karlsbad.

Surber, E., Amiet, R. and Kobert, H. (1973) *Das Brachlandproblem in der Schweiz*, Bericht Nr. 112 der Eidg, Anstalt für das forstliche Versuchswesen, Birmensdorf, pp. 1–138.

Sutton, R. (1993) Some 19th century records of the Large Copper, *Lycaena dispar* (Haworth) and Scarce Copper, *Lycaena viguareae* (L.) in Somerset. *Butterfly Conservation News*, no. 52, 18–20.

Sutton, S.L. and Beaumont, H.E. (eds) (1989) *Butterflies and Moths of Yorkshire, Distribution and Conservation*, Yorkshire Naturalists' Union, Doncaster.

Svendsen, P. and Fibiger, M. (1992) *The Distribution of European Macrolepidoptera Noctuidae*, Vol. 1, European Faunistical Press, Copenhagen.

Swaay, C.A.M. van (1990) An assessment of the changes in butterfly abundance in The Netherlands during the 20th century. *Biological Conservation*, 52, 287–302.

Swaay, C.A.M. van (1991) *Bibliografie van de Nederlandse Dagvlinders*, De Vlinderstichting, Wageningen.

Tarmann, G. (1984) Psodos (Trepidina) burmanni n. sp. eine neue Geometride aus den Tiroler Alpen (Lep. Geometridae). *Zeitschrift der Arbeitsgemeinschaft österreichischer Entomologen*, 36, 1–7.

Tax, M.H. (1989) *Atlas van de Nederlandse Dagvlinders*, De Vlinderstichting, Wageningen and Natuurmonumenten, s'-Graveland.

Thomas, C.D. (1983) The ecology and status of *Plebejus argus* L. in North West Britain. M.Sc. Thesis, University of Wales, Bangor.

Thomas, C.D. (1985) The status and conservation of *Plebejus argus* (Lepidoptera: Lycaenidae) in north-west Britain. *Biological Conservation*, 33, 29–51.

Thomas, C.D. (1992) The establishment of rare insects in vacant habitats. *Antenna*, 16, 89–93.

Thomas, C.D. (1993) Local extinctions, colonizations and distributions: habitat tracking by British butterflies, in *Individuals, Populations and Patterns in Ecology*, (eds S.R. Leather, A.D. Watt, N.J. Mills and K.F.A. Walters), Intercept, Andover.

Thomas, C.D. and Harrison, S. (1992) Spatial dynamics of a patchily-distributed butterfly species. *Journal of Animal Ecology*, 61, 437–46.

Thomas, C.D. and Jones, T.M. (1993) Partial recovery of a skipper butterfly (*Hesperia comma*) from population refuges: lessons for conservation in a fragmented landscape. *Journal of Animal Ecology*, 62, 472–81.

Thomas, C.D., Thomas, J.A. and Warren, M.S. (1992) Distributions of occupied and vacant butterfly habitats in fragmented landscapes. *Oecologia*, 92, 563–7.

Thomas, J.A. (1974) Factors influencing the numbers and distribution of the brown hairstreak, *Thecla betulae*, and the black hairstreak, *Strymonidia pruni*. Ph.D. thesis, Leicester University, UK.

Thomas, J.A. (1975) The black hairstreak, conservation report. Unpublished report ITE/NCC.

Thomas, J.A. (1976) *The biology and conservation of the Large Blue butterfly* Maculinea arion L., ITE, Monks Wood, Abbots Ripton, UK.

Thomas, J.A. (1977) *Ecology and Conservation of the Large Blue butterfly – second report*, ITE, Monks Wood, Abbots Ripton, UK.

Thomas, J.A. (1980) Why did the large blue become extinct in Britain? *Oryx*, 15, 243–7.

Thomas, J.A. (1983a) The ecology and status of *Thymelicus acteon* (Lepidoptera: Hesperiidae) in Britain. *Ecological Entomology*, 8, 427–35.

Thomas, J.A. (1983b) The ecology and conservation of *Lysandra bellargus* (Lepidoptera: Lycaenidae) in Britain. *Journal of Applied Ecology*, 20, 59–83.

Thomas, J.A. (1983c) A WATCH census of common British butterflies. *Journal of Biological Education*, 17, 333–8.

Thomas, J.A. (1984a) The conservation of butterflies in temperate countries: past efforts and lessons for the future, in *The Biology of Butterflies*, (eds R.I. Vane-Wright and P.R.Ackery), Symposium of the Royal Entomological Society, London, no. 11, Academic Press, London, pp. 327–32.

Thomas, J.A. (1984b) The behaviour and habitat requirements of *Maculinea nausithous* (the dusky large blue butterfly) and *M. teleius* (the scarce large blue) in France. *Biological Conservation*, 28, 325–47.

Thomas, J.A. (1987) The return of the large blue. *British Butterfly Conservation Society News*, 38, 22–6.

Thomas, J.A. (1989) The return of the large blue butterfly. *British Wildlife*, 1, 2–13.

Thomas, J.A. (1991) Rare species conservation: case studies of European butterflies, in *The Scientific Management of Temperate Communities for Conservation*, (eds I.F. Spellerberg, F.B. Goldsmith and M.G. Morris), Blackwell Scientific Publications, Oxford, pp. 149–97.

Thomas, J.A. (1992a) Adaptations to living near ants, in *The Ecology of British Butterflies*, (ed. R.L.H. Dennis), Oxford University Press, Oxford, pp. 109–15.

Thomas, J.A. (1992b) Relationships between butterflies and ants, in *The Ecology of British Butterflies*, (ed. R.L.H. Dennis), Oxford University Press, Oxford, pp. 149–54

Thomas, J.A. (1993) Holocene climate change and warm man-made refugia may explain why a sixth of British butterflies inhabit unnatural early-successional habitats. *Ecography*, 16, 278–84.

Thomas, J.A. and Elmes, G.W. (1992) The ecology and conservation of *Maculinea* butterflies and their Ichneumon parasitoids, in *Future of Butterflies in Europe: Strategies for Survival*, (eds T. Pavlicek-van Beek, A.H. Ovaa, and J. van der Made), Agricultural University, Wageningen, pp. 116–23.

Thomas, J.A. and Elmes G.W. (1993) Specialised searching and the hostile use of allomones by a parasitoid whose host, the butterfly *Maculinea rebeli*, inhabits ant nests. *Animal Behaviour*, 45, 593–602.

Thomas, J.A. and Lewington, R. (1991) *The Butterflies of Britain and Ireland*, Dorling Kindersley, London.

Thomas, J.A. and Wardlaw, J.C. (1990) The effect of queen ants on the survival of *Maculinea arion* in *Myrmica* ant nests. *Oecologia*, 85, 87–91.

Thomas, J.A. and Wardlaw, J.C. (1992) The capacity of a *Myrmica* ant nest to support a predacious species of *Maculinea* butterfly. *Oecologia*, 91, 101–9.

Thomas, J.A. and Webb, N. (1984) *Butterflies of Dorset*, Dorset Natural History and Archaeological Society, Dorchester.

Thomas, J.A., Elmes, G.W. and Wardlaw, J.C. (1993) Contest competition among *Maculinea rebeli* larvae in ant nests. *Ecological Entomology*, 18, 73–6.

Thomas, J.A., Snazell, R.G. and Moy, I. (1994) *The Conservation of Violet-feeding Fritillaries in the British Isles*, English Nature, Peterborough (in press).

Thomas, J.A., Thomas, C.D., Simcox, D.J. and Clarke, R.T. (1986) The ecology and declining status of the silver-spotted skipper butterfly (*Hesperia comma*) in Britain. *Journal of Applied Ecology*, 23, 365–80.

Thomas, J.A., Elmes, G.W., Wardlaw, J.C. and Woyciechowsky, M. (1989) Host plant specificity of Maculinea butterflies in Myrmica ants nests. *Oecologia*, 79, 452–7.

Thomas, J.A., Munguira, M.L., Martin, J. and Elmes, G.W. (1991) Basal hatching by *Maculinea* butterfly eggs: a consequence of advanced myrmecophily? *Biological Journal of the Linnaean Society*, 44, 175–84.

Thompson, P. (1992) Field Officer's 1991 report. *The Game Conservancy Reviews of 1991*, No. 23, The Game Conservancy, Fordingbridge.

Thomson, G. (1980) *The Butterflies of Scotland, a Natural History*, Croom Helm, London.

Thomson, G. (1987) Enzyme variation at morphological boundaries in *Maniola* and related genera (Lepidoptera: Nymphalidae: Satyrinae). Ph.D. Thesis, University of Stirling.

Toso, G.G. and Balletto, E. (1976) Una nuova specie del genera Agrodiaetus Huebn. *Annali del Museo Civico di Storia naturale Giacomo Doria*, 81, 124–30.

Tubbs, C.R. (1986) *The New Forest*, Collins New Naturalist, London.

Turner, J.R.G., Gatehouse, C.M. and Corey, C.A. (1987) Does solar energy control organic diversity? Butterflies, moths and the British climate. *Oikos*, 48, 195–205.

Vane-Wright, R.I., Humphries, C.J. and Williams, P.H. (1991) What to protect? – Systematics and the agony of choice. *Biological Conservation*, 55, 235–54.

Veling, K. and Swaay, C.A.M. van (1992) How to Achieve More with Butterfly Inventory Data. Proceedings of the 8th International colloquium of the European Invertebrate Survey, Brussels, 9–10 September 1991, Koninkliik Belgisch Instituut voor Natuurwetenschappen, Brussel.

Verrall, G.H. (1909) The large copper butterfly (*Chrysophanus dispar*). *Entomologist*, 42, 183.

Verspui, K. and Visser, S. (1992) Ecological research on a population of the Heath Fritillary (*M. athalia*), in *Future of Butterflies in Europe: Strategies for Survival*, (eds T. Pavlicek-van Beek, A.H. Ova and J.G. van der Made), Agricultural University, Wageningen, pp. 172–6.

Vickerman, G.P. (1991) The effects of different pesticide regimes on the invertebrate fauna of winter wheat, in *The Boxworth Project: Cereal Farming and the Environment*, (eds P.W. Greig-Smith, G.K. Frampton and A.R. Hardy), HMSO, London.

Vickerman, G.P. and Sunderland, K.D. (1977) Some effects of dimethoate on arthropods in winter wheat. *Journal of Applied Ecology*, **14**, 767–77.

Vickery, M. (1991a) National garden butterfly survey, part 1. *Butterfly Conservation News*, **47**, 32–8.

Vickery, M. (1991b) National garden butterfly survey, part 2. *Butterfly Conservation News*, **48**, 26–30.

Vickery, M. (1992a) Garden butterfly survey 1991 report. *Butterfly Conservation News*, **50**, 42–9.

Vickery, M. (1992b) Garden stinging nettle survey. *Butterfly Conservation News*, **50**, 21–3.

Vickery, M. (1993) Garden butterfly survey 1992. *Butterfly Conservation News*, **53**, 50–9.

Viedma, M.G. and Gómez Bustillo, M.R. (1976) *Libro Rojo de los Lepidópteros Ibéricos*, ICONA, Madrid.

Viedma, M. G. and Gómez Bustillo, M.R. (1985) *Revisión del Libro Rojo de los Lepidópteros Ibéricos*, Monografías ICONA (No. 42), Madrid.

Viedma, M.G., Escribano, R., Gómez Bustillo, M.R. and Mattoni, R.H.T. (1985) The first attempt to establish a Nature Reserve for the conservation of Lepidoptera in Spain. *Biological Conservation*, **32**, 255–76.

Viejo, J.L. (1986) Diversity and species richness of butterflies and skippers in Central Spain habitats. *Journal Research Lepidoptera*, **24**, 364–71.

Viejo, J.L., Viedma, M.G. and Martínez, E. (1989) The importance of woodlands in the conservation of butterflies (Lep.: Papilionoidea and Hesperioidea) in the Centre of the Iberian Peninsula. *Biological Conservation*, **48**, 101–14.

Viejo, J.L., Munguira, M.L., Ibero, C. and Alvarez, C. (1992) Aspectos legales sobre la conservación de las mariposas en España. *Shilpa Revista Lepidopterología*, **20**, 355–65.

Wagener, P.S., Kinkler, H., Löser, S. and Rehnelt, K. (1979) Rote Liste der in Nordrhein-Westfalen gefährdeten Grossschmetterlinge (Macrolepidoptera), 2. Fassung (Stand 1. 9. 1978). *Schriftenreihe der Landesanstalt für Ökologie, Landschaftsentwicklung und Forstplanung Nordrhein-Westfalen*, **4**, 51–64.

Wallis, E. (1986) Europe's problem regions. *Geofile*, issue no. 67, Mary Glasgow Publications, London.

Warren, B.C.S. (1949) A note on the Central European races of *Papilio machaon* and their nomenclature. *Entomologist*, **82**, 150–3.

Warren, M.S. (1981) The ecology of the wood white butterfly. Ph.D. thesis, University of Cambridge.

Warren, M.S. (1984) The biology and status of the wood white butterfly *L. sinapis* in the British Isles. *Entomologist Gazette*, **35**, 207–23.

Warren, M.S. (1985) The influence of shade on butterfly numbers in woodland rides, with special reference to the Wood White (*L. sinapis*). *Biological Conservation*, **33**, 147–64.

Warren, M.S. (1987a) The ecology and conservation of the heath fritillary butterfly *Mellicta athalia*. I. Host selection and phenology. *Journal of Applied Ecology*, **24**, 467–82.

Warren, M.S. (1987b) The ecology and conservation of the heath fritillary butterfly, *Mellicta athalia*. II. Adult population structure and mobility. *Journal of Applied Ecology*, **24**, 483–98.

Warren, M.S. (1987c) The ecology and conservation of the heath fritillary butterfly *Mellicta athalia* III: Population dynamics and the effects of habitat management. *Journal of Animal Ecology*, **24**, 499–513.

Warren, M.S. (1989) Pheasants and fritillaries: is there really any evidence that pheasant rearing may have caused butterfly decline? *British Journal of Entomology and Natural History*, **2**, 169–75.

Warren, M.S. (1990a) The conservation of *Eurodryas aurinia* in the United Kingdom, in *Colloquy on the Berne Convention on invertebrates and their conservation*, Environment Encounters series, no. 10, Council of Europe, Strasbourg, pp. 71–4.

Warren, M.S. (1990b) The chequered skipper *Carterocephalus palaemon* in northern Europe. Unpublished report to British Butterfly Conservation Society.

Warren, M.S. (1991) The successful conservation of an endangered species, the heath fritillary butterfly (*M. athalia*) in Britain. *Biological Conservation*, **55**, 37–56.

Warren, M.S. (1992a) Butterfly populations, in *The Ecology of Butterflies in Britain*, (ed. R.L.H. Dennis), Oxford University Press, Oxford, pp. 73–92.

Warren, M.S. (1992b) The conservation of British butterflies, in *The Ecology of Butterflies in Britain*, (ed. R.L.H Dennis), Oxford University Press, Oxford. pp. 246–74.

Warren, M.S. (1992c) The high brown fritillary – Britain's most endangered butterfly? *Butterfly Conservation News*, **50**, 26–30.

Warren, M.S. (1992d) Britain's vanishing fritillaries. *British Wildlife Magazine*, **3**, 282–96.

Warren, M.S. (1993a) A review of butterfly conservation in central southern Britain. I. Protection, evaluation and extinction in prime sites. *Biological Conservation*, **64**, 25–35.

Warren, M.S. (1993b) A review of butterfly conservation in central southern Britain: II. Site management and habitat selection by key species. *Biological Conservation*, **64**, 37–49.

Warren, M.S. (1993c) Bracken: hot-spots for invertebrates. *Enact*, **2**, 6.

Warren, M.S. (1994) The UK status and metapopulation structure of a threatened European butterfly, *Eurodryas aurinia* (the marsh fritillary). *Biological Conservation* (in press).

Warren, M.S. and Fuller, R.J. (1990) *Woodland Rides and Glades: Their Management for Wildlife*, Nature Conservancy Council, Peterborough.

Warren, M.S. and Key, R.S. (1991) Woodlands: past, present and potential for insects, in *The Conservation of Insects and Their Habitats*, (eds N.M. Collins and J.A. Thomas), Academic Press, London, pp. 155–211.

Warren, M.S. and Stephens, D.E.A. (1989) Habitat design and management for butterflies. *The Entomologist*, **108**, 123–34.

Warren, M.S. and Thomas, J.A. (1992). Butterfly responses to coppicing, in *The Ecological Effects of Coppicing*, (ed. G.P. Buckley), Chapman & Hall, London, pp. 249–70.

Warren, M.S. and Thomas, J.A. (1994). *Management Options for the Silver Spotted Skipper Butterfly; a Study of the Timing of Grazing at Beacon Hill NNR, Hants*, English Nature, Peterborough, (in press).

Warren, M.S., Pollard, E. and Bibby, T.J. (1986) Annual and long-term changes in a population of the wood white butterfly *Leptidea sinapis*, *Journal of Animal Ecology*, **55**, 707–19.

Warren, M.S., Thomas, C.D. and Thomas, J.A. (1984) The status of the heath fritillary butterfly *Mellicta athalia* Rott, in Britain. *Biological Conservation*, **29**, 287–305.

Warren, R.G. (1984) *Atlas of the Lepidoptera of Staffordshire*, Staffordshire Biological Recording Scheme Publication No. 11, Stoke-on-Trent.

Warrick, R.A. and Barrow, E.M, (1991) Climate change scenarios for the UK. *Transactions of the Institute of British Geographers, NS*, **16**, 387–99.

Watt, T.A., Smith, H. and Macdonald, D.W. (1990) *The Control of Annual Grass Weeds in Fallowed Field Margins Managed to Encourage Wildlife*. Proceedings of the European Weed Research Society, Symposium Helsinki 1990: Integrated Weed Management in Cereals, pp. 187–95.

Watt, W.B., Hoch, P.C. and Mills, S. (1974) Nectar resource use by *Colias* butterflies: chemical and visual aspects. *Oecologia*, **90**, 581–5.

Weidemann, H.-J. (1988) *Tagfalter*, Vol. 2: *Biologie, Okologie, Biotopschutz*, Neuman-Neudamm, W. Germany.

Weiss, S. J., Murphy, D.D. and White, R.R. (1988) Sun, slope and butterflies: topographic determinants of habitat quality in *Euphydryas editha*. *Ecology*, **69**, 1486–96.

Wiklund, C. (1973) Host plant suitability and the mechanism of host selection in larvae of *Papilio machaon*. *Entomologia experimentalis et applicata*, **16**, 232–42.

Wiklund, C. (1974) Oviposition preferences in *Papilio machaon* in relation to the host plants of the larvae. *Entomologia experimentalis et applicata*, **17**, 189–98.

Wiklund, C. (1975) The evolutionary relationship between adult oviposition preferences and larval host plant range in *Papilio machaon*. *Oecologia*, **18**, 185–97.

Wiklund, C. (1977) Oviposition, feeding and spatial separation of breeding and foraging habitats in a population of *Leptidia sinapsis*. *Oikos*, **28**, 56–68.

Wiklund, C. (1981) Generalist vs. specialist oviposition behaviour in *Papilio machaon* (Lepidoptera) and functional aspects of the hierachy of oviposition preferences. *Oikos*, **36**, 163–70.

Wiklund, C. and Ahberg, C. (1978) Host plants, nectar source plants and habitat selection of males and females of *Anthocharis cardamines*. *Oikos*, **31**, 169–83.

Wiklund, C. and Persson, A. (1983) Fecundity, and the relation of egg weight variation to offspring fitness in the speckled wood butterfly, *Pararge aegeria*, or why don't butterfly females lay more eggs? *Oikos*, **40**, 53–63.

Wilkinson, R. S. (1981) The first records of *Papilio machaon* in England. *Entomologists Record and Journal of Variation*, **93**, 4–6.

Wilkinson, R. and Beecroft, R. (1988) Kagoro Forest conservation study (Nigeria). *International Council for Bird Preservation, Study Report no. 28*, Cambridge.

Williams, L. and Barker, A. (eds) (1993) *Hampshire and Isle of Wight Butterfly Report 1992*, Hampshire Branch of Butterfly Conservation.

Willmott, K.J. (1988) An appraisal of the Chiddingfold forest and its current butterfly population. Unpublished report to the Nature Conservancy Council.

Wilson, P.J. (1992) The natural regeneration of vegetation under set-aside in southern England, in *BCPC Monograph No. 50: Set-aside*, (ed. J.H. Clarke), BCPC, Farnham, Surrey, pp. 73–8.

Wood, H. (1908) Notes on the life-history of *Cyclopides palaemon*. *Entomologist's Record*, **20**, 65–6.

Wood, P. and Samways, M.J. (1991) Landscape element pattern and continuity of butterfly flight paths in an ecologically landscaped botanic garden, Natal, South Africa. *Biological Conservation*, **58**, 149–66.

Woodwell, G.M. (1989) The warming of the industrialized middle latitudes 1985–2050: causes and consequences. *Climatic Change*, **15**, 31–50.

Woolley, D. (1993) Rhos Llaw-cwrt NNR: Marsh Fritillary studies 1992. Unpublished report, Countryside Council for Wales, Aberystwyth.

Wynhoff, I. (1992) Micro-distribution and flower preference of the meadow brown (*Maniola jurtina*) and the ringlet (*Aphantopus hyperantus*), in *Future of Butterflies in Europe: Strategies for Survival*, (eds T. Pavlicek-van Beek, A. H. Ovaa and J.G. van der Made), Agricultural University, Wageningen, pp. 177–85.

Wynhoff, I., Made, J. van der and Swaay, C.A.M. van (1990) *Dagvlinders van de Benelux*, De Vlinderstichting, Wageningen.

Young, M.R. and Ravenscroft, N.O.M. (1991) Conservation of the chequered skipper butterfly (*Carterocephalus palaemon* Pallas) in Scotland. Confidential Report to Nature Conservancy Council.

Zoller, H. and Bischof, N. (1980) Stufen der Kulturintensität und ihr Einfluss auf Artenzahl und Artengefüge der Vegetation. *Phytocoenologia*, **7**, 35–51.

Butterfly species index

Subject index